POLYMER REVIEWS

H. F. Mark *and* E. H. Immergut, *Editors*

Volume 1: **The Effects of Ionizing Radiation on Natural and Synthetic High Polymers.** By Frank A. Bovey

Volume 2: **Linear and Stereoregular Addition Polymers: Polymerization with Controlled Propagation.** By Norman G. Gaylord and Herman F. Mark

Volume 3: **Polymerization of Aldehydes and Oxides.** By Junji Furukawa and Takeo Saegusa

Volume 4: **Autohesion and Adhesion of High Polymers.** By S. S. Voyutskii

Volume 5: **Polymer Single Crystals.** By Philip H. Geil

Volume 6: **Newer Methods of Polymer Characterization.** Edited by B. Ke

Volume 7: **Thermal Degradation of Organic Polymers.** By S. L. Madorsky

Volume 8: **Metalorganic Polymers.** By K. A. Andrianov

Volume 9: **Chemistry and Physics of Polycarbonates.** By H. Schnell

Volume 10: **Condensation Polymers: By Interfacial and Solution Methods.** By P. W. Morgan

Volume 11: **Oxidation-Reduction Polymers (Redox Polymers).** By H. G. Cassidy and K. A. Kun

Volume 12: **Polymerization by Organometallic Compounds.** By L. Reich and A. Schindler

Volume 13: **Polymeric Sulfur and Related Polymers.** By Arthur V. Tobolsky and William J. MacKnight

Volume 14: **Infrared Spectra of Polymers: In the Medium and Long Wavelength Regions.** By Dieter O. Hummel

Volume 15: **Stress-Strain Behavior of Elastic Materials: Selected Problem of Large Deformations.** By O. H. Varga

Volume 16: **Graft Copolymers.** By H. A. J. Battaerd and G. W. Tregear

Volume 17: **High Temperature Resistant Polymers.** By A. H. Frazer

Additional volumes in preparation

HIGH TEMPERATURE RESISTANT POLYMERS

A. H. FRAZER

E. I. du Pont de Nemours & Company
Pioneering Research Laboratory
Wilmington, Delaware

Interscience Publishers, A division of John Wiley & Sons

New York   London   Sydney   Toronto

Copyright © 1968 by John Wiley & Sons, Inc.

All rights reserved. No part of this book may be reproduced by any means, nor transmitted, nor translated into a machine language without the written permission of the publisher.

Library of Congress Catalog Card Number 68-21491
SBN 470   276509
Printed in the United States of America

This book is dedicated to my wife, Christine, and my children, Tina, Marilee, and Jeff, for their understanding and encouragement.

INTRODUCTION TO THE SERIES

This series was initiated to permit the review of a field of current interest to polymer chemists and physicists *while* the field was still in a state of development. Each author is encouraged to speculate, to present his own opinions and theories, and, in general, to give his work a more "personal flavor" than is customary in the usual reference book or review article. Whenever background material was required to explain a new development in the light of existing and well-known data, the authors have included them, and, as a result, some of the volumes are lengthier than one would expect of a "review."

We hope that the books in this series will generate as much new research as they attempt to review.

H. F. MARK
E. H. IMMERGUT

PREFACE

As with preceding volumes in this series, the purpose of this book is to review a field of current interest in polymer chemistry. Since the thrust for intensive research in this area in the past several years is the need for such polymers in specific end uses, in this volume, unlike previous volumes, the utilization of these polymers in various applications requiring high temperature resistance is discussed.

The subject matter is arranged so as to define what is meant by thermal stability, what it implies and how it is evaluated before discussing particular polymer systems. The arrangement of the discussion of these systems is almost chronological, covering those polymers resulting from research aimed (1) at improving existing polymers by structural modification and (2) at devising new polymer systems tailored to be heat resistant.

As in any volume of this type, complete coverage of the literature cannot be claimed. However, to the best of the author's knowledge, the majority of pertinent recent references have been consulted.

The reader will note the absence of any detailed and specific discussion of the nature, mechanism, kinetics, or products of the degradation of the particular polymers covered. This is not an oversight on the part of the author, but a reflection of the limited knowledge in this area and, in turn, an indication of the need for information and understanding of the processes which lead to high temperature resistance in polymers.

As has been said by many previous authors, no book of this type is ever written without the assistance of many people. At this time, I wish to thank the Management of Textile Fibers Department for permission to write this book, to express my appreciation to Ilya Sarasohn, John Schaefgen, Ralph Beaman, Wilfred Sweeny, Paul Morgan, Fred Knobloch, Roe Blume, and Eugene Magat for their critical reviews of all or parts of the manuscript, and to acknowledge the pertinent comments and suggestions of countless people in the du Pont Company.

Thanks is also due the secretarial staff, in particular, Helen Fulmer, Doris Springfield, Theresa Monell, Arlene Cicconi, Lucille Primaldi,

Audrey Johnson, Maureen Fitzharris, and Evelyn Whiteside, and the library staff of Helen Rowley, Rachel Morrison, Linda Ringer, and Joanne Germak.

Finally, a special thanks is extended to Walter Holley whose tireless devotion and assistance made this undertaking possible.

A. H. Frazer

April, 1968

CONTENTS

I. Thermal Stability of Polymers **1**
   A. Introduction 1
   B. Thermal Stability 1
   C. Reversible Property Changes 2
      1. Glass Transition Temperature (T_g) 3
      2. Melting Point 7
      3. Nature of Property Changes Below the Melting Point . 9
   D. Irreversible Property Changes (Decomposition) 14
      1. Factors Contributing to Stability 14
      2. Types of Decomposition Reactions 18
      3. Nature of Property Changes Caused by Decomposition . 22
   E. Determination of Thermal Stability 22
      1. Methods Related to Softening Behavior 23
      2. The Isoteniscope 25
      3. Differential Thermal Analysis (DTA) 25
      4. Torsional Braid Analysis (TBA) 26
      5. Thermogravimetric Analysis (TGA) 27
      6. Application Methods 28
   F. Structure and Thermal Stability 32
   References 34

II. Aromatic Polymers **38**
   A. Introduction 38
   B. Poly-*p*-Phenylenes 38
      1. Preparation 38
      2. Properties 43
   C. Poly(Methylene Phenylenes) 49
   D. Poly-*p*-Xylylenes 53
      1. Preparation 53
      2. Properties 56
      3. Poly(Alkylene Phenylenes) 69

4. Poly-*p*-Xylylidenes.	69
5. Polycyanoterephthalylidene	72
E. Other Aromatic Polymers	73
References.	75

III. Aromatic Polymers Containing Functional Groups in the Chain — 78

A. Aromatic Polyamides.	78
1. Aromatic Polyamides from Short-Chain Aliphatic Primary and Secondary Diamines and Cycloaliphatic Secondary Diamines	79
2. Aromatic Polyhydrazides.	85
3. Polyamides from Primary Aromatic Diamines and Aromatic Dicarboxylic Acids.	90
B. Aromatic Polyanhydrides.	97
C. Aromatic Polycarbonates.	100
D. Aromatic Polyesters	106
E. Aromatic Polyethers (Polyphenylene Oxides)	114
F. Aromatic Polysulfides (Polyphenylene Sulfides)	123
G. Aromatic Polysulfonates and Aromatic Sulfonamides.	127
H. Aromatic Polysulfones (Polyphenylene Sulfones)	129
I. Aromatic Polyureas and Aromatic Polyurethanes.	132
References.	132

IV. Aromatic Polymers Containing Heterocyclic Rings — 138

A. Polybenzimidazoles	138
B. Polybenzoxazoles.	151
C. Polybenzothiazoles.	155
D. Aromatic Polyimides.	159
E. Aromatic Polyoxadiazoles.	176
1. Aromatic Poly-1,2,4-Oxadiazoles.	176
2. Aromatic Poly-1,3,4-Oxadiazoles.	178
F. Aromatic Polypyrazoles	188
G. Polyquinoxalines.	192
H. Aromatic Poly-1,3,4-Thiadiazoles	197
I. Aromatic Polythiazoles.	199
J. Aromatic Polytetraazopyrenes.	202
K. Aromatic Poly-4-Phenyl-1,2,4-Triazoles.	204
L. Poly(Quinazolinediones) and Poly(Benzoxazinones)	207
References.	210

V. Inorganic Polymers — 214

A. Boron-Containing Polymers.	214
1. Addition Polymers of Borozens.	215

CONTENTS

- 2. Addition Polymers of Borizons (Condensed Systems) . . 215
- 3. Addition Polymers of Borozins (Fused Systems) . . . 216
- 4. Polymeric Boron Complexes 216
- 5. Polymers Having Only Boron and Nitrogen in the Main Chain 216
- 6. Polymers Having Only Boron, Nitrogen, and Elements Other than Carbon in the Main Chain 217
- 7. Polymers Having Boron, Nitrogen, and Carbon in the Main Chain 218
- 8. Polymers Derived from Isocyanates 220
- 9. Polymers Containing Boron, Oxygen, and Carbon in the Main Polymer Chain 220
- 10. *p*-Vinylphenylboronic Acid and Related Polymers . . . 221
- 11. Polymers Containing Boron–Phosphorus Bonds . . . 221
- B. Phosphorus-Containing Polymers 223
  - 1. Phosphorus–Nitrogen Polymers 223
  - 2. Phosphoryl Amides and Derivatives 225
  - 3. Phosphonic Acids, Phosphonic Amides, and Derivatives . 226
- C. Silicon-Containing Polymers 228
  - 1. Silicones or Polysiloxanes 228
  - 2. Modified Silicones or Polymetallosiloxanes 234
- D. Other Organometallic Polymers 241
  - 1. Aluminum–Oxygen Polymers 241
  - 2. Aluminum–Nitrogen Polymers 242
  - 3. Germanium–Oxygen Polymers 243
  - 4. Tin–Oxygen Polymers 243
  - 5. Polymers Containing Group IV Atoms and Carbon Chains 243
  - 6. Polymers Containing Arsenic and Carbon Chains . . . 247
- E. Metal Chelate Polymers 249
  - 1. Linking of Ligands with Ions 249
  - 2. Polymer Formation in the Presence of Metals 253
  - 3. Incorporation of Metal Ions in Preformed Polymers . . 255
  - 4. Reaction with Chelates Containing Functional Groups . 258
  - 5. Ferrocenes 261
- References 263

VI. Ladder Polymers **268**

- A. Pyrolyzed Polyacrylonitrile and Polymethacrylonitrile . . 270
- B. Polyphenylsilsesquioxanes 271
- C. Cyclized Poly-3,4-Isoprene 271
- D. Cyclized Polybutadiene, Cyclized Polyisoprene, and Cyclized Polychloroprene 272

E. Ladder Polymers from Vinyl Isocyanate 273
F. Polyquinone Ethers, Polyquinone Thioethers, and Polyquinone Imines 274
G. Polyquinoxaline Ladder Polymers 275
H. Polydioxins. 277
I. Ladder Polymers from Vinylbutadiene with Cyclic Bisdienophiles 277
J. Ladder Polymers Derived from 4,4-Dimtheyl-1,6-Heptadiene-3,5-Dione 279
K. Poly(Imidazopyrrolones) 279
L. Poly(Benzimidazo-Benzophenanthrolines) 282
References 283

VII. Applications **285**

A. Adhesives 285
B. Coatings 287
C. Composite Structure (Glass-Reinforced Laminates) . . . 289
D. Fibers 293
E. Films 302
F. Papers 304
G. Resins 314
References 317

Author Index **319**

Subject Index **332**

I. THERMAL STABILITY OF POLYMERS

A. INTRODUCTION

Nearly all polymers of present commercial interest which are fabricable into films, fibers, laminates, etc., are relatively low melting and not appreciably stable above their melting points. Even those polymers now employed at temperatures up to 250°C are not suitable for more elevated temperatures, because they undergo thermal and chemical decomposition.

Recent advances in space and other technologies have produced a continuing and growing need for materials that will withstand prolonged exposure at temperatures in excess of 300°C and/or relatively short exposures at much higher temperatures (1–3). The need has become particularly acute for high temperature materials used in electrical and thermal insulation, gaskets, flexible tubing, and many other items currently made from fibers, films, plastics, and rubber. Since this need has emerged in both civilian and military items, both industrial concerns and military agencies have initiated and sponsored research in this area. With additional impetus from aerospace requirements, considerable progress has been made in the last few years, and a firm information base has been established for the ultimate development of polymers—organic, organic-inorganic (semiorganic), and inorganic—which are much more thermally stable than those commercially available today. Before discussing these developments, it is necessary to consider what is meant by thermal stability, its measurement, and the factors affecting it.

B. THERMAL STABILITY

The term "thermal stability" in relation to polymeric materials is often misused and more often misinterpreted. In most of these cases the thermal stability of a material is identified by a specific temperature, without reference to test method, surrounding atmosphere, and time involved.

One of the main reasons for the confusion about thermal stability is that two basically different mechanisms of property loss (deterioration) exist, both of which are temperature dependent. The first mechanism is a reversible process and represents a "softening" of the material with increasing temperature. This process is a function of temperature only. Although temperature alone determines the magnitude of a reversible property

---

At the glass transition temperature, the characteristics of the material change upon cooling from a rubbery state to a glassy rigid state. The molecular mechanism associated with this process is a decrease in the vibration of segments of the polymer chain. It is generally assumed that at the glass transition temperature the microbrownian movement of chain segments ceases completely. However, smaller atomic groupings and atoms themselves exhibit considerable motion well below the glass transition temperature (7–14); nuclear magnetic resonance transitions have been found in the 77–300°K region (14).

The glass transition temperature is not an equilibrium phenomenon but, rather, a rate effect (15). It depends on the time scale of the experiment. Shorter times result in higher values; in extremely long tests a transition might not be observed at all (16). Spencer and Boyer (15) noticed, when measuring the expansion of polystyrene in a dilatometer, that upon change of temperature the sample expanded to its new volume almost instantaneously, after which there was a slow isothermal expansion until time equilibrium was reached. The plot of the true equilibrium volume versus temperature does not indicate a transition. Their conclusion (17), based on reaction rate equations, was that the logarithm of the test time varies linearly with the reciprocal transition temperature.

TABLE I
Glass Transition Temperatures (T_g) of Polystyrene and Polyvinyl Acetate in Relation to the Molecular Weight

Polyvinyl acetate (24)		Polystyrene (18)	
Visc. mol. wt.	T_g(°C)	Visc. mol. wt.	T_g(°C)
15,000	15	2,970	40
104,000	26	4,300	63
570,000	28	5,600	76
1,120,000	29	85,000	100

The glass transition temperature depends also upon such variables as the forces applied during the test, the molecular weight of the material, crosslinking, and plasticizers. It has been shown (17) that the glass transition temperature of Saran* copolymers determined by the discontinuity of the change in elongation with temperature, changes from about −5°C for 5000 psi to −40°C for 20,000 psi. The molecular weight dependence of the glass transition temperature (Table I) has been established by various authors (18–26). Low molecular weight or the presence of plasticizers,

*Trademark for Dow Chemical polyvinylidene chloride copolymers.

change, the *determination* of such a property may involve the time factor because of the relaxation behavior of polymeric materials. This happens when two physical constants which are characteristic of thermoplastics and, to a certain degree, of thermosetting materials, are determined: the glass transition temperature and the melting point (softening range).

The second mechanism is the irreversible decomposition of the substance caused by heat. This process is both temperature and time dependent, but is also affected by other factors such as the surrounding atmosphere. Decomposition under an inert atmosphere or under vacuum can be considerably different from decomposition under ambient air or oxygen. It is also different in a sealed system, where the products of decomposition cannot be removed and can affect the course of the reaction.

Most thermoplastic materials decompose to a considerable degree and at a noticeable rate only above their melting points. The decomposition of these materials, although extensively studied, is seldom of importance for the conventional application because the polymers will melt before they decompose. Thermal decomposition of thermosetting materials, i.e., below the melting point, however, has to be taken into consideration when these materials are to be used at higher temperatures and over prolonged periods of time.

The fact that decomposition reactions may occur during the determination of reversible property changes should be taken into consideration in such a way that the material to be tested should be exposed to the test temperature for the shortest time possible. The effect of decomposition on the properties of plastics will be treated later.

The type of decomposition reaction has considerable influence on the direction of the change of a property. For example, chain rupture will lower tensile and flexural strength, while crosslinking, within limits, increases strength and makes the material insoluble and infusible.

C. REVERSIBLE PROPERTY CHANGES

When heating or cooling a polymeric material, the changes in a number of properties with temperature are different, depending upon whether the material is crystalline or amorphous. A crystalline material, will, at a certain temperature, undergo an abrupt change of the property itself (Fig. 1a) while an amorphous material will show a variation with temperature in the rate of change of the property in question (Fig. 1b).

The first type is a first-order transition, and the crystalline melting point is a typical example. This transition point is thermodynamically defined, although it can be influenced by exterior forces, such as pressure (4). A transition which involves the rate at which a property changes is called a

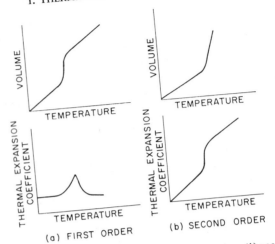

Fig. 1. First and second order transitions: (a) first order; (b) second ord[er]

second-order transition. Second-order transitions are typical of high[poly]mers especially those containing large amorphous regions. [Poly]property changes occurring at specific transition temperatures, the of the properties change continuously over a temperature range. Inc[rease] temperature causes increased vibration of molecules or chain seg[ments] This motion disturbs and weakens the cohesive forces between the [mole]cules. Decreased cohesion results in softening of the material, and [corre]spondingly affects other physical, optical, mechanical, and el[ectrical] properties.

1. Glass Transition Temperature (T_g)

The most prominent of several possible second-order transitio[ns in a] polymeric material is the glass transition temperature. The glass tra[nsition] temperature occurs at considerably lower temperatures than the cry[stalline] melting point (softening range) and can be considered an "internal [melting] point" (5). Beaman (6) established an empirical relationship betwe[en the] transition temperature (T_g) and melting point (T_m), so that

$$T_g/T_m \text{ (in °K)} = 0.58\text{--}0.78$$

Exceptions are highly crystalline polymers which are symm[etrically] substituted along the chain in pairs, such as polyethylene and po[lyvinyl]idene fluoride; here T_g/T_m ratios of 0.53 and 0.49, respectively, ha[ve been] found.

which reduce viscosity, lower the temperature necessary to freeze the molecular movements and lower T_g (26). Crosslinking increases the glass transition temperature, as nuclear magnetic resonance studies of styrene-divinylbenzene copolymers have indicated (61,87).

Comparison of the glass transition of various polymeric materials allows the influence of the molecular structure on the micromolecular mobility to be determined. The glass transition temperature can therefore serve as a means to determine chain stiffness and rigidity of the molecules. Flexible groups in the polymer chain, i.e., those which allow the chain segments to rotate freely, such as the oxygen atom, lower the glass transition temperature; rigid groups, such as p-phenylene groups, increase it (27). Increasing the cohesive forces between chains also raises the transition temperature. Orientation and crystallinity have a marked effect on the glass transition, while stereospecificity seems to be only of negligible influence (28). Polar and sterically hindering groups increase the transition temperature, while aliphatic sidechains decrease it (29).

Numerous methods have been used to determine the glass transition temperature of polymers. The results are not always in close agreement, but most fall into the same range. Not in all cases, however, can it be safely assumed that the basic molecular process of motion is the same. This is especially true when dielectric and mechanical dispersions are compared (30). It has to be considered in this case, since the difference in relaxation times (frequencies of the dielectric measurements) and the exclusive dependence on the dielectric dispersion from dipoles cannot possibly reflect the same vibration mechanism.

The method most widely used for the determination of the glass transition temperature is thermal expansion or determination of the specific volume by use of dilatometers (18,31–35). Dannis (36,37) used a differential expansion apparatus coupled with an X–Y recorder, which records expansion versus temperature. Figure 2 shows the curve obtained for uncured polybutadiene. The determination of elongation of films and its temperature dependence has been used by Jensen (38). Other mechanical determination methods include the use of penetrometers (27,39), torsional apparatus (28,40), vibrating reed (28,41), and thermal distortion tester (28). Specific heat, thermal conductivity (42), refractive index (23,43–45) have also been applied. Differential thermal analysis (see Section D-3) has also been used successfully to detect the glass transition and the melting point of powdered polyethylene terephthalate (52) (Fig. 3). By amplifying the sensitivity of the differential thermocouples, Keavney and Eberlin (53) could detect temperature changes of 0.01°C and determine accurately the glass transition temperatures of polyvinyl chloride, polystyrene, polymethyl methacrylate, polyvinylidene chloride, and polyacrylonitrile.

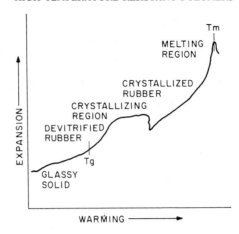

Fig. 2. Expansion versus temperature of uncured polybutadiene (37).

Nuclear magnetic resonance (46–51) is especially useful in detecting transitions below the glass transition temperature.

The glass transition temperature is often identified with the "brittle temperature," the temperature at which a material breaks upon cooling coupled with the sudden application of a load. Although, for high molecular weight polymers, glass transition temperature and brittle temperature are very similar, the brittle point test is an empirical test and may vary with the test conditions (54).

Table II lists glass transition temperatures and melting points for a number of crystalline polymers.

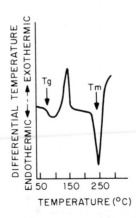

Fig. 3. Differential thermal analysis of polyethylene terephthalate (52).

TABLE II[a]
Glass Transition Temperatures and Melting Points
of Crystalline Polymers (4)

Polymers	$T_g(°C)$	$T_m(°C)$
Polypropylene	−18	165
Polybutene-1	−25	120
Polypentene-1	−24	70
Poly-3-methyl-1-butene	50	300
Poly-4-methyl-1-pentene	29	240
Polystyrene	90	230
Polyvinylcyclohexane	90	372
Polyallylbenzene	70	185
Isotatic polymethyl methacrylate	45	160
Polymethylyl isopropenyl ketone	114	240
Isotatic polyisopropyl acrylate	−11	162
Polybisphenol A carbonate	150	265
Polytetrafluoroethylene	−150, 127	327
Linear polyethylene	−85, −30	137
Polychlorotrifluoroethylene	50	220
Polyformaldehyde	−76, −73	179
trans-Polyisoprene	−67	62
cis-Polyisoprene	−75	30
cis-Polybutadiene	−108	2
trans-1,4-Polybutadiene	−14	145
Polydimethylsiloxane	−123	−80

[a] T_g = glass transition temperature; T_m = crystalline melting point.

2. Melting Point

The melting mechanism of high polymers has been generally considered to be a first-order transition. While sharp melting points have been observed for highly crystalline polymers, such as polyethylene, polyvinylidene chloride, polyamides, cellulose esters, and polytetrafluoroethylene (54), some authors (55) feel that a first-order transition requires the coexistence of several phases, and that a partially crystalline high polymer has to be considered a homogeneous system rather than a two-phase system (56). The work of Muenster (57) strongly argues that melting and crystallization of high polymers can be represented as a second-order transition. The fact that no long chain polymer is completely crystalline leads necessarily to a melting range rather than a melting point. The width of this range depends on the amount of crystallinity, the chain length, and the method used. Even within the crystalline regions, it has been observed by x-ray diffraction that the size of these regions does not decrease equally upon melting, but that certain parts melt considerably earlier than others (58).

It seems that the melting point of high polymers, as determined by various methods indicated later, represents the upper end of a melting range and is fairly well-defined as such. The melting point T_m can be defined as

$$T_m = \Delta H/\Delta S$$

where ΔH is the enthalpy of fusion and ΔS the entropy of fusion (59). The enthalpy can be identified with the heat of fusion, the entropy is related to the molecular order, symmetry, and flexibility of the chains.

Some structural elements of the polymer chain affect the melting point in the same direction as the glass transition temperature. For example, increased flexibility decreases both melting point and glass transition temperature. The higher melting point of polytetrafluorethylene as compared with that of polyethylene is believed to be caused by the stiffer chain of the first material, because of the higher potential barrier to free rotation in the CF_2 chain. Aliphatic polyesters melt below polyethylene, because of the flexibility of the in-chain oxygen atom. Benzene rings in the structure increase the rigidity and tendency to crystallize, and increase the melting point (60).

Other factors have opposite effects on melting point and glass transition: bulky sidegroups and branching decrease the melting point and increase the glass transition temperature. Symmetry increases the tendency for crystallization and therefore increases polymer melt temperatures. Polar groups, which increase the interchain attraction (cohesion), also increase the melting point; hydrogen bonding affects the melting point in the same direction.

The effect of the molecular weight on the melting point is of importance for the estimation of the softening behavior of a novel polymeric structure from the melting point of a low molecular weight model compound. The melting point increases first with increasing molecular weight and approaches asymptotically an endpoint (61). Billmeyer (62) expresses this by the empirical equation:

$$1/T_m = (a + b)/x$$

where x represents the chain length. Thus, the melting point depends only very slightly, above certain limits, on the molecular weight (63).

The melting phenomenon is accompanied by considerable changes in the value of a variety of properties. Measurement of all these properties can be used to determine the melting point. Edgar, Hill, and Ellery (60,64) used a penetrometer for determination of the melting points of polyethylene terephthalates, sebacates, and adipates. Other mechanical properties, such as Young's modulus and viscosity can also be used. Scott (52) determined the melting point of polyethylene terephthalate by the use of differential thermal analysis (see Fig. 3). Optical methods such as changes in birefringence with temperature have also been applied (65). Another method consists in determining the heating and cooling curves (temperature of the sample versus time) (66). When the melt of a polymer is cooled slowly, the exothermic heat of fusion at the melting point delays the cooling of the sample. Although cooling curves normally result in a melting *point*, heating curves rather indicate a temperature *range* of melting. Hysteresis between both curves results in differences between melting and freezing points of approximately 12°C for most polymers (67); not in all of the cases can this difference be reduced or eliminated by a highly reduced rate of cooling or heating. Measurement of the specific heat indicates a considerable increase at the melting point.

3. Nature of Property Changes Below the Melting Point

The properties of polymeric materials below their melting points determine their practical applicability. While a considerable amount of data is available on the properties of polymeric materials at room temperature, much less information has been published (in the literature) on the temperature dependence of the properties of various plastics, and often the data are not comparable because of variations in test conditions.

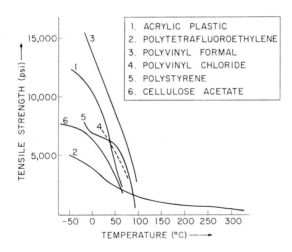

Fig. 4. Temperature dependence of tensile strength: *1*, acrylic plastic; *2*, polytetrafluoroethylene; *3*, polyvinyl formal; *4*, polyvinyl chloride; *5*, polystyrene; *6*, cellulose acetate (68a).

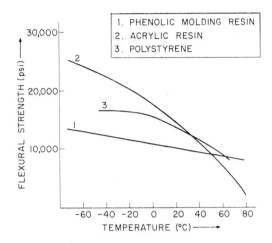

Fig. 5. Temperature dependence of flexural strength: *1*, phenolic molding resin; *2*, acrylic resin; *3*, polystyrene (68*a*).

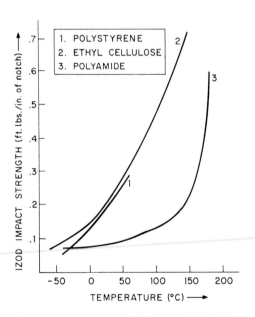

Fig. 6. Temperature dependence of Izod impact strength: *1*, polystyrene; *2*, ethyl cellulose; *3*, polyamide (68*a*).

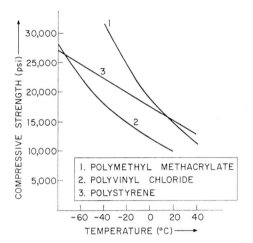

Fig. 7. Temperature dependence of compressive strength: *1*, polymethyl methacrylate; *2*, polyvinyl chloride; *3*, polystyrene (68*b*).

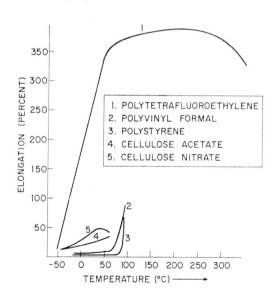

Fig. 8. Temperature dependence of elongation: *1*, polytetrafluoroethylene; *2*, polyvinyl formal; *3*, polystyrene; *4*, cellulose acetate; *5*, cellulose nitrate (68*a*).

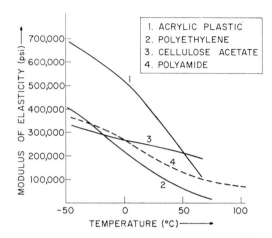

Fig. 9. Temperature dependence of modulus elasticity: *1*, acrylic plastic; *2*, polyethylene; *3*, cellulose acetate; *4*, polyamide (68*a*).

Figures 4–17 show the temperature dependence of some properties of plastics. These figures are presented only for the purpose of giving a general picture of the direction and tendency of the changes of some properties and the magnitude of their changes. The absolute values may vary, depending upon the test method used, the test conditions, and the specific composition of the material.

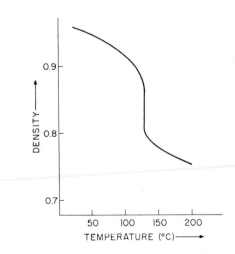

Fig. 10. Temperature dependence of the density of polyethylene (68*a*).

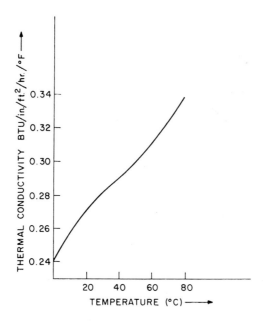

Fig. 11. Temperature dependence of the thermal conductivity of foamed polyurethane (68a).

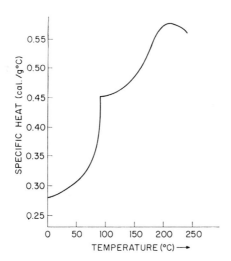

Fig. 12. Temperature dependence of the specific heat of polystyrene (68b).

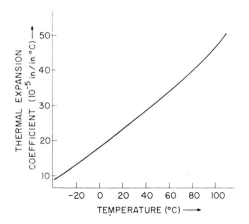

Fig. 13. Temperature dependence of the thermal expansion coefficient of polyethylene (68b).

D. IRREVERSIBLE PROPERTY CHANGES (DECOMPOSITION)

1. Factors Contributing to Stability

The strength of the chemical bond puts an upper limit on the vibrational energy that molecules may possess without bond rupture. Because the vibrational energy is increased by heat, thermal stability is related to the dissociation energy of the various bonds. The bond dissociation energies

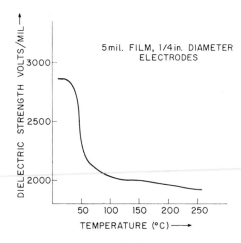

Fig. 14. Temperature dependence of the dielectric strength of polytetrafluoroethylene (68a).

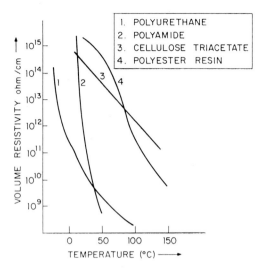

Fig. 15. Temperature dependence of volume resistivity: *1*, polyurethane; *2*, polyamide; *3*, cellulose triacetate; *4*, polyester resin (68a).

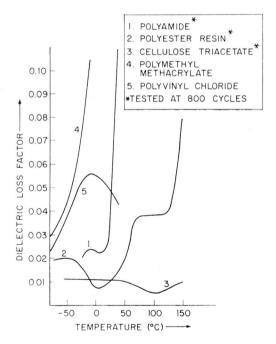

Fig. 16. Temperature dependence of the dielectric loss factor: *1*, polyamide*; *2*, polyester resin*; *3*, cellulose triacetate*; *4*, polymethyl methacrylate; *5*, polyvinyl chloride. *Tested at 800 cycles (68b).

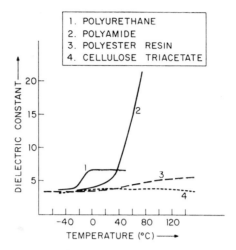

Fig. 17. Temperature dependence of dielectric constant: *1*, polyurethane; *2*, polyamide; *3*, polyester resin; *4*, cellulose triacetate (68*b*).

between two elements can be determined directly by measuring the energy of dissociation into atoms, or the heat of formation from the elements. More often, however, they are calculated from known dissociation energies, based on the postulate of the additivity of normal covalent bonds (69). Since a covalent bond A—B is similar in character to the bonds A—A and B—B, it is expected that the value of the bond energy A—B will be intermediate between the values for A—A and B—B. The arithmetic mean of the two bond energy values D(A—A) and D(B—B) would be

$$D(A-B) = \tfrac{1}{2}[D(A-A) + D(B-B)]$$

while the geometric mean is

$$D(A-B) = [D(A-A) \cdot D(B-B)]^{1/2}$$

The geometric mean leads to somewhat more satisfactory values than the arithmetic mean (69).

For simple organic compounds it has been found that the heat of formation can be calculated by summing the bond energies. These calculated values agree with the experimental values within a few kcal/mole for nearly all molecules.

Among the various investigators, there is considerable disagreement about the absolute values of bond energies. For example, bond energies in the following ranges have been reported for a few representative bonds (in kcal/mole):

$$\begin{array}{ll} \text{C—C:} \ 74\text{–}124 & \text{C—H:} \ 83\text{–}121 \\ \text{C}{=}\text{C:} \ 100\text{–}145 & \text{C—F:} \ 94\text{–}119 \\ \text{C—N:} \ 49\text{–}82 & \text{N—N:} \ 20\text{–}37 \\ \text{C}{=}\text{N:} \ 94\text{–}138 & \end{array}$$

However, at least, a general order of stability of the various bonds can be derived.

While the primary bond energies provide the greatest contribution to the stability of the compound, polymers may obtain additional strength and stability by secondary or van der Waals bonding forces. Secondary bonding forces determine the cohesion, the forces of attraction, between the chain molecules; these, in turn, affect the magnitude of the glass transition temperature and the melting point, but also, to a certain degree, the stability against thermal decomposition. The magnitude and efficiency of the secondary bonding forces depend upon the average chain length, the polarity, symmetry, and degree of orientation (54). Secondary bonding forces can be due to the attraction between unlike dipoles (dipole effect, up to 8.7 kcal/mole); they can be due to forces between permanent and induced dipoles (induction effect, up to 0.5 kcal/mole); and they can be caused by temporary displacements of nuclei and electrons during the vibrations, which cause attractive forces (dispersion effect, in the order of 2–6 kcal/mole). Finally, hydrogen bonding, the attraction of a hydrogen atom by rather strong forces to *two* atoms (preferably F, O, N) provides forces in the order of 6–10 kcal/mole and can be considered a strong secondary bond or a weak primary bond.

Another factor, which may contribute to the bond strength and therefore to the stability against decomposition, is the resonance stabilization of certain cyclic structures, such as benzene, pyridine, quinoline, etc. The resonance energy of these systems is in the order of 40–70 kcal/mole. Isoteniscope studies have indicated that these ring structures are in fact quite stable; they do not decompose noticeably below 1100–1200°F (70). Similarly, crosslinking and multiple bonding in a polymer also decrease or, at least, delay decomposition. Rearrangement to more stable structures may be another mechanism whereby polymer resistance to decomposition may be improved. On the other hand, sterically crowded structures decrease stability.

Unfortunately, irregularities in the structure, such as branching, chain ends, impurities, peroxidated groups, hydrogen in a position where it can be abstracted—in general, so-called weak links—are sites where, in most cases, the degradation reaction is initiated. The energy necessary for breaking such a weak link is often considerably lower than the energy required to break the primary bonds in question.

Another paramcter which has to be considered is the type of reaction which occurs after bond cleavage. If the activation energy of the next reaction step is high, recombination might occur rather than continued cleavage. Bond strength can therefore serve as a guide for the synthesis of polymers of higher thermal stability, but other factors, as indicated above, have also to be taken into consideration.

2. Types of Decomposition Reactions

The pathways of pyrolytic decomposition of organic polymers are so varied and complex as to render generalizations practically useless. In the polymer field, the random scission versus zip-down depolymerization has received some attention, with the conclusion that, if there are possibilities for a resonance stabilization of the intermediate fission product, chain depolymerization will be likely. Extensive studies by Madorsky, Wall, and co-workers have also led to interesting conclusions concerning the decomposition of rubbers, halogen-containing polymers, crosslinked polymers and polyamides. These workers have shown that isolated allylic groups, branch points and halogenated or oxygenated compounds break down more easily than aromatic substituted hydrocarbons. Compounds which allow five- or six-membered ring formation of split-off products are understandably unstable. Sensitivity to hydrogen abstraction, because of electronegativity of neighboring atoms or groups, or resonance stabilization by neighboring groups, also destabilizes a polymer. Substitution of labile hydrogens by more stable groups such as fluorine, methyl, phenyl, etc., offers an excellent way to improve pyrolytic stability. Mechanisms which lead to the formation of conjugated double bonds along the chain, e.g., the condensation of cyano groups in polyacrylonitrile or the fission of halogen acids from halo polymers, also enhance stability. In addition, polymeric materials may decompose by loss of side groups, may crosslink, especially under the influence of oxygen, or may rearrange to more stable or to less stable structures than the starting material.

According to Winslow and co-workers (71), polymer pyrolysis consists of two competitive processes, scission reactions to form low molecular weight fragments and gases, and recombination reactions resulting in the formation of polymeric carbon. While the first type of reaction occurs at relatively low temperatures, up to about 500°C, the formation of high molecular weight polymer carbon from fragmcnts left during the first process requires temperatures between 600 and 1000°C.

The scission reactions have been explored extensively for vinyl polymers,

I. THERMAL STABILITY OF POLYMERS   19

while little is yet known about the very complicated mechanism for the recombination reactions yielding polymeric carbon and graphite-like structures.

According to Grassie (72), one can distinguish between three different mechanisms of the depolymerization reaction of vinyl polymers: (*1*) random degradation, where chain scission occurs at random points along the chain leaving relatively large fragments; (*2*) stepwise depolymerization where an unzipping reaction starts from the chain end, producing exclusively monomer; and (*3*) reverse polymerization reaction where chain scissions at random points or at weak links yield free radicals which rapidly lose monomer until they are completely degraded, provided that the depolymerization reaction is not interrupted by a chain transfer. After scission at weak links occurs, the reverse polymerization process actually is identical to the stepwise depolymerization mentioned before. Determination of molecular weight of the residue and amount of monomer produced as a function of time gives an indication of which process prevails. During the first process, the molecular weight will decrease considerably before any monomer appears. In the second process, decrease in molecular weight and monomer production will be proportional, or no appreciable change in molecular weight will occur, depending upon whether the initiation reaction (see next paragraph) is fast or slow compared to the propagation reaction. The third process will give monomer without appreciable loss in molecular weight of the remaining product.

Many of the addition polymers, such as vinyl polymers, seem to decompose according to a reverse polymerization mechanism. The reaction scheme of such a mechanism follows:

$$\begin{array}{c} H\ H\ H\ H \\ \sim\!C\!-\!C\!-\!C\!-\!C \\ H\ X\ H\ X \end{array} \longrightarrow \begin{array}{c} H\ H \\ \sim\!C\!-\!C\cdot \\ H\ X \end{array} + \begin{array}{c} H\ H \\ \sim\!C\!-\!C\cdot \\ X\ H \end{array} \quad \text{Initiation}$$

$$\begin{array}{c} H\ H\ H\ H \\ \sim\!C\!-\!C\!-\!C\!-\!C\cdot \\ H\ X\ H\ X \end{array} \longrightarrow \begin{array}{c} H\ H \\ \sim\!C\!-\!C\cdot \\ H\ X \end{array} + \begin{array}{c} H\ H \\ C\!=\!C \\ H\ X \end{array} \quad \begin{array}{l}\text{Propagation}\\\text{(primary radical)}\end{array}$$

$$\begin{array}{c} H\ H \\ \sim\!C\!-\!C\cdot H \\ H\ X \end{array} + \begin{array}{c} H\ \ \ H\ H \\ \sim\!C\!-\!C\!-\!C\!-\!C\sim \\ X\ H\ X\ H \end{array}$$
$$\longrightarrow \begin{array}{c} H\ H \\ \sim\!C\!-\!CH \\ H\ X \end{array} + \begin{array}{c} H\ \ \ \cdot\ \ H\ H \\ \sim\!C\!-\!C\!-\!C\!-\!C\sim \\ X\ H\ X\ H \end{array} \quad \begin{array}{l}\text{Transfer}\\\text{(intermolecular)}\end{array}$$

$$\begin{array}{c} H\ H\ \ \ H\ H \\ \sim\!C\!-\!C\!-\!C\!-\!C\!-\!C\cdot H \\ X\ H\ X\ H\ X \end{array} \longrightarrow \begin{array}{c} H\ H\ \cdot\ H\ H \\ \sim\!C\!-\!C\!-\!C\!-\!C\!-\!CH \\ X\ H\ X\ H\ X \end{array} \quad \begin{array}{l}\text{Transfer}\\\text{(intramolecular)}\end{array}$$

$$\begin{array}{c} H\ \cdot\ H\ H \\ \sim\!C\!-\!C\!-\!C\!-\!C\sim \\ X\ H\ X\ H \end{array} \longrightarrow \begin{array}{c} H \\ \sim\!C\!=\!C \\ X\ H \end{array} + \begin{array}{c} H\ H \\ \cdot C\!-\!C\sim \\ X\ H \end{array} \quad \begin{array}{l}\text{Propagation}\\\text{(secondary radical)}\end{array}$$

```
                              Disproportionation      H     H  H
                         ┌─────────────────────→  ~C=C  +  HC—C~
  H  H      H  H         │                          X  H     H  X
 ~C—C·  +  ·C—C~         │                                            Termination
  X  H      H  X         │                        H  H  H  H
                         └─────────────────────→  ~C—C—C—C~
                              Combination          X  H  H  X
```

Transfer is much more important in decomposition reactions than in polymerizations, because the higher temperatures involved favor transfer. Because transfer reactions provide opportunities to form nonmonomeric material, they tend to reduce monomer yields (73,74). Tertiary hydrogen and chlorine atoms are very readily removed in transfer reactions. Quaternary carbon atoms are, as are tertiary carbons, more susceptible toward degradation than secondary carbon atoms. However, they do not contain hydrogen for transfer reactions; polymers which contain quaternary carbon are more likely to give large amounts of monomer than do those with tertiary carbon atoms. Polymethyl methacrylate produces more monomer than polymethyl acrylate and polymethacrylonitrile more than polyacrylonitrile.

Polymeric materials prepared by a condensation process, for example, polyesters, polyamides, etc., decompose according to random chain scission mechanism but seldom form products which can be considered as true monomers. Although the decomposition of particular condensation polymers has been studied, the mechanisms are so complex that generalizations such as those suggested for vinyl polymers are impossible.

However, the rate of the decomposition reaction is

$$\frac{dc}{dt} = -KC^n$$

where C is the concentration of the chemical species undergoing transformation, K is the rate constant, and n determines the order of reaction. It has been shown (75) that a mathematical function of the concentration can be found, regardless of the order of reaction, such that

$$f(C) = -Kt$$

When a property P can be found which is proportional to the concentration of the chemical constituent undergoing decomposition, then

$$P = f_0(C)$$

From the above two equations,

$$f'_0(P) = -Kt$$

If, therefore, the function of the property, $f'_0(P)$, determined at constant temperature over a period of time, is plotted versus time for several tem-

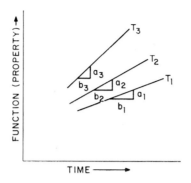

Fig. 18. Calculation of the rate constant.

peratures, the rate constant for the decomposition reactions at these temperatures will be

$$K = a/b \quad \text{(see Fig. 18)}$$

The chief factor determining the rate of a decomposition process is the activation energy, which is the energy necessary for the molecule to reach the activated state required for the dissociation (76). From the Arrhenius equation, the activation energy for the process can be calculated. From this equation

$$K = A \cdot e^{-E/RT}$$

$$\log K = \log A - \frac{E \log e}{R} \cdot \frac{1}{T}$$

where A is a constant, the frequency factor of the reaction, E is the activation energy for the specific decomposition reaction, and T is the absolute temperature, it follows that the expression $(E \log e)/R$ represents the slope of a plot of $\log K$ versus $1/T$ (Fig. 19):

$$E \log e / R = x/y$$

so that

$$E = \frac{(x/y)R}{\log e}$$

The activation energy so computed is seldom meaningful, however, because often the specific decomposition reaction is not known, or several reactions with different rates proceed simultaneously or successively in an overlapping fashion over the temperature range. In many cases the activation energy has meaning only as an average activation energy, valid only over a limited temperature range.

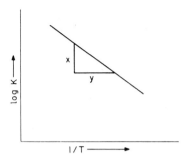

Fig. 19. Calculation of the activation energy.

3. Nature of Property Changes Caused by Decomposition

The nature of the temperature dependence of many reversible property changes can be predicted. It depends mainly upon the flexibility of molecular chains, which in turn increases with increasing temperature.

The situation is considerably more complex, when decomposition affects the properties. The two most important reactions, which may occur, are chain scission and crosslinking; both have adverse effects. In a decomposition process, such as the air aging of natural rubber, chain scission results in softening, whereas crosslinking causes embrittlement and, within narrow limits, in increased mechanical strength. A study of mechanical properties over an extended period of time at constant temperature or over an increasing temperature scale may reveal that the material first loses strength because of chain scission, later regains strength caused by crosslinking of the fragments, and finally, loses strength again because of complete disintegration or rearrangement. Continuous high temperature pyrolysis may cause carbonization, which would be expected to increase dielectric loss and constant, and decrease the dielectric strength. Normally, this is the case. However, the polysiloxanes show a decrease of loss and dielectric constant upon heat aging, possibly because the structure approaches that of silica as the organic components are lost by volatilization.

A prediction of the course of a property change of a material during aging is therefore difficult, unless information on related materials is already available.

E. DETERMINATION OF THERMAL STABILITY

The thermal stability of a material can be expressed either by temperature or by a temperature–time limit, within which a material can be used. Which one of these limits is used depends upon the application. Below a temperature where no noticeable decomposition occurs over any length of time, and where a reversible property determines the temperature limit, temperature alone can be used to define thermal stability. However, if the material is to be used within a temperature range where decomposition occurs, a temperature–time limit should be given. For example, the thermal stability of polystyrene will be somewhere in the range of 70–110°C, depending upon the application. Here, softening determines the endpoint of its use, before any decomposition occurs. The thermal stability of a polysiloxane, on the other hand, may be 1000 hr at 250°C, or 10 years at 180°C. When these aging periods are exceeded, the material becomes brittle and develops cracks because of decomposition.

A number of rapid test methods are available, in which the property change can be observed while the temperature is increased at a constant rate. Several methods related to the softening behavior of the material are in use. The methods of differential thermal analysis (DTA), thermogravimetric analysis (TGA), torsional braid analysis (TBA), along with the isoteniscope have been used to record the degree of decomposition of a material.

Since the definitive determination of thermal stability of a polymeric material depends on the specific application, application test methods have been developed. Since several property requirements determine utility in many applications, physical, mechanical, and electrical properties as time–temperature functions are of major importance.

1. Methods Related to Softening Behavior

A number of methods for determining the glass transition temperature and the melting point as intrinsic properties characterizing the reversible softening behavior have already been mentioned earlier in this chapter. The conventional methods of determining melting points of organic compounds (visual inspection of a sample in a capillary or in a Fisher-Johns melting point block) can also be used for polymeric materials, if these are available in the proper form. A softening range rather than a melting point will normally be obtained. The brittle point, a method of practical importance which is related to the glass transition temperature, has also been mentioned.

A great number of methods have been developed to determine, in well-defined temperature limits, the end point of the usability of a material due to softening. Although some methods determine a "flow" temperature, the majority of the methods are designed to determine a temperature limit of structural stability, which is (sometimes considerably) below the melting point of the polymeric material. Only a few of the more important methods are indicated in the following.

The ring and ball method belongs to the first type of methods mentioned above. According to the American Specification ASTM Spec E 28-51T (77), a disc of a sample held within a horizontal ring is forced downward a distance of one inch under the weight of a steel ball in water or glycerine heated at a constant rate.

The deflection temperature of plastics under load (ASTM D 648-56) (78a) is of the second type, determining the "heat distortion" of a material. A test bar, $\frac{1}{2}$-in. thick is placed on two supports 4 in. apart and is center-loaded to a final stress of 264 psi. The apparatus is immersed in an oil bath, the temperature of which is raised at 2°C/min. The temperature at which the bar has deflected 0.010 in. will be reported as the deflection temperature. Table III shows the ASTM deflection temperatures for a number of conventional plastics.

Another widely used heat distortion method is the *Martens-Method* (German Specifications DIN 53458 and 53462) (79). A bending moment of 711 psi is loaded at the top of a vertical bar clamped at the bottom end. The temperature is raised at a rate of 50°C/hr, and the temperature is noted at which a deflection of 0.24 in. is observed.

The *Vicat softening point* of plastics (American Specification ASTM D1525-85T) (78a) uses samples of a minimum width of $\frac{3}{4}$ in., and a thickness of $\frac{1}{8}$ in. These samples are subjected to a load of 1000 g by a flat ended needle of 1 mm^2 cross-sectional area. The sample is heated in an immersion bath at a rate of 50°C/hr. The temperature is recorded when the indicator reads 1 mm penetration. The Vicat method is advantageous because the shape of the test specimens is not critical.

The original German specification (VDE 0302/III.43) (79) for the Vicat method uses a 5-kg load. Even with this load, a penetration of 1 mm will

TABLE III
Deflection Temperatures of Polymers under Load (78b)
(ASTM D648)

Polymers	$T(°C)$
Polyethylene (66 psi)	41–82
Polystyrene	66–113
Polyvinyl acetate	38
Polyvinyl chloride (rigid)	54–74
Polyvinylidene chloride	54–66
Polymethyl methacrylate	66–99
Cellulose acetate (66 psi)	43–99
Polyester cast resin (filled)	116–>204
Epoxy cast resin	46–288
Phenol-formaldehyde resin	116–127

not be obtained with many thermosetting materials, especially laminates, because the material will break rather than soften to a sufficient degree. A modification has been suggested (80) whereby a penetration of 0.25 mm will be used as the criterion in conjunction with a 5-kg load.

ASTM deflection temperatures under load are of the order of 40–110°C for thermoplastics (unfilled), and up to 290°C for unfilled thermosetting resins such as phenolics, polyesters, and epoxies. Vicat temperatures normally are between 0 and 40°C higher. Both temperatures differ considerably for certain resilient materials such as polyamides.

2. The Isoteniscope

The principle of an isoteniscope involves heating the material in a sealed system under vacuum and the determination of the resultant vapor pressure. When a material volatilizes without decomposition, the instrument can be used to determine the temperature dependence of the vapor pressure. Decomposition of a compound can be measured by observing the isothermal pressure change. The temperature at which the first noticeable decomposition occurs, i.e., where the vapor pressure increases with increasing time, can be used as the criterion for stability. It has been suggested, however, that a specific rate of pressure increase, such as 1% per hr (81), be considered as the decomposition temperature because specific rates are easier to define and recognize than a determination of when the onset of decomposition occurs.

The isoteniscope method is very useful for determining the thermal stability of materials of low molecular weight and of liquid materials. Simple organic model compounds can be tested and will indicate the order of stability to be expected for a polymer of a similar structure, so that the isoteniscope can be used to guide research on new polymers. It should be realized, however, that the criterion for the thermal stability has been arbitrarily selected, and that the method is based on the assumption that the rate of production of gaseous products is directly proportional to the decomposition rate. This method does allow the determination of orders of stability and, of greater importance, of comparative stabilities of organic compounds. For example, by using this criterion, the stability of benzene, diphenyl, naphthalene, and diphenyl ether were established in the region of 540–600°C (70).

3. Differential Thermal Analysis (DTA)

Differential thermal analysis is a method for detecting transitions which are associated with heat loss or gain, or more precisely for detecting endo- or exothermic processes while heating the material at a constant rate of temperature increase.

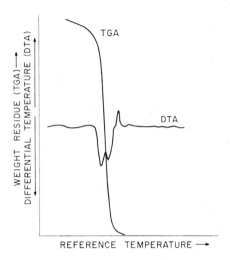

Fig. 20. TGA and DTA of polymethyl methacrylate (81).

In practice, a sample, often diluted with an inert material such as aluminum oxide, is heated in a furnace side by side with an inert reference material; the temperature is increased at a constant rate. The temperature difference between sample and reference material is then measured by a differential thermocouple and plotted versus time or versus temperature. The resulting curve indicates exo- and endothermic peaks, as shown in Figure 20 (81).

This method has been widely used to characterize inorganic materials (79). It has also been successfully used in a number of cases to determine transitions, such as the glass transition temperature and the melting point of polymers, as mentioned earlier in this chapter. Recently, DTA also has been used to indicate endo- or exothermic processes which can be related to degradation (81,82). Quite generally, sensitivity and reproducibility are highly affected by variables such as heating rate, dilution of the sample, design of the sample holder, and others.

4. Torsional Braid Analysis (TBA)

Torsional braid analysis (TBA) is a dynamic mechanical method for the investigation of polymers (83–87). The technique is based on the fact that changes in dynamic mechanical behavior with temperature (or other variables) of a composite made up of a mechanically inert substrate (the braid) and a polymer can be attributed to changes in the polymer. In contrast with classical methods of dynamic mechanical spectroscopy, for which geometrical specimens are fabricated, this technique permits the investigation of any type of polymer (or chemical compound) which can be deposited or synthesized on the braid. In some cases TBA has been utilized to determine first and second order transitions along with those related to decomposition. It has been suggested that the greatest utility of TBA may be in its use in conjunction with DTA and TGA.

5. Thermogravimetric Analysis (TGA)

Determination of the weight loss of a material as an indication of its decomposition is the most widely used, simple, and rapid method. Very often this method is performed in ordinary laboratory ovens in ambient air and over temperature and time ranges which are of interest for the application of a specific material. Sometimes, considerably higher aging temperatures are used and the results extrapolated to obtain the life expectancy at lower temperatures.

Normally, a certain percentage of weight loss is taken as a criterion, and the time necessary to reach this weight limit at a certain temperature or the temperature necessary to reach it after a certain time defines the stability. However, a certain rate of degradation or a sharp increase in rate can also serve to indicate the failure of a material, because the degradation rates of many materials change abruptly in certain temperature ranges (88,89). These rates can be obtained from a plot of weight loss versus aging temperature, such as in Figure 20; the "knee" in the curves indicates a sudden increase in rate.

Curves very similar to the weight loss versus aging-temperature plot are obtained conveniently in the thermobalance. In the thermobalance, the weight of a sample is recorded continuously while it is heated at a controlled rate, either in ambient air, controlled atmosphere, or in a vacuum. The thermobalances are either of the spring (88) or beam type (90–92). The weight change is recorded either photographically, or by using a servomechanism or a differential transformer. The final plot is one of residual weight versus the temperature of air close to the sample. It has been empirically found (93) that the TGA curve obtained at a heating rate of 150°C/hr corresponds roughly to aging data obtained from isothermal aging over 10–60 min periods; this means that the weight loss after reaching a certain temperature in the thermobalance would be the same as if the material had been exposed to this temperature alone over a period of 10–60 min.

Although the shape alone of the weight loss versus temperature curve is of considerable interest, attempts have also been made to establish criteria for defining thermal stability and to evaluate the curves on a quantitative basis. A major breakdown temperature, if available, can be used for an endpoint definition, or the selection of a certain rate of decomposition, such as 10% weight loss per hour. By integrating the area under the weight loss curve a value is obtained which can be used as a quantitative basis for comparing several materials. Doyle (94) established an "integral procedural decomposition temperature" on the basis of the area under the curve and the residual weight at 900°C. A differential plot of the decomposition rates versus temperature yields curves with one or several peaks, indicating the temperatures where the most rapid breakdown occurs.

Thermogravimetric analysis is an easy, rapid, and very reproducible method for screening the stability characteristics of polymers up to temperatures well above 1000°C (82,95). Using this method Ehlers (96) has studied over 450 experimental polymers and suggested relative thermal stability indices along with correlations between structure and stability.

In all of the thermal stability indices in which weight loss is the criterion of stability, more emphasis is placed on the thermal stability of the reaction products rather than on the starting material itself. Bloomfield (97) has argued against this emphasis by indicating that the actual percentage weight loss is not important unless it can be related to the degradation process, e.g., a dimethylsilicone polymer can be completely degraded to give silica with only a 20% weight loss.

It is apparent that preliminary thermal analysis of a polymer should involve several complementary techniques. No one technique responds to all the transitions and transformations which occur in a polymer system. Furthermore, the magnitude of response of a technique to a change will differ with technique and type of change. Thus, the most significant route to relative thermal stability indices for polymers would be the use of several of these methods, especially TBA, DTA, and TGA.

6. Application Methods

It should be emphasized, however, that no temperature or temperature–time limit of general validity can be given even for a very well-defined plastic material with known history, because the specific application in mind always determines the kind and magnitude of the required properties. The most common methods used to determine the thermal stability for a specific application are determinations of physical, mechanical, and electrical properties in temperature-dependent (reversible) and temperature/time dependent (irreversible) instances. Criteria are established in the form of property limits, which are derived from practical experience. The temperature limits or the temperature–time limits corresponding to these criteria are determined.

The fact that the tests are often destructive, or that repeated testing at several temperatures would incorporate an uncontrolled aging effect, requires in most cases that a new sample be used for each temperature to be

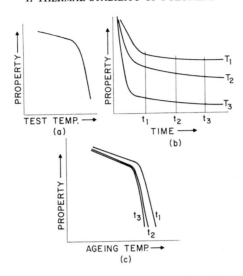

Fig. 21. Property versus test temperature, time, and aging temperature.

tested. If tests are carried out at temperatures higher than room temperature, an oven is needed to keep the sample at constant temperature for the test. Normally, a half-hour exposure to the temperature should be sufficient to bring the sample to a constant temperature and to perform the measurement. The property often will be determined at several temperatures of interest, and the data plotted as property versus test temperature (Fig. 21a).

When the effect of aging is to be determined, a number of approaches are possible. Aging can be achieved at one or at several temperatures, and for these tests, conventional laboratory ovens are usually used. The aged samples, in turn, can be tested at room temperature, at one or several higher temperatures (very often the aging temperature is used as the testing temperature) or over a temperature range of interest. The test result should clearly indicate this, for example:

flexural strength (psi) at 25°C
flexural strength (psi) at 200°C
flexural strength (psi) at 25°C after aging (750 hr 200°C)
flexural strength (psi) at 200°C after aging (750 hr 200°C)

Measurements of this kind will result in a plot of property versus time for various aging temperatures (T_1–T_3 in Fig. 21b), determined at a certain temperature. Data from these curves, taken after several aging times (t_1, t_2, t_3) can be plotted as property versus aging temperature (Fig. 21c).

The difficulties involved with data scattering of thermal aging data, can be avoided by superimposing data for several aging temperatures onto one master aging curve (98), thereby gaining a larger number of data points and giving a better representation of the true curve.

For specific end use applications, practical considerations have dictated the particular parameters of physical, chemical, mechanical, and/or electrical properties which are utilized as measurements of thermal stability of the polymeric material as a time–temperature function. We shall now consider these tests as related to a particular application.

In adhesive applications (99,100), lap shear strength of the adhesive joint over a temperature range both before and after heat aging is the most commonly used measurement. Not only is high initial tensile shear strength over a temperature range important but retention of high strength over a temperature range after long exposure at elevated temperature is required.

In coating applications such as wire enamels and varnishes (101,102), retention of tensile and electrical properties over a temperature range after long-term exposure to elevated temperature is the most used parameter. Such properties of the coating as dielectric strength, dielectric constant, dissipation factor, abrasion resistance, and cut-through resistance, must be retained, after long-term exposure to elevated temperature.

For composite structures (101,102), initial flexural strengths and electrical properties over a temperature range both before and after heat aging are the most commonly utilized tests. In these applications, the laminate must possess good tensile and tear strength which along with such electrical properties as dielectric strength, dielectric constant, power factor, dissipation factor and volume resistivity must be maintained at elevated temperatures even after long-term exposure at high temperatures.

In fiber applications (103,104), tensile properties of the fiber over a temperature range both before and after heat aging are the most widely used measurements. The retention of initial tenacity, elongation, and initial modulus of the fiber, both straight and loop, at elevated temperatures even after long-term exposure to elevated temperature are the most important requirements.

For film applications (101,105,106), mechanical and electrical properties of the film over a temperature range both before and after exposure to elevated temperature are considered to be the important parameters. Such mechanical properties of the film as tensile strength, elongation, tensile modulus, and creasibility along with such electrical properties as dielectric constant, dissipation factor, volume resistivity, and dielectric strength must be retained even at elevated temperature after heat aging.

For paper applications (103,104), as for film applications, mechanical and electrical properties of the paper over a temperature range both before and after exposure to heat aging are the most commonly used measurements. Here again, such mechanical properties of the paper as tensile strength, elongation, tear strength, edge tear strength coupled with such electrical properties as enumerated in the previous paragraph must be retained even at elevated temperature after long-term high temperature exposure.

Once it had been established that decomposition is a function of time and temperature, some estimate of the life expectancy at lower temperatures from high temperature aging data can be made since the function of a property, $f(P)$, can be expressed as

$$f(P) = -Kt$$

where K is the rate constant and t is time. If a certain property value such as 50% of the original flexural strength is taken as the criterion for failure,

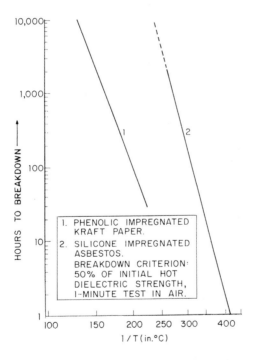

Fig. 22. Life expectancy as a function of temperature: *1*, phenolic impregnated kraft paper; *2*, silicone impregnated asbestos. Breakdown criterion: 50% of initial hot dielectric strength, 1-min test in air (107).

$f(P)$ can be considered a constant and the time to reach this value takes the place of the rate constant K in the Arrhenius equation:

$$f(P)/t = Ae^{-E/RT}$$

$$\log \frac{f(P)}{t} = \log A - (E \log e/RT)$$

It follows that the logarithm of the times necessary to reach the limiting value at various temperatures plotted versus the reciprocal of the corresponding temperatures (in °K) should give a straight line. Extrapolation of these lines to lower temperature gives an approximation of the life expectancy at these temperatures. Figure 22 shows an example of this.

Although not highly accurate, this exponential law indicates that, at higher temperatures, differences in temperature have a relatively larger effect on rate constant and life expectancy than at lower temperatures. This is actually the case. The Arrhenius law has been used very extensively and successfully in estimating the life of electrical insulations for electrical machinery and transformers.

F. STRUCTURE AND THERMAL STABILITY

In the intensive research initiated in the past several years in an attempt to synthesize polymer systems capable of withstanding prolonged exposure to elevated temperatures, the effort has been three-pronged:

1. Improve existing polymers by introducing structural modifications.
2. Devise new organic systems tailored to resist the effects of heat.
3. Synthesize an entirely new class of inorganic and organic-inorganic (semiorganic) polymers which would be thermally stable.

In the first approach, (*1*) conventional organic polymers consisting of inherently flexible chains were modified such that the practical use-temperature ceiling was raised or (*2*) intractable, inherently inflexible polymers were modified to yield more tractable systems. In the former case, polymers such as polyamides, polyesters, etc., were synthesized containing carbocyclic rings, preferably *p*-phenylene, alternating in the polymer chain with connecting amide, ether, sulfide, ester, carbonate, or other flexible units such that the use-temperatures were raised to 400°C or above. In the latter case, intractable polymers such as poly-*p*-phenylenes were modified by the insertion of methylene units or short chains of methylene units between the phenylene rings such that the resulting polymers were then fabricable.

In the second method of attack, the most fruitful of these three lines of research, a whole new generation of carbon-based polymers with inherently rigid chains has been developed. These materials were prepared by incorporating highly stable, rigid, aromatic carbocyclic or heterocyclic ring systems directly into the polymer chain. As with organic polymers in general, an infinite variety of such structures is possible, and many have been synthesized. For the most part, the synthetic routes to such new polyaromatic heterocyclic polymers involved syntheses via a novel two-step process in which soluble high molecular weight prepolymers were first synthesized and then rigid stable rings were formed by thermally or chemically induced condensation of reactive groups on the polymer chains.

The third line of attack, the synthesis of inorganic and organic-inorganic (semiorganic) polymers, was prompted largely by the unusual success of the silicones developed in the early 1940's, and the fact that many bond types are stronger than the carbon–carbon bond. Research efforts have been aimed at the synthesis of stable, inorganic polymeric materials having linear chains consisting of such typical repeating units as silicon–nitrogen, boron–nitrogen and phosphorus–nitrogen. In addition, the inorganic-organic (semiorganic) polymers having inorganic chains framed by organic substituents, such as in the silicones, or organic units as members of the chains themselves, have been thoroughly investigated. Much of the effort in this field has been aimed at the replacement of silicon in silicone-like structures by elements such as aluminum, titanium, tin, and boron. Also, considerable effort has been expended on the synthesis of metal chelate polymers. Such polymers have been prepared either via the polymerization of difunctional or tetrafunctional intermediates with suitable metal ions or by the postreaction of high molecular weight organic polymers containing suitable functional groups with chelating metal ions.

Falling in the classes of both organic and inorganic polymers are specialized types of rigid-chain polymers, ladder polymers. These polymers are double strand structures which consist of two polymer chains periodically bound together by chemical bonds. In principle these materials should show superior thermal stability because the polymer chains cannot be severed by a single bond-breaking reaction. In this general area of research both inorganic and organic ladder polymers have been prepared.

In the remainder of this book these recent developments will be examined in detail. The aforementioned polymeric compositions, their preparation and properties, will be described and discussed. Since the ultimate test of the thermal stability of any of these polymers is in a use or application, the final chapter will be devoted to these new polymeric structures in applications and in use tests.

References

1. R. Bartholomew, G. R. Eykamp, and W. E. Gibbs, *Rubber Chem. Technol.*, **32**, 1587 (1959).
2. L. D. Jaffe, *Chem. Eng. Progr. Symp.*, **59** (40), 81 (1963).
3. J. H. Ross, *Textile Res. J.*, **32**, 768 (1962).
4. (*a*) H. A. Stuart, *Die Physik der Hochpolymeren*, Vol. 3, Springer, Berlin, 1955, p. 468. (*b*) R. F. Boyer, *Rubber Chem. Technol.*, **36**, 1303 (1963). (*c*) S. D. Burck, *J. Polymer Sci. B*, **4**, 933 (1966). (*d*) S. Krause, J. J. Gormby, N. Roman, J. A. Shetter, and W. H. Watanabe, *J. Polymer Sci. A*, **3**, 3573 (1965). (*e*) S. Krause and N. Roman, *J. Polymer Sci. A*, **3**, 1631 (1965). (*f*) A. Eisenberg, H. Farb, and L. G. Cool, *J. Polymer Sci. A-2*, **4**, 855 (1966). (*g*) T. J. Dudek and J. J. Lohr, *J. Appl. Polymer Sci.*, **9**, 2489 (1965). (*h*) S. C. Temin, *J. Appl. Polymer Sci.*, **9**, 471 (1965).
5. K. Ueberreiter, *Angew. Chem.*, **53**, 247 (1940).
6. R. G. Beaman, *J. Polymer Sci.*, **9**, 470 (1952).
7. R. L. Miller, *Polymer*, **1**, 135 (1960).
8. A. Nishioka, Y. Koike, M. Owaki, T. Naraba, and Y. Kato, *J. Phys. Soc. (Japan)*, **15**, 416 (1960).
9. J. G. Powles, *J. Polymer Sci.*, **22**, 79 (1956).
10. J. A. Sauer, R. A. Wall, N. Fushchillo, and A. E. Woodward, *J. Appl. Phys.*, **29**, 1385 (1958).
11. K. M. Sinnot, *J. Polymer Sci.*, **42**, 3 (1960).
12. W. P. Slichter and E. R. Mandell, *J. Appl. Phys.*, **30**, 1473 (1959).
13. W. P. Slichter and E. R. Mandell, *J. Appl. Phys.*, **29**, 1438 (1958).
14. A. E. Woodward, Paper presented at the ACS Meeting, Chicago, September, 1961.
15. R. S. Spencer and R. F. Boyer, *J. Appl. Phys.*, **17**, 398 (1946).
16. F. W. Billmeyer, Jr., *Textbook of Polymer Chemistry*, Interscience, New York, 1957, p. 42.
17. R. F. Boyer and R. S. Spencer, *J. Appl. Phys.*, **16**, 594 (1945).
18. T. G. Fox and P. J. Flory, *J. Appl. Phys.*, **21**, 581 (1950).
19. E. Jenckel and K. Ueberreiter, *Z. Physik. Chem. (Leipzig)*, **A182**, 361 (1938).
20. E. H. Merz, L. E. Nielsen, and R. Buchdahl, *Ind. Eng. Chem.*, **43**, 1936 (1951).
21. K. Ueberreiter, *Z. Physik. Chem.*, **B45**, 25 (1950).
22. Ueberreiter, *Z. Physik. Chem.*, **B45**, 361 (1940).
23. R. H. Wiley and G. M. Brauer, *J. Polymer Sci.*, **3**, 647 (1948).
24. R. H. Wiley and G. M. Brauer, *J. Polymer Sci.*, **11**, 221 (1953).
25. R. H. Wiley, G. M. Brauer, and A. R. Bennett, *J. Polymer Sci.*, **5**, 609 (1950).
26. (*a*) Ref. 4*a*, p. 630. (*b*) J. A. Faucher, *J. Polymer Sci. B*, **3**, 143 (1965).
27. O. B. Edgar, *J. Chem. Soc.*, **1952**, 2638.
28. S. Newman and W. P. Cox, *J. Polymer Sci.*, **46**, 29 (1960).
29. (*a*) Ref. 4*a*, p. 654. (*b*) Ref. 4*a*, p. 643.
30. C. J. Aloiso, S. Matsurka, and B. Maxwell, *J. Polymer Sci. A-2*, **4**, 113 (1966).
31. N. Bekkendahl, *J. Res. Natl. Bur. Std.*, **3**, 411 (1934).
32. R. F. Boyer, and R. S. Spencer, *J. Appl. Phys.*, **15**, 398 (1944).
33. J. D. Ferry and G. S. Parks, *J. Chem. Phys.*, **4**, 70 (1936).
34. L. Mandelkern, G. M. Martin, and F. A. Quinn, Jr., *Bull. Am. Phys. Soc.*, **1**, 123 (1956).
35. F. E. Wiley, *Ind. Eng. Chem.*, **34**, 1052 (1942).

36. M. L. Dannis, *J. Appl. Polymer Sci.*, **1**, 121 (1959).
37. M. L. Dannis, *J. Appl. Polymer Sci.*, **4**, 249 (1960).
38. P. W. Jensen, *J. Polymer Sci.*, **28**, 635 (1958).
39. F. Rybnikar, *J. Polymer Sci.*, **28**, 633 (1958).
40. L. E. Nielsen, R. E. Pollard, and E. McIntyre, *J. Polymer Sci.*, **5**, 661 (1951).
41. M. Baccaredda and E. Butta, *J. Polymer Sci.*, **22**, 217 (1956).
42. K. Ueberreiter and S, Nems, *Kolloid-Z.*, **123**, 92 (1951).
43. R. H. Wiley, *J. Polymer Sci.*, **2**, 10 (1947).
44. R. H. Wiley and G. M. Brauer, *J. Polymer Sci.*, **3**, 455 (1948).
45. R. H. Wiley and G. M. Brauer, *J. Polymer Sci.*, **3**, 704 (1948).
46. C. M. Huggins and D. R. Carpenter, WADD Tech. Rept. 61-225, Aeronautical Systems Division, Wright-Patterson AFB, Ohio, 1961.
47. A. Odajima, J. Sohma, and M. Koike, *J. Phys. Soc., Japan*, **12**, 272 (1957).
48. J. G. Powles, *Polymer*, **1**, 219 (1960).
49. J. A. Sauer and A. E. Woodward, *Rev. Mod. Phys.*, **32**, 88 (1960).
50. W. P. Slichter, *Fortschr. Hochpolymer. Forsch.*, **1**, 35 (1958).
51. (*a*) A. E. Woodward, J. A. Sauer, and A. Odajima, *Bull. Am. Phys. Soc.*, **6**, 133 (1961). (*b*) K. Ito and Y. Yamashita, *J. Polymer Sci. B*, **3**, 625 (1965). (*c*) Y. Abe, M. Tasumi, T. Shimanouchi, S. Satok, and R. Chujo, *J. Polymer Sci. A-1*, **4**, 1413 (1966).
52. N. D. Scott, *Polymer*, **1**, 114 (1960).
53. J. J. Kearney and E. C. Eberlin, *J. Appl. Polymer Sci.*, **3**, 47 (1960).
54. A. X. Schmidt and C. A. Marlies, *Principles of High Polymer Theory and Practice*, p. 182, McGraw-Hill, New York, 1948.
55. Ref. 4*a*, p. 557.
56. E. Jenckel, *Kunstoff*, **43**, 454 (1953).
57. A. Muenster, *Z. Physik. Chem. (Leipzig)*, **1**, 259 (1954).
58. Ref. 4*a*, p. 454.
59. Ref. 4*a*, p. 579.
60. O. B. Edgar and E. Ellery, *J. Chem. Soc.*, **1952**, 2633.
61. Ref. 4*a*, p. 586.
62. Ref. 16, p. 27.
63. H. F. Mark, ASTM Bull. No. 245, 31 (April, 1960).
64. O. B. Edgar and R. Hill, *J. Polymer Sci.*, **3**, 1 (1952).
65. Ref. 4*a*, p. 44.
66. Ref. 4*a*, p. 441.
67. P. W. Jensen, *J. Polymer Sci.*, **28**, 635 (1958).
68. (*a*) *Technical Data on Plastics*, Manufacturing Chemists Association, Inc., Washington, D.C., 1957. (*b*) H. K. Nason, T. S. Carsivell, and C. H. Adams, *Mod. Plastics*, **29**, 4, 127 (1951).
69. L. Pauling, *The Nature of the Chemical Bond*, Cornell University Press, Ithaca, N.Y., 1948.
70. J. W. Dale, J. B. John, E. A. McElhill, and J. O. Smith, WADC Tech. Rept. 59–95, Aeronautical Systems Division, Wright-Patterson AFB, Ohio, 1959.
71. F. H. Winslow, W. O. Baker, N. R. Pope, and K. W. Matreyek, *J. Polymer Sci.*, **16**, 101 (1955).
72. N. Grassie, *The Chemistry of High Polymer Degradation Processes*, Butterworths, London, 1955, p. 25.
73. B. D. Achhammer, M. Tyron, and G. M. Kline, *Mod. Plastics*, **36**, 4, 131 (1959).
74. L. A. Wall and R. E. Plorin, *J. Res. Natl. Bur. Std.*, **60**, 455 (1958).

75. T. W. Dakin, *AIEE Trans.*, **67**, 113 (1948).
76. Ref. 54, p. 51.
77. A. E. Lever and J. Rhys, *The Properties and Testing of Plastic Materials*, Chemical Publishing Co., New York, 1958.
78. (*a*) *ASTM Standards on Plastics*, Philadelphia, Pa., American Society for Testing Materials (1959). (*b*) *Modern Plastics Encyclopedia*, Plastics Properties Chart, New York, Plastic Corp., 1965.
79. DIN Tacshenbuch 21, *Kunststoffnormen*, Berlin, Beuth Vertrieb GmbH, 1955.
80. G. F. L. Ehlers, *ASTM Bull.*, **236**, 54 (1959).
81. C. D. Doyle, WADD Tech. Rep. 60-283, Aeronautical Systems Division, Wright-Patterson AFB, Ohio, 1960.
82. (*a*) W. J. Smothers and Y. Chiang, *Differential Thermal Analysis*, Chemical Publishing Co., New York, 1958. (*b*) B. Ke, *Newer Methods of Polymer Characterization*, Interscience, New York, 1964. (*c*) W. W. Wendlant, "Thermal Methods of Analysis," *Chemical Analysis*, Vol. 19, Interscience, New York, 1965. (*d*) P. E. Slade, Jr. and L. T. Jenkins, *Thermal Analysis*, Dekker, New York, 1966. (*e*) J. D. Matlack and A. P. Metzger, *J. Polymer Sci. B*, **4**, 875 (1966). (*f*) L. Reich, *J. Appl. Polymer Sci.*, **10**, 465, 813, 1033, 1801 (1966). (*g*) R. M. Perkins, G. L. Drake, and W. A. Reeves, *J. Appl. Polymer Sci.*, **10**, 1041 (1966). (*h*) R. H. Still and C. J. Keattch, *J. Appl. Polymer Sci.*, **10**, 193 (1966).
83. A. F. Lewis and J. K. Gillham, *J. Appl. Polymer Sci.*, **6**, 422 (1962).
84. A. F. Lewis and J. K. Gillham, *J. Appl. Polymer Sci.*, **7**, 685 (1963).
85. J. K. Gillham and A. F. Lewis, *Nature*, **195**, 1199 (1962).
86. J. K. Gillham and A. F. Lewis, *J. Appl. Polymer Sci.*, **7**, 2293 (1963).
87. (*a*) J. K. Gillham and A. F. Lewis, *Science*, **139**, 494 (1963). (*b*) A. Adicoff and A. A. Yukelson, *J. Appl. Polymer Sci.*, **10**, 159 (1966).
88. G. F. L. Ehlers, *Elektrotech. Z.* **75**, 469 (1954).
89. L. E. Sieffert and E. M. Schoenborn, *Ind. Eng. Chem.*, **42**, 496 (1959).
90. Cooke, Troughton and Simms, Inc., "Instructions for the Assembly and Use of the Chevenard Thermobalance."
91. C. Duval, *Inorganic Thermogravimetric Analysis*, Elsevier, New York, 1953.
92. E. L. Simons, A. E. Newkirk, and J. Aliferis, *Anal. Chem.*, **59**, 48 (1957).
93. E. Ehlers, WADD Tech. Rep. 620261, Aeronautical Systems Division, Wright-Patterson AFB, Ohio, 1961.
94. C. D. Doyle, *Anal. Chem.*, **33**, 77 (1961).
95. (*a*) J. Mitchell, I. M. Rolthoff, and A. Weissberger, *Organic Analysis*, Vol. 4, Interscience, New York, 1960. (*b*) L. Reich, J. T. Lee, and D. W. Levi, *J. Appl. Polymer Sci.*, **9**, 351 (1965). (*c*) L. Reich, *J. Appl. Polymer Sci.*, **9**, 3033 (1965).
96. G. F. L. Ehlers, Paper presented at Twelfth Canadian High Polymer Forum, May, 1964.
97. P. R. Bloomfield, *Thermal Degradation of Polymers*, SCI Monograph No. 13, Society of Chemical Industry (1961), p. 89.
98. C. D. Doyle, *Mod. Plastics*, **34** (II), 141 (1957).
99. H. H. Levine, *J. Appl. Polymer Sci.*, **6**, 184 (1962).
100. *Mater. Design Eng.*, **57**, 92 (1963).
101. L. W. Frost and G. M. Bower, *Annual Report, 1962, Conf. Elec. Insul.*, NAS-NRD, 45 (1963).
102. J. H. Freeman, E. J. Traymor, J. Miglarese, and R. H. Lunn, *SPE Trans.*, **2**, 216 (1962).
103. L. K. McCune, *Textile Res. J.*, **32**, 762 (1962).

104. W. R. Clay, EIEET-136-12, Electrical Insulation Conference, Washington, Feb., 1962.
105. H. C. Stewart, L. C. Whitman, and A. L. Scheidler, *Power Apparatus and Systems*, No. 5, 267, 1953.
106. L. E. Amborski. *Am. Chem. Soc., Div. Polymer Chem., Preprints*, **4**, No. 1, p. 175 (1963).
107. S. V. Abramo, C. E. Berr, W. M. Edwards, A. L. Endrey, K. L. Oliver, and C. S. Sroog, *J. Polymer Sci. A*, **3**, 1373 (1965).

II. AROMATIC POLYMERS

A. INTRODUCTION

As discussed in Chapter I, ring structures stabilized by resonance have been shown by isoteniscopic measurements to be stable to over 600°C (1). For p-terphenyl, the amount of pyrolysis products at 400°C amounts to only 3% (2,3). Since the melting point of p-phenylenes are strikingly high—namely, p-tetraphenyl, 320°C, p-pentaphenyl, 395°C, p-hexaphenyl, 475°C—it has been suggested that high molecular weight poly-p-phenylene would be infusible and should be stable up to temperatures of 800–900°C (2–4).

In this chapter, poly-p-phenylene and related polymers of the general formula, $+AR-R+$, where R is a hydrocarbon radical such as methylene, ethylene, etc., will be discussed. Emphasis in this discussion will be placed on the preparation and properties of these polymers, along with an intercomparison of thermal stability.

B. POLY-p-PHENYLENES

1. Preparation

The first reported synthesis and characterization of poly-p-phenylene with molecular weight in excess of 1000 was that of Goldfinger and co-workers (5–7). Using the Wurtz-Fittig reaction of 1,4-dichlorobenzene with metallic sodium or a liquid potassium–sodium alloy in dioxane solution, these workers obtained a benzene-soluble fraction which did not melt up to 550°C.

$$Cl-\langle\bigcirc\rangle-Cl \xrightarrow[\text{or K-Na}]{\text{Na}} +\langle\bigcirc\rangle+_x \qquad (1)$$

By cryoscopic measurements and chlorine analysis this material was shown to have an average degree of polymerization (DP) of approximately 34 which corresponds to a molecular weight of between 2200 and 2800.

Using a similar reaction scheme, Hellmann and co-workers (8,9) prepared polyperfluoro-p-phenylene. In this work, 1,4-dibromo- or 1,4-diiodo-2,3,5,6-tetrafluorobenzene was reacted with copper at 200–250°C for up to 80 hr (see reaction 2) where X = Br or I. The insoluble portion of the product, which did not melt at 555°C, was shown to have an average DP of 4–5 or a molecular weight of about 850 by bromine or iodine analysis, respectively.

$$X-\underset{F\ F}{\overset{F\ F}{\bigcirc}}-X \longrightarrow \left[\underset{F\ F}{\overset{F\ F}{\bigcirc}}\right]_n \qquad (2)$$

Similar polyaromatics have been prepared by the reaction of 4,4'-biphenylyl bisdiazonium salts or their derivatives with cuprous salts (10,11). From the bisdiazonium salt of benzidine an infusible, nitrogen-containing, reddish-brown product was obtained which had the composition **1**.

$$\left[\left[\bigcirc\right]_9\left[\bigcirc\right]-N=N\right]_1$$
1

The decomposition of the 4,4'-bisdiazonium benzidine-2,2'-dicarboxylic acid, gave a polymer of structure **2**.

$$\left[\underset{CO_2H\ CO_2H}{\bigcirc-\bigcirc}\right]_x$$
2

Structure **2** was an infusible reddish-brown powder, soluble in N,N-dimethylformamide. Heating **2** at 330° for $1\frac{1}{2}$ hr, resulted in a 33% weight loss with formation of an insoluble product presumed to be a poly-p-phenylene.

Polymerizations to yield polyphenylenes have also been reported to occur (*1*) on heating p-dibromobenzene with activated copper (12), (*2*) in the catalytic hydrogenation of dibromobenzene in the presence of methanol (13), (*3*) on heating benzene and biphenyl in the presence of sulfur at 650–950°C (14), (*4*) in the reaction of dilithiobenzene with heavy metal halides (15,16), and (*5*) in the electrolysis of phenylmagnesium bromide (17).

One of the more successful routes to poly-p-phenylenes is via the dehydrogenation of polycyclohexadienes derived from the polymerization of 1,3-cyclohexadiene.

$$\left(\underset{x}{\bigcirc} \right) \xrightarrow{-H} \left(\underset{x}{\bigcirc} \right) \quad (3)$$

Hoffmann and Damm (18) first reported having obtained a polymer by heating 1,3-cyclohexadiene in a sealed tube. Subsequently many studies have dealt with the polymerization of 1,3-cyclohexadiene in the presence of peroxides (19–23), diisobutylene oxide (24), phosphoric anhydride (25), aluminum chloride (26), and azobisisobutyronitrile (27). However, Marvel and Hartzell (28) first reported the preparation of high molecular weight poly-1,3-cyclohexadiene.

These latter workers reacted 1,3-cyclohexadiene with a Ziegler catalyst derived from triisobutylaluminum and titanium tetrachloride in normal heptane or cyclohexane at reaction temperatures of 25°C or lower, to prepare essentially linear polymers containing recurring cyclohexene units.

$$\bigcirc \longrightarrow \left(\underset{x}{\bigcirc} \right) \quad (4)$$

These polymers were amorphous, waxy powders melting at 170–180°C with inherent viscosities in the range of 0.11–0.16 which corresponded to a molecular weight of between 5000 and 10,000.

In subsequent work, the polymerization of 1,3-cyclohexadiene was studied in the presence of titanium tetrachloride (29,30), boron trifluoride (31), metallic aluminum with a titanium derivative (32), aluminum chloride in benzene (32), and by use of high energy radiation polymerization in thiourea canal complexes (33). The highest molecular weight poly-1,3-cyclohexadiene to date, based on inherent viscosity data, was reported by Dawans (34). Using butyllithium in noncomplexing media of low polarity (n-heptane, cyclohexane, or benzene) poly-1,3-cyclohexadiene with inherent viscosities as high as 0.35 at high conversion rates (Fig. 1 and 2) were obtained. The molecular weight was found to vary inversely as the concentration of the n-butyllithium, and the molecular weight distribution was monodisperse.

The conversion of poly-1,3-cyclohexadiene to poly-p-phenylene by various methods has been carried out by several investigators (28,34,35). Although poly-1,3-cyclohexadiene has been dehydrogenated with palladium, chloranil, sulfur, and N-bromosuccinimide, and by bromination and/or chlorination followed by dehydrohalogenation, the best results were obtained by use of chloranil (28). Treatment of poly-1,3-cyclohexadiene with chloranil in refluxing xylene for 48 hr gave a 60% yield of a tan-brown insoluble polymer. Analysis of the material showed the presence

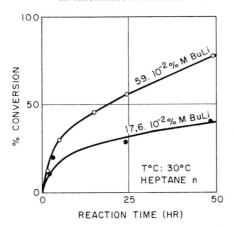

Fig. 1. Influence of the reaction time on the percent conversion in polymerization of 1,3-cyclohexadiene with n-BuLi (34).

of 8.9% chlorine which was assumed to be due to unextracted tetrachloro-p-hydroquinone and could correspond to 15.6% of this impurity in the product mixture. On further heating of this impure and incompletely aromatized polymer at 450°C for 5 hr, a black powder was obtained which did not contain chlorine and gave an analysis corresponding to the

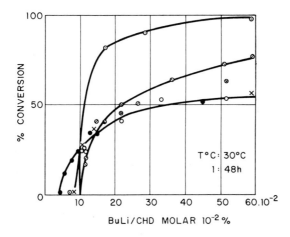

Fig. 2. Influence of the initiator concentration on the percent conversion in polymerization of 1,3-cyclohexadiene with n-BuLi in various reaction media: (●) in bulk; (○) tetrahydrofuran; (×) dioxane; (⊗) diethyl ether; (⊘) n-heptane; (⊙) cyclohexane; (◎) benzene (34).

empirical formula for poly-*p*-phenylene containing approximately 10% carbon as an impurity.

The most recently reported synthesis of poly-*p*-phenylene was via the polymerization of benzene. Kovacic and co-workers (36,39) have found that benzene readily polymerizes in a system consisting of a Lewis acid catalyst–cocatalyst–oxidizing agent. For example, in the presence of aluminum chloride–cupric chloride, benzene was polymerized under remarkably mild reaction conditions—temperatures of 35–50°C, times of 15 min—to a brown solid in about 60% yield.

$$\bigcirc + 2MCl_y \longrightarrow +(\bigcirc)_x + 2MCl_{y-1} + 2HCl \quad (5)$$

It has been proposed that the reaction sequence was that of an oxidative cationic polymerization as shown in the diagram below. The initiation would entail formation of a sigma complex (benzenonium ion) which then undergoes propagation as illustrated in reaction 6. It is significant that

$$MCl_y + H_2O \rightleftharpoons H_2O \cdots \rightarrow MCl_y \rightleftharpoons H^+MCl_y(OH)^-$$

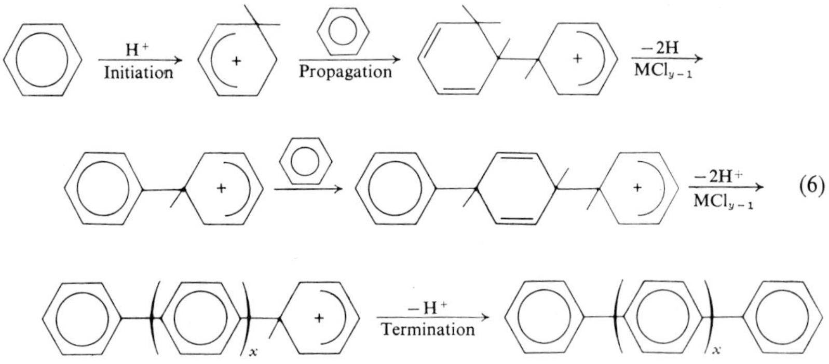

whereas cationic polymerization of olefins can occur at −100°C within a few seconds, temperatures in the vicinity of room temperature were required in order for the benzene polymerization to take place at any appreciable rate.

In the case of the Lewis acids, ferric chloride (38) and molybdenum pentachloride (39), it was presumed that they act both as catalysts and as oxidants. In order to test this hypothesis, the dehydrogenation of 1,4-cyclohexadiene by ferric chloride was investigated. This reaction was found to occur readily with the formation of benzene. An alternative hypothesis for the initiating species with these particular Lewis acids would require the one electron reduction of metal chloride to yield a radical carbonium ion such as **3**.

3

2. Properties

Although these various workers in all probability have prepared poly-*p*-phenylene, their products, however, were markedly different in a number of properties, the poly-*p*-phenylene prepared by the Wurtz reaction being quite different from that prepared by the dehydrogenation of poly-1,3-cyclohexadiene or the polymerization of benzene.

The Wurtz product was the only poly-*p*-phenylene which was soluble in benzene, and the only product on which direct determination of molecular weights has been carried out. Since the molecular weights were about 2700–2800, one is tempted to conclude that the other poly-*p*-phenylenes must have a higher molecular weight or that the Wurtz product might contain other than *p*-substituted phenylene groupings.

The poly-*p*-phenylenes prepared by the other aforementioned routes have been described as dark-brown to black powders, not melting up to 530°C. These polymers were crystalline, insoluble in such polymer solvents as *N,N*-dimethylacetamide, *N,N*-dimethylformamide, dimethyl sulfoxide, boiling biphenyl, and hot *p*-terphenyl, but slowly dissolved, probably with reaction, in hot sulfuric acid and fuming nitric acid. For the poly-*p*-phenylenes derived from poly-1,3-cyclohexadiene and from the polymerization of benzene, the infrared absorption spectra were similar to the spectra of *p*-terphenyl and *p*-tetraphenyl (28). The absorption maxima corresponding to aromatic stretching (3035 cm^{-1}) and C=C skeletal in-plane vibrations (1603, 1575, and 1486 cm^{-1}) were at wavenumbers nearly identical with those observed in the aforementioned *p*-phenylenes. The maxima at 1006 cm^{-1} characteristic of *p*-aromatic substitution also corresponded very closely. The most intense absorption maximum in the infrared spectra of the poly-*p*-phenylenes from poly-1,3-cyclohexadiene was that occurring at 811 cm^{-1} which can be assigned to *p*-phenylene substitution. In the case of the poly-*p*-phenylene from the polymerization of benzene this principal absorption band occurred at 805–807 cm^{-1}. It has been argued (37) that since this absorption band shifts to longer wavelengths with increasing molecular weight in the *p*-phenylene series (*p*-terphenyl, 837 cm^{-1}; *p*-tetraphenyl, 826 cm^{-1}; *p*-pentaphenyl, 818 cm^{-1}), this indicates that the poly-p-phenylenes from the benzene polymerization were of higher molecular weight than those from poly-1,3-cyclohexadiene.

Absorption maximum at 695 cm^{-1} in the spectra of the poly-p-phenylene corresponded to the out-of-plane deformation vibration of five adjacent carbon–hydrogen bonds. The presence of these monosubstituted aromatic absorption maxima could be explained as the endgroups of the polyphenylene molecule.

The polyphenylenes derived from poly-1,3-cyclohexadiene have absorption bands at 885 and 760 cm^{-1} (characteristic of o-aromatic substitution) which could arise from 1,2-polymerization of the cyclohexadiene (34). These absorption bands indicative of o-aromatic substitution were not observed in the spectra of the polyphenylenes obtained from benzene polymerization.

Ultraviolet absorption spectra of cyclohexane solutions of low molecular weight polyphenylenes synthesized either from poly-1,3-cyclohexadiene or from benzene showed either absorption bands located at 273 and 286 mμ or only one more intense band whose maximum is at about 300 mμ (34). These observations were in agreement with the comparative measurements made on p-terphenyl (absorption maxima 276 mμ) and on p-tetraphenyl (maximum at 235 mμ) and confirmed the hypothesis that the ultraviolet absorption frequency undergoes displacement depending on the number of phenyl rings (40).

The x-ray diffraction patterns demonstrated that the polyphenylenes were quite crystalline with diffractions at 4.6, 4.4, and 3.25 Å, and d^{hkl} distance of 4.6, the most intense diffraction being very close to the length of a phenyl unit (4.4–4.5 Å). The diffraction patterns of the poly-p-phenylenes were in close agreement with those of p-terphenyl and p-tetraphenyl, the d^{hkl} values of p-terphenyl being 12.6, 6.5, 4.5, 3.78, 3.14, 2.97, and 2.37 Å; and for p-tetraphenyl, the d^{hkl} values were 8.85, 5.99, 4.61, 4.4, 3.86, 3.20, and 2.38 Å (40).

Of major importance is, of course, the thermal stability of these poly-p-phenylenes. The poly-p-phenylene derived from dichlorobenzene, which did not melt up to 550°C, lost hydrogen at 500°C (4). The thermal stability of the polyphenylenes prepared from the biphenylyl bisdiazonium salts was indicated by weight loss measurements at elevated temperatures. For polymer prepared from benzidine containing ca. 2.5% residual —N=N— linkages, the weight loss after 1.5 hr at 300°C was 3.9%; at 350°C for the same time period, 17%; extensive decomposition ensued at 400°C. For polymer prepared from benzidine 3,3'-dicarboxylic acid containing ca. 1.5% residual —N=N— linkages, the weight loss after 1 hr at 450°C was 28–42%. After heating for 1.5 hr at this temperature, the product lost no more weight on continued heating (10,11).

Similarly, the thermal stability of these polyphenylenes from the polymerization of benzene (36–39) or the dehydrogenation of poly-1,3-cyclohexadiene (28), was demonstrated by weight loss measurements at elevated temperatures. For the poly-*p*-phenylene from benzene polymerization, the weight loss after 30 min *in air* at 350°C was 0.5%; at 400°C, ca. 1.0%; at 450°C, 16%; at 500°C, 55%; complete volatilization occurred in the same time period at 550°C (35). The poly-*p*-phenylene from poly-1,3-cyclohexadiene was reported to possess similar stability and, in addition, withstood heating for 72 hr at 230–240°C without any change in properties (28).

The thermal degradation of poly-*p*-phenylene *in vacuo* at 750–800°C yielded benzene as a major product, along with a sublimate and a residue. The sublimate was shown to be a mixture containing biphenyl and lower molecular weight *p*-polyphenyls, such as *p*-terphenyl, *p*-tetraphenyl, and *p*-pentaphenyl; no evidence for the presence of products other than poly-*p*-phenyls was found. The residue based on infrared spectra appeared to be graphite (37).

The thermogravimetric analyses of both the poly-*p*-phenylenes from poly-1,3-cyclohexadiene and from benzene has been carried out in both air and nitrogen. The polymer from poly-1,3-cyclohexadiene in nitrogen showed an initial break in the TGA curve at 550°C, with 86% of the original weight remaining even up to 817°C. For the same polymer in air, the initial break occurred approximately 100°C lower (462°C) with recovery at 703°C and only 10% of the original weight remaining at 812°C. For the poly-*p*-phenylene derived from benzene, the TGA curve in nitrogen was quite similar to that of the poly-*p*-phenylene from poly-1,3-cyclohexadiene showing an initial break at 555°C with 91% of the original weight remaining even at 817°C. Here again, in air the TGA curve of this polymer and the previously described poly-*p*-phenylene were quite similar, having an initial break at 508°C with only 4% of the original weight remaining at 817°C (Figs. 3–6).

As in the case of the TGA, the differential thermal analysis of these two poly-*p*-phenylenes in air and in nitrogen were quite similar. At the rate of heating used (10°C per min), no transitions of the first or second order, as evidenced by endotherms, were observed for either polymer both in air and/or nitrogen up to 825°C. For the poly-*p*-phenylene derived from poly-1,3-cyclohexadiene, in nitrogen—an exotherm with a peak at 725°C, and, in air—exotherms with peaks at 560 and 745°C were observed. For the polymer from benzene, similar behavior was observed, in nitrogen—an exotherm at 743°C and, in air—exotherms at 548 and 743°C (Figs. 7–10).

The polyperfluoro-*p*-phenylenes were reported to withstand heating for 1 hr *in vacuo* at 500°C without any structural change and only some darkening being observed (9). Due to the intractability and limited

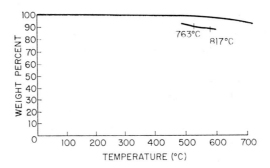

Fig. 3. TGA in nitrogen of poly-*p*-phenylene from poly-1-3-cyclohexadiene (heating rate, 10°C/min).

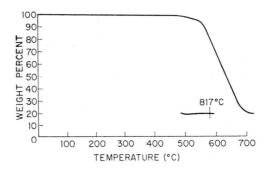

Fig. 4. TGA in air of poly-*p*-phenylene from poly-1,3-cyclohexadiene (heating rate, 10°C/min).

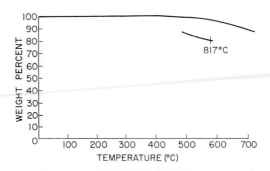

Fig. 5. TGA in nitrogen of poly-*p*-phenylene from benzene (heating rate, 10°C/min).

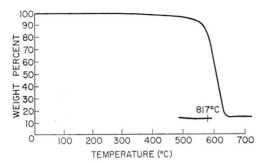

Fig. 6. TGA in air of poly-*p*-phenylene from benzene (heating rate, 10°C/min).

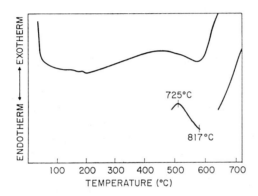

Fig. 7. DTA in nitrogen of poly-*p*-phenylene from poly-1,3-cyclohexadiene (heating rate, 10°C/min).

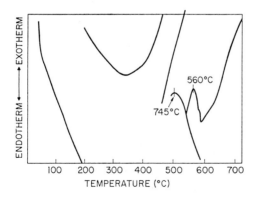

Fig. 8. DTA in air of poly-*p*-phenylene from poly-1,3-cyclohexadiene (heating rate, 10°C/min).

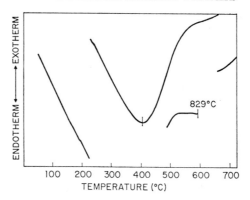

Fig. 9. DTA in nitrogen of poly-*p*-phenylene from benzene (heating rate, 10°C/min).

descriptive data, no true assessment of thermal stability of these fluorinated poly-*p*-phenylenes can be made.

The chlorinated polyphenyls derived from the polymerization of suitable benzene derivatives, however, showed significantly poorer stability than the unsubstituted poly-*p*-phenylenes. Whereas the unsubstituted poly-*p*-phenylenes in air showed no significant loss in weight up to temperatures of 450–500°C, the chlorinated products lost over 50% of the original weight at this temperature (39).

Because of the high melting point and absence of suitable solvents and plasticizers, fabrication of poly-*p*-phenylenes to useful products has not been accomplished, and meaningful thermal data on poly-*p*-phenylenes in

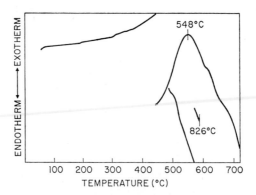

Fig. 10. DTA in air of poly-*p*-phenylene from benzene (heating rate, 10°C/min).

applications or in end use tests are not available. Thus, the definitive assessment of the thermal stability of poly-*p*-phenylenes must await the results of such tests.

C. POLY(METHYLENE PHENYLENES)

As indicated in the previous section, the high polymer melt temperature and intractability of the poly-*p*-phenylene have not permitted its fabrication into useful articles. Molecular models show that poly-*p*-phenylene is a rigid linear molecule and that shaped articles from this polymer probably would be unusually brittle and almost inextensible since the molecule cannot be bent or coiled. Since the insertion of methylene groups or their derivatives between phenylene groups along the polymer chain should give more flexible and more tractable polymers, considerable research effort has been expended in this area. The first member of such a homologous series would be the poly(methylene phenylenes).

The preparation of poly(methylene phenylenes) or polybenzyls has been extensively studied. In one of the earliest papers on the subject, Ingold and Ingold (41) described the preparation of benzyl fluoride and reported that it polymerized violently to a glasslike solid, $(C_7H_6)_n$, when treated with a catalytic amount of concentrated sulfuric acid or hydrogen fluoride. In a study of polymerization of benzyl alcohol and benzyl halides, Shriner and Berger (42) found that the polymers from these benzyl derivatives were all structurally similar and arose from a Friedel-Crafts reaction which yielded predominantly the linear *para*-substituted polymer. These workers found the ratio of *para* to *ortho* linkages within the polymer to be about 6 to 1, and the structure assigned to the polymer was described approximately by **4**.

4

One would expect such a polymer to be high melting crystalline and more stable to high temperatures than poly-*p*-xylylene (see Section D) since it does not contain the relatively weak dibenzyl linkages.

Haas and co-workers (43) have found that the polymer prepared from benzyl fluoride by the addition of a catalytic amount of concentrated sulfuric acid had a softening range of 75–80°C, and was completely liquid at 95°C. X-ray diffraction studies showed the polymer to be amorphous.

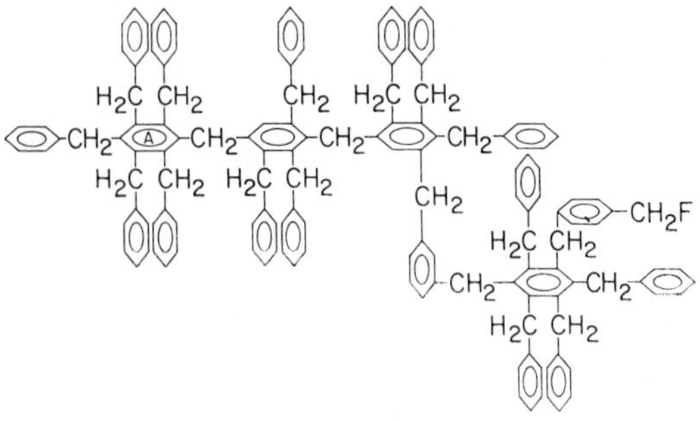

Fig. 11. Structure of polybenzyl (43).

In solubility studies, the polymer was found to be soluble in such solvents as benzene, dioxane, carbon tetrachloride, and chloroform and to be insoluble in acetone, methanol, and petroleum ether. Similarly, these workers found that the polymer formed from benzyl alcohol by treatment with sulfuric acid was also soluble in dioxane and had a softening range between 75 and 80°C. In addition, polymers obtained from *p*-isopropylbenzyl chloride, *p*-methoxybenzyl chloride, *p*-methylbenzyl chloride, and *p*-methylbenzyl bromide had similar solubilities and softened in the 75–80°C range. Based on infrared spectra, oxidation, and degradation studies, Haas and co-workers (43) concluded that the structure of the polybenzyls could best be described as a nonrandomly substituted, highly branched molecule which contained predominantly two types of benzene rings, a small number of highly substituted rings, and a considerably larger number of monosubstituted benzenes existing in the form of pendant benzyl groups as indicated in Figure 11.

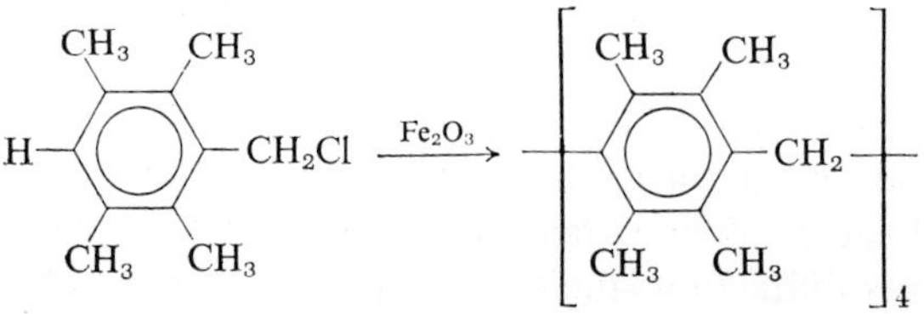

These workers did, however, prepare what surely must have been a linear poly(methylene phenylene) structure via the polymerization of durylmethyl chloride. When durylmethyl chloride was heated with a trace of iron oxide at 100°C, a small amount of hydrogen chloride was evolved and solidification occurred. This solid melted at 260–280° range, and, based on analysis, was the tetramer with the structure shown in equation 7. Similarly, the preparation of poly(methylene arylenes), [$ArCH_2$] $_x$, has been reported by Vansheidt and co-workers. In the original work (44) aromatic hydrocarbons such as *p*-xylene, durene, and isodurene were heated with formaldehyde in anhydrous acetic acid, containing sulfuric acid, to yield high melting, crystalline, difficulty soluble polymers (see reaction 8). The properties of these polymers suggested that the polymer

$$\underset{R}{\text{Ar}} + CH_2O \xrightarrow[H_2SO_4]{CH_3-C(=O)-OH} \left[\underset{R}{\text{Ar}} - CH_2 \right]_x \quad (8)$$

chains consisted of identical regular repeating units with the above formula. In subsequent work (45), using the bisacetoxy and bischloromethyl derivatives, these workers reported the preparation of similar polymeric derivatives of the aromatic structures shown in Table I. In addition, the

TABLE I
Composition and Properties of Poly(arylene methylenes) Prepared by Polycondensation of *p*-Xylene and Durene with Their Bisacetoxy and Bischloromethyl Derivatives (45)

Reagents		Polymeric reaction products					
Hydro-carbon	Second component	Yield (%)	mp (°C)	Functional groups	Found (%)	Soluble in	Remarks
p-Xylene	Paraform	98	245–265	CH_3COO	1.7	Anisole	
p-Xylene	Bisacetoxy-methyl-*p*-xylene	90		CH_3COO	2.1	Above 130°C	Diffraction patterns identical
p-Xylene	Bischlorome-thyl-*p*-xylene	74	240–245	Cl	1.4	Above 130°C	
Durene	Paraform	96	340–350	CH_3COO	2.8	Diphenyl-methane	
Durene	Bisacetoxy-methyldurene	88	315–325	CH_3COO	2.7	Above 200°C	Diffraction patterns identical
Durene	Bischlorome-thyldurene	79	310–325	Cl	2.0	Above 200°C	

preparation of ordered copoly(methylene arylenes) $-\!\!\!+\!ArCH_2\!-\!BCH_2\!+\!\!\!-_x$, the chains of which consist of a regular sequence of different aromatic units A and B was also reported. Such polymers from the condensation of durene with bis(chloromethyl)xylene would lead to the formation of a strictly regular alternating copolymer as shown in reaction 9. The copolymeric poly(methylene arylenes) which were prepared are summarized in Table II.

$$\text{[durene]} + \text{ClCH}_2\text{-[trimethylbenzene]-CH}_2\text{Cl} \xrightarrow{-\text{HCl}}$$

$$\left[\text{-[durene]-CH}_2\text{-[trimethylbenzene]-CH}_2\text{-} \right]_x \quad (9)$$

TABLE II
Properties of Pol*y*(arylene methylenes)
Prepared by Polycondensation of *p*-Xylene and Durene
with Bischloromethyl Derivatives of other Hydrocarbons (45)

Reagents			Polymers				
Hydro-carbon	Bischloro-methyl derivative	Structure of polymer chain[a]	Yield (%)	mp (°C)	Chlorine content	Solubility	Crystal-linity
p-Xylene	*p*-Xylene	(—ACH$_2$—)$_n$	74	240–245	1.4	Anisole (130°C)	+
p-Xylene	Durene	(—ACH$_2$OCH$_2$—)$_n$	65	280–285	6.3	Anisole (130°C)	+
			80	255–275	2.3	Anisole (130°C)	+
Durene	Durene	(—DCH$_2$—)$_n$	79	310–325	2.0	Diphenyl-methane	+
Durene	*p*-Xylene	(—DCH$_2$ACH$_2$—)$_n$	80	310–325	1.5	(>200°C)	+
Durene	Benzene (·*p*)	(—DCH$_2$BCH$_2$—)$_n$	75	305–310	1.1	(<200°C)	+

[a] A = 2,5-dimethylphenylene; B = *p*-phenylene; D = 2,3,5,6-tetramethylphenylene.

Another route to the poly(methylene phenylenes) which has been explored was via dehydrogenation of poly(methylene cyclohexanes) (46).

$$\left[\text{-CH}_2\text{-cyclohexylene-} \right]_n \xrightarrow{-\text{H}} \left[\text{-CH}_2\text{-phenylene-} \right] \quad (10)$$

These cyclohexane derivatives have been prepared via the intermolecular–intramolecular polymerization of 1,6-hexadiene.

The poly(methylene phenylene) was obtained by catalytic dehydrogenation with palladium and platinum and chemical dehydrogenation with potassium perchlorate. Using the latter method, a benzene-soluble material was obtained which showed infrared absorption maxima for aromatic CH (3040 cm^{-1}), aromatic C=C (1607 and 1505 cm^{-1}), and m-aromatic substitution (815 and 748 cm^{-1}). The ultraviolet spectra of an ethanolic solution of this benzene-soluble product showed a λ_{max} in the expected 225 mμ region. Because of the small amount of product isolated, no further polymer characterization could be carried out (46).

Thus, even though many attempts have been made to prepare poly(methylene phenylenes), the successful preparation of high molecular weight linear poly(methylene phenylenes) has not been achieved, and the evaluation of the thermal stability of these structures must await such an achievement.

D. POLY-p-XYLYLENES

The next member of the homologous series of polymers containing methylene and phenylene groupings is the poly(xylylenes) in which two methylene groups are inserted between the phenylene groups. All of the isomeric poly(xylylenes) have been prepared in reasonable molecular weights and have been characterized. However, since the poly-o-xylylene has a polymer melt temperature of 110°C (47) and the poly-m-xylylene has a softening temperature of only 60°C (48), the p-isomer, poly-p-xylylene, which melts in excess of 400°C (49), is the only poly(xylylene) of interest within the framework of this discussion.

1. Preparation

Although poly-p-xylylene has been synthesized by numerous methods, the first reported synthesis of this polymer in film-forming molecular weights was that of Szwarc (49) by the pyrolysis of p-xylene.

$$CH_3-\langle\bigcirc\rangle-CH_3 \longrightarrow \left[CH_2-\langle\bigcirc\rangle-CH_2 \right]_x \quad (11)$$

Using high temperatures (700–1000°C) and reduced pressures (1–5 mm), Szwarc (50–52) and other workers (53–56) have prepared poly-p-xylylene and polymers of analogous structure by the pyrolysis of suitably substituted aromatic monomers such as 1,2,4-trimethylbenzene, 1,4-dimethylnaphthalene, 2,5-dimethylpyrazine, 5,8-dimethylquinoline and copolymers of *p*-xylylene with such comonomers as maleic anhydride, chloroprene, chloro-*p*-xylene, and pseudocumene.

Recently Errede and co-workers (57–60), by carrying out the pyrolysis at 1000°C at <10 mm pressure have been able to produce the pseudoradical, *p*-xylylene, and trap it in solvent at temperatures of −78°C. At this temperature, the radical had a half-life of 22 hr, whereas at −36°C *p*-xylylene was stable for only 12 min. When polymerization did occur, the poly-*p*-xylylene formed had molecular weights in excess of 200,000.

Using these long-lived radicals in hexane at −78°C, Errede has been able to prepare copolymers of *p*-xylylene with oxygen, a peroxide polymer, **5**, which decomposed explosively at 100°C (**5**) with sulfur dioxide, an

$$\left[-O-CH_2-\underset{5}{\underset{}{\bigcirc}}-CH_2-O- \right]_x$$

insoluble polysulfone, **6**, which fused at 200°C under 500 atm pressure

$$\left[-CH_2-\underset{6}{\underset{}{\bigcirc}}-CH_2-SO_2- \right]_x$$

with the formation of translucent films, with various nitroso compounds (nitrosobenzene, *p*-nitroso-*N*,*N*-dimethylaniline, α-nitroso-β-naphthol), soluble copolymers which melted or softened at 200°C with the structure, **7**,

$$\left[-CH_2-\underset{7}{\underset{}{\bigcirc}}-CH_2-\underset{R}{N}-O- \right]_x$$

and with such vinyl monomers as maleic anhydride, diethyl maleate, diethyl fumarate, acrylonitrile, butyl acrylate, and styrene, one to one alternating copolymers with this typical structure, **8**.

$$\left[-CH_2-\underset{8}{\underset{}{\bigcirc}}-CH_2-HC-CH- \atop O=C\diagdown_O\diagup C=O \right]$$

Poly-*p*-xylylene has also been prepared by the Wurtz reaction with bischloromethyl derivatives of aromatic hydrocarbons (61–64), by the Hoffman degradation of *p*-methylbenzyltrimethylammonium halides (65), by hydrogenation of poly-*p*-xylylidene and its analogs (66), by dehydrohalogenation or reduction of *p*-halogenated aromatics (67) and by polymerization of *p*-xylylene (67). Of these various methods, the Hoffman degradation is the most intriguing because the poly-*p*-xylylene obtained was much more soluble (see reaction 12). By the slow addition of a 50%

$$\left[CH_3-\langle\bigcirc\rangle-CH_2-\overset{+}{N}(CH_3)_3 \right] X^- \xrightarrow{N(CH_3)_3 H\overset{+}{C}l^-} \left(CH_2-\langle\bigcirc\rangle-CH_2 \right)_n \quad (12)$$

aqueous solution of *p*-methyl benzyl trimethylammonium bromide or chloride to a refluxing rapidly stirred solution of aqueous 50% sodium hydroxide, polymer with inherent viscosities as high as 2 was prepared (65).

Although all of the mechanisms of the various routes to poly-*p*-xylylenes have not been investigated, the pyrolysis route has been thoroughly investigated, and the mechanism appears to be well worked out. Szwarc (49) in his initial work suggested that the polymer was formed by a series of gas-phase reactions whereby CH bonds of *p*-xylene were cleaved thermally to give *p*-xylyl radicals that disproportionate on collision to produce *p*-xylene and *p*-xylylene which subsequently polymerizes (see reactions 13). In addition to the kinetic arguments of Szwarc, Farthing, and

$$CH_3-\langle\bigcirc\rangle-CH_3 \longrightarrow CH_3-\langle\bigcirc\rangle-CH_2^\cdot + H\cdot$$

$$H\cdot + CH_3-\langle\bigcirc\rangle-CH_3 \longrightarrow CH_3-\langle\bigcirc\rangle-CH_2\cdot + H_2 \quad (13)$$

$$2CH_3-\langle\bigcirc\rangle-CH_2\cdot \longrightarrow CH_2=\langle\bigcirc\rangle=CH_2 + CH_3-\langle\bigcirc\rangle-CH_3$$

$$CH_2=\langle\bigcirc\rangle=CH_2 \longrightarrow \left[CH_2-\langle\bigcirc\rangle-CH_2 \right]_x$$

Brown (61,68) have indicated the presence of *p*-xylylene by careful studies of the by-products of the pyrolysis of *p*-xylene and the isolation of a cyclic dimer of *p*-xylylene, **9**.

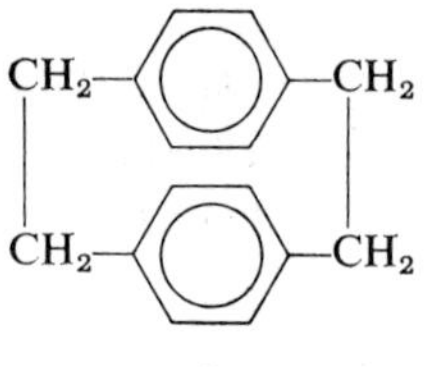

$$\text{CH}_2\text{—}\langle\bigcirc\rangle\text{—CH}_2$$
$$|\qquad\qquad|$$
$$\text{CH}_2\text{—}\langle\bigcirc\rangle\text{—CH}_2$$

9

Definite proof for the disproportionation of the *p*-xylyl radicals was provided by Schaefgen's (69) studies of the pyrolysis of the linear dimer, **10**.

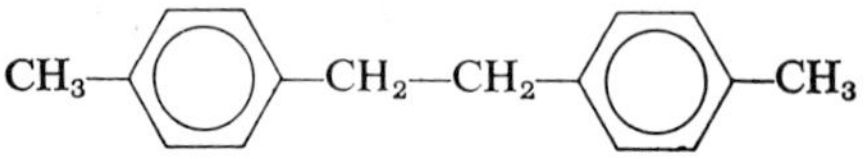

$$\text{CH}_3\text{—}\langle\bigcirc\rangle\text{—CH}_2\text{—CH}_2\text{—}\langle\bigcirc\rangle\text{—CH}_3$$

10

In this work equivalent amounts of *p*-xylene and poly-*p*-xylylene were obtained. The formation of the former was conclusive evidence of the disproportionation mechanism.

The calculations of Coulson (70) and others (71) indicate an energy difference of only 12 kcal between the singlet and triplet states of *p*-xylylene. This, plus the fact that the molecule in the ground state has a calculated resonance energy of 38 kcal/mole, implies that the molecules at room temperature are virtually all in the quinoidal form. Thus, it would be expected that even in the nonpyrolytic reactions such as the Hoffman degradation, *p*-xylylene was the intermediate species in these polymerizations.

2. Properties

The poly-*p*-xylylene obtained from the pyrolysis of *p*-xylene has been extensively characterized. Briefly, this polymer was an insoluble, crystalline material which did not melt below 425°C and yielded optically anisotropic films (72). The structure, **11**, of this poly-*p*-xylylene appeared to consist of

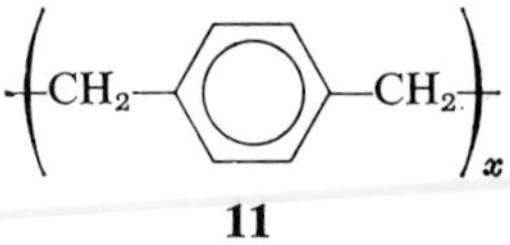

$$\left(\text{CH}_2\text{—}\langle\bigcirc\rangle\text{—CH}_2\right)_x$$

11

linear units as shown below as evidenced by chromic acid oxidation to terephthalic acid in 99% yields. The ultraviolet and the infrared spectra were consistent only with this structure (63), and the x-ray diffraction patterns (51,56,63) indicated a highly crystalline structure which excludes extensive irregular branching. X-ray fiber diagrams (53), on oriented

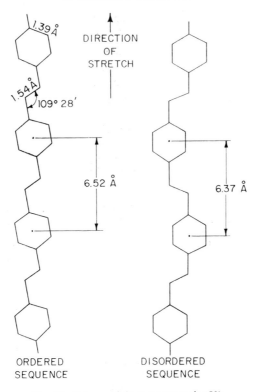

Fig. 12. Poly-*p*-xylylene repeat unit (53).

specimens, showed an identity period corresponding to one *p*-xylylene repeat unit in an ordered sequence (see Fig. 12). The poly-*p*-xylylene crystal structure has been thoroughly investigated and the existence of two modifications, classified as α and β forms, has been shown (63). It has been suggested that the configuration of the α form was similar to that of the molecule of 4,4'-dimethyl dibenzyl—the rings lying parallel to each other in the same molecule; however, they were not coplanar but arranged stepwise. On the other hand, in the β form the benzene rings were both parallel and coplanar as in the molecule of diphenyl. The α form has been converted to the β form by heat, preferentially near the melting point and this suggests that these modifications are polymorphic (73).

Over 250 materials of all chemical types were tried as solvents for these polyhydrocarbons (56). The best solvents were the chlorinated diphenyls of which Aroclor* 1248 was the most acceptable. But even in these preferred solvents, the polymer did not dissolve below 300°C. The viscosity of dilute

*Trademark for Monsanto Corp. chlorinated aromatic hydrocarbon.

solutions of poly-*p*-xylylene was measured at high temperatures (302°C), and it was found that all "soluble" polymers, i.e., materials which could be dissolved at that temperature in less than 5 min, had low inherent viscosities (56). The viscosity increased, however, as the rate of solution of polymer decreased, and the relevant results which are collected in Table III indicate that the molecular weight of some samples was at least 20,000.

TABLE III
Viscosities of Poly-*p*-Xylylene (56)

Method of preparation	Time to solution (min)	Pyrolysis conditions			Concn. (g/100 cc)	ln $\eta_{r/c}$
		Temp. (°C)	Pressure (mm)	Contact time (sec)		
Pyrolysis	40	—	—	—	0.768	0.27
Pyrolysis	25	—	—	—	0.673	0.325
Pyrolysis	6	815	7.8	1.8	0.504	0.153[a]
Wurtz reaction[b]	10 (at 259°)	—	—	—	0.838	0.095
Wurtz reaction[c]	11 (at 259°)	—	—	—	0.865	0.15
Pyrolysis	5	900	4.5	<0.1	0.261	0.35
Pyrolysis[d]	30	—	—	—	0.618	0.73

[a] Polymer was drawable at 150°C before dissolving.

[b] Sodium naphthalene in glycol dimethyl ether plus *p*-chloromethylbenzyl chloride. Run at −30°C. Polymer melted at 390–400°C when heated rapidly. Degradation was quite slow at 259°C. Calculated for C_8H_8: C, 92.3; H, 7.7. Found: C, 90.4; H, 7.8; Cl, nil.

[c] Wurtz reaction[b] at −60°C. Viscosity at 259°C (more soluble polymer).

[d] Swollen gel present after 5 min. After 30 min solution resulted, and solution had turned from colorless to amber.

The poly-*p*-xylylenes prepared by the Wurtz reaction or by the Hoffman degradation were much more soluble than those prepared from *p*-xylene pyrolysis. The Wurtz product which melted at 395–405°C was soluble in diphenyl, α-bromonaphthalene, and benzyl benzoate. Viscosity measurements on these polymers indicated that they were of lower molecular weight than those prepared from the *p*-xylene pyrolysis (see Table III). The poly-*p*-xylylene from the Hoffman degradation formed highly viscous solutions at 20 wt% of polymer in such solvents as Aroclor 1248 and benzyl benzoate at 300°C. The inherent viscosity of these polymers was in the range of 1.0–2.0 (69).

In Tables IV and V are summarized the general properties of this type of polyhydrocarbon polymer (53,56).

TABLE IV
General Characteristics of Poly-*p*-Xylylene (53,56)

Characteristic	Observation
Visual appearance	Clear, essentially colorless films becoming whitish opaque on standing
Odor	Practically odorless after extraction of natural plasticizing by-products
Spinnability	Poor quality fibers spun only with considerable difficulty
Drawability	Easily drawn 3–8× at temperatures varying from 25 to 250°C
Toughness	Reasonably tough when prepared—becoming brittle and friable on standing if not oriented by drawing
Solubility	Soluble only in a few solvents, all boiling over 300°C
Flammability	Burn fiercely, as do *p*-xylene and related hydrocarbons
Crystallinity	No truly amorphous polymers of this type have ever been isolated even at −190°C. Poly-*p*-xylylene has been found having two different crystal forms; a low temperature form for polymer collected at 0°C or below, and a high temperature form for polymer collected at 75°C or above. Collection at room temperature resulted in a mixture of the two forms. The low temperature form can be irreversibly transformed into the high temperature form by heating at or above 225°C
Birefringence	Transparent film exhibits uniaxial, negative birefringence

TABLE V
Physical Properties of Poly-*p*-Xylylene (53,56)

Cryst. mp	375–425°C
Stick or flow temp.	~400°C
(Apparent) second-order transition temp.[a]	~55°C
Moisture absorption (max.)	1%
Inherent viscosity (at 302°C) in Aroclor 1248	
Initial	0.2–0.7
After 2 hr	0.14
Density	1.14
Electrical properties at 1000 cps[b]	
Dissipation factor	0.00078–0.0017
Dielectric const.	3.30–4.10[c]
Resistivity (ohm cm)	>1.5–1.9×10^{18}

[a] Method of J. W. Ballou and J. C. Smith, *J. Appl. Phys.*, **20**, 493 (1949).

[b] These electrical properties place poly-*p*-xylylene slightly below polyethylene. Teflon tetrafluoroethylene resin and mica, but in the same class as Mylar polyester film and far superior to most polymeric materials.

[c] The dielectric constant values are higher than expected, probably because of surface contamination of the films.

In spite of the high melting point and intractability of poly-*p*-xylylenes, films and coherent fibers have been fabricated (53,56). Films were prepared either in the pyrolysis apparatus by deposition or by isolation of the polymer as a dispersion in Aroclor 1248, followed by hot pressing. Films prepared in either manner could be oriented by drawing to yield highly oriented structures with typical properties listed in Table VI.

TABLE VI
Textile Properties[a] of Poly-*p*-Xylylene
(53,56)

Tenacity (g/denier)	3.5
Elongation (%)	9
Initial modulus (g/denier)	36
Compliance ratio[b]	0.05
Work recovery[c]	
1	54
3	50
5	44

[a] Drawn 6× at 235–250°C, heat set in boiling water for 30 min, no shrinkage.
[b] R. M. Hoffman and L. F. Beste, *Textile Res. J.*, **20**, 441 (1950).
[c] From area under stress-strain curve from the respective 1, 3, and 5% elongations.

Although the polymer was not attacked appreciably (74) by concentrated sulfuric acid at 150°C, sulfonation occurred readily in the presence of traces of silver ion (53,54,56) and the product contained one sulfonic acid per benzene ring (54). Interestingly enough this sulfonated product swelled about 100-fold in water and about 400-fold in dilute aqueous base without dissolving (54). This swelling was anisotropic; that is, a sulfonated film extended in length and width but not in thickness.

Similarly, the polymer could be chlorinated (53) by treatment with sulfuryl chloride in the presence of ultraviolet light (53). The chlorination occurred predominantly at the ethylene groups and the chlorinated product, which was fairly soluble in low-boiling solvents, displayed anisotropic swelling. On treatment with concentrated nitric acid (56) at 50°C, nitration occurred, yielding a poly(dinitro-*p*-xylylene) which was a high explosive with the sensitivity of pentaerythritol tetranitrate and the power of trinitrotoluene. On boiling in 5*N* nitric acid (63), no effect was noted. However, the polymer was readily oxidized on treatment with chromic oxide in boiling acetic acid (63). Similarly, polymer was fully oxidized at high temperature by atmospheric oxygen, and the material burned fiercely when ignited (56).

The thermal stability of poly-*p*-xylylene has been extensively studied. Severe degradation of poly-*p*-xylylene occurred on melting, even in vacuum (54). The material fused much lower on remelting and after prolonged fusion in vacuum at 520°C became sublimable and completely soluble in boiling benzene. The decomposition of poly-*p*-xylylene in vacuum (75) and in solution (75,76) has been studied. Madorsky and Strauss (75) found that at temperatures of 415–430°C (Fig. 13) the loss of weight as a function of time was a first-order reaction up to 60–80% decomposition with an activation energy of 76 kcal/mole. Benzene, toluene, *p*-xylene, *p*-methylstyrene and *p*-ethyltoluene were identified as volatile products of the decomposition. The amount of monomer that could be present was estimated to be less than 10%. Auspos and co-workers (55) studied the

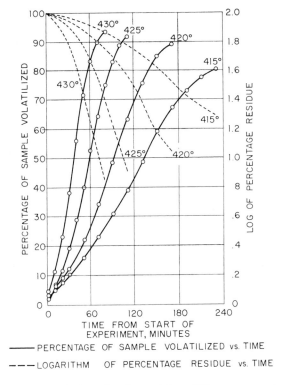

Fig. 13. Thermal degradation of poly-*p*-xylylene (75): ——— percentage of sample volatilized versus time; ----- $\log_{10}$ of percentage residue versus time temperatures expressed in degrees centigrade.

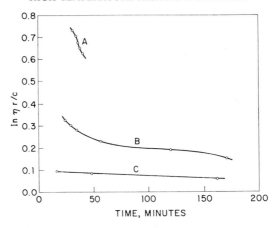

Fig. 14. Degradation of poly-*p*-xylylene in Arochlor 1248 solution, *A* and *B*, pyrolysis polymer at 302°C; *C*, Wurtz polymer at 259°C (56).

TABLE VII
Rates of Degradation of Poly-*p*-Xylylene (96)

Polymer no.[a]	Solvent[b]	Polymer concn. (g/100 ml)	Temp. (°C)	Time to dissolve (min)	η_0[c]	$M_0 \times 10^{-3}$	$k_{\text{initial}} \times 10^8$ (sec^{-1})	$k \times 10^8$ (sec^{-1})
1	A	0.673	304.6	25	0.32	9.8	—	330
2	A	0.618	304.6	32	0.73	28	—	165
3	B	0.845	304.6	6	0.82	34	12	7.3
3	A	0.708	305.6	3.5	0.81	34	16	7.7
3	A	0.734	305.6	3.5	0.84	35	20	7.8
3[d]	A	0.718	305.9	3.5	0.90	37	14	4.7
3[e]	A	0.720	305.9	3	0.88	36	8.5	3.02
3[f]	A	0.688	305.8	3	0.88	36	42	4.9
3[g]	A	0.692	306.3	3	0.75	30	31	8.2
3[h]	A	0.714	305.9	3	0.82	33	21	6.0
3	A	0.722	285.2	10	0.97	42	8.8	0.75
3[d]	A	0.712	285.6	6	0.93	39	12	0.74
3[d]	A	0.715	321.5	3	0.84	35	—	17.5
3[e]	A	0.669	321	3	0.78	32	—	11.0

[a] Polymers 1 and 2 prepared by pyrolysis of *p*-xylene; polymer 3 prepared by Hoffman reaction.
[b] Solvents: A = Aroclor 1248; B = benzyl benzoate.
[c] Inherent viscosity at zero time.
[d] Carefully degassed sample.
[e] Anthracene, ~50%, based on polymer, added.
[f] Dilauryl disulfide, 53%, based on polymer, added.
[g] Lauryl mercaptan, 68%, based on polymer, added.
[h] Hydroquinone, 48%, based on polymer, added.

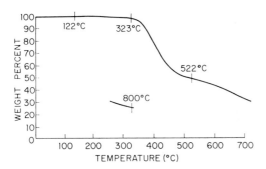

Fig. 15. TGA in nitrogen of poly-*p*-xylylene from pyrolysis (heating rate, 10°C/min).

decomposition of poly-*p*-xylylene at 302°C in Aroclor 1248 and found that decomposition occurred even at this temperature and continued on prolonged heating (Fig. 14).

Schaefgen (76) has compared the thermal stability of poly-*p*-xylylene prepared by the Hoffman reaction with that prepared from the pyrolysis of *p*-xylene. These measurements were made by following the decrease in inherent viscosity of polymer in solution at temperatures of 285, 304, and 321°C. The decomposition was found to be a first-order random cleavage reaction with an activation energy of 58 kcal/mole. Based on a comparison of specific velocity constants, this poly-*p*-xylylene was an order of magnitude more stable than the pyrolysis polymer (Table VII).

The thermogravimetric analysis of the poly-*p*-xylylenes (*1*) from the pyrolysis of *p*-xylene, (*2*) from the Wurtz reaction of α,α'-dichloro-*p*-xylene and sodium, and (*3*) the Hoffman degradation of *p*-methylbenzyltrimethylammonium chloride has been carried out in both air and nitrogen. The TGA curves are shown in Figures 15–20 and the pertinent information

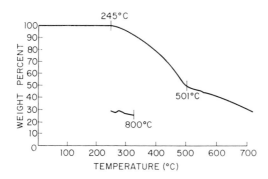

Fig. 16. TGA in nitrogen of poly-*p*-xylylene from Wurtz reaction (heating rate, 10°C/min).

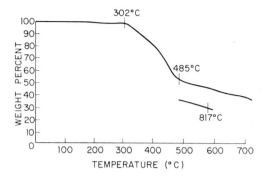

Fig. 17. TGA in nitrogen of poly-*p*-xylylene from Hoffman reaction (heating rate, 10°C/min).

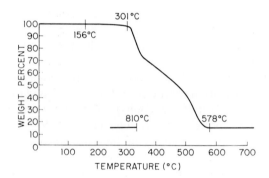

Fig. 18. TGA in air of poly-*p*-xylylene from pyrolysis (heating rate, 10°C/min).

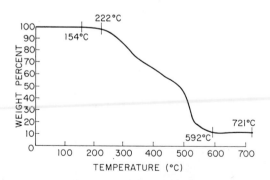

Fig. 19. TGA in air of poly-*p*-xylylene from Wurtz reaction (heating rate, 10°C/min).

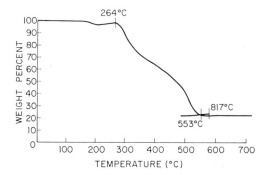

Fig. 20. TGA in air of poly-*p*-xylylene from Hoffman reaction (heating rate, 10°C/min).

summarized in Table VIII. From an inspection of these curves and these data, it is apparent that in nitrogen the TGA of the pyrolysis polymer and the polymer prepared from the Hoffman reaction were quite similar, with the Wurtz product showing an initial break in the curve 50–75°C lower. Similarly, in air, although there were differences in the behavior of the pyrolysis and Hoffman products, the initial break in the curves of these polymers here again was 40–80°C higher than that of the Wurtz product. In all probability, the apparent greater instability of the Wurtz product reflects the low molecular weight and greater number of ends or sites for initiating degradation.

TABLE VIII
Thermogravimetric Analysis of Poly-*p*-Xylylenes

Polymer source	1st break (°C)	Recovery (°C)	Wt loss (%)
	In Nitrogen		
From pyrolysis	323	522	84
From Wurtz reaction	245	501	81
From Hoffman reaction (low temp.)	302	485	85
	In Air		
From pyrolysis	301	577	96
From Wurtz reaction	222	592	100
From Hoffman reaction (low temp.)	264	553	93

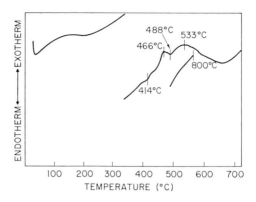

Fig. 21. DTA in nitrogen of poly-*p*-xylylene from pyrolysis (heating rate, 10°C/min).

The differential thermal analysis of these three polymers has also been carried out in both air and nitrogen. The DTA curves are shown in Figures 21–26 and the pertinent information summarized in Table IX. An inspection of these curves and these data clearly show that the pyrolysis product was quite different from both the Wurtz and Hoffman products. At the rate of heating used (10°C/min), the pyrolysis polymer showed no transitions, as evidenced by endotherms, up to temperatures of 400°C. The endotherms at 414 and 480°C reflect melting and decomposition of the polymer.

Both the Wurtz and Hoffman polymers show endotherms in the 233–238°C region. These common endotherms must reflect second-order transitions and may well represent polymorphic changes. It is tempting to

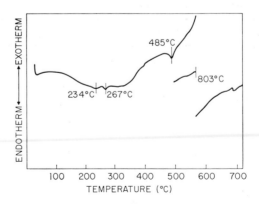

Fig. 22. DTA in nitrogen of poly-*p*-xylylene from Wurtz reaction (heating rate, 10°C/min).

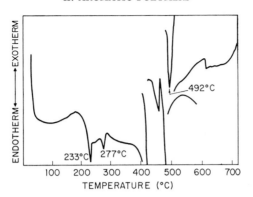

Fig. 23. DTA in nitrogen of poly-*p*-xylylene from Hoffman reaction (heating rate, 10°C/min).

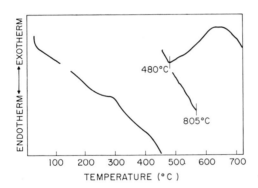

Fig. 24. DTA in air of poly-*p*-xylylene from pyrolysis (heating rate, 10°C/min).

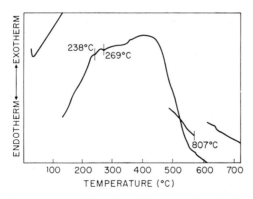

Fig. 25. DTA in air of poly-*p*-xylylene from Wurtz reaction (heating rate, 10°C/min).

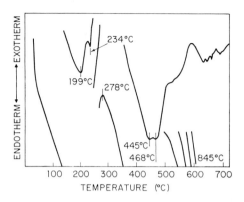

Fig. 26. DTA in air of poly-*p*-xylene from Hoffman reaction (heating rate, 10°C/min).

suggest that these endotherms indicate the conversion of the α form of poly-*p*-xylylene to the β form. Under nitrogen, both Wurtz and Hoffman poly-*p*-xylylenes showed the melting-decomposition endotherm above 400°C.

Poly-*p*-xylylene on heating at 300°C in air lost 6.25% by weight after 5 hr (54), and the oriented films on heating in air at 105°C for $2\frac{1}{2}$ days lost 5% of their weight and were embrittled (56). Even the poly-*p*-xylylene films which were drawn at temperatures as low as 260°C no longer maintained their solubility. In fact, repeated drawing at this temperature rapidly embrittled the films (56).

Thus, by any standard of stability, poly-*p*-xylylene is considerably less stable than poly-*p*-phenylene. It must then be concluded that even though the poly-*p*-xylylenes are much more tractable polymers yielding orientable

TABLE IX
Differential Thermal Analysis of Poly-*p*-Xylylenes

Polymer source	Endotherms (°C)
In Nitrogen	
From pyrolysis	414, 487
From Wurtz reaction	234, 267, 485
From Hoffman reaction (low temp.)	233, 277, 426, 492
In Air	
From pyrolysis	480
From Wurtz reaction	238, 269
From Hoffman reaction (low temp.)	199, 234, 445, 468

films and coherent fibers, the relatively weak dibenzyl linkages in the polymer chain limit their usefulness in high temperature applications.

3. Poly(Alkylene Phenylenes)

Polymers which might be considered as derivatives of poly-*p*-xylylene are those having in the polymer chain two alkylene groups between phenylene rings $-\!(AR-CR_2-CR_2)_x$. Such polymers have been prepared by Korshak and co-workers (77) by a polyrecombination reaction. In this synthesis which involves the reaction of aromatic hydrocarbons such as *p*-diisopropylbenzene with peroxides such as *t*-butyl peroxide at 170–200°C, a polymer was formed having repeating units with the structure **12**. The product was a difficulty soluble, crystalline material melting at

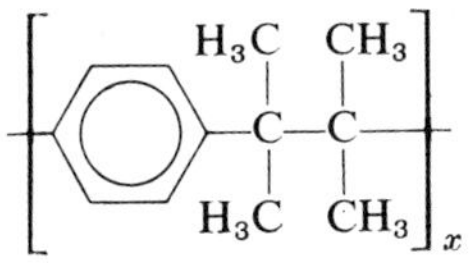

12

about 300°C. It has been suggested that the polymer was formed as a result of multiple recombinations of radicals of secondary origin (in the particular case, hydrocarbon radicals). Thus, the reaction can be represented schematically as shown in reaction 14 as a polyrecombination reaction.

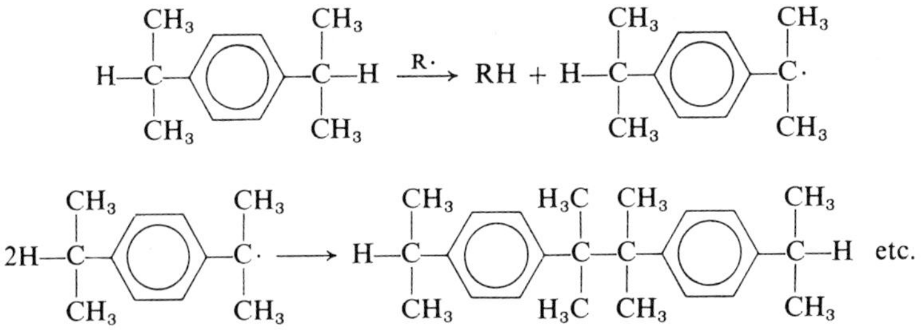

(14)

Although only limited characterizations of this polymer have been carried out, based on the relatively low melting point and considering the greater instability of the alkylene group over the methylene group (78) in the phenylene-R polymers, it would be supposed that such polymers would be even less thermally stable than the poly-*p*-xylylenes.

4. Poly-*p*-Xylylidenes

Another class of polymers which might be considered as derivatives of poly-*p*-xylylene are the poly-*p*-xylylidenes, **13**. These polymers are of interest not only from the expectation that such structures would possess high thermal stability because of the conjugated unsaturation in the main chain but would also show useful electrical and optical properties.

$$\left[\underset{}{\underline{\bigcirc}} -CH=CH \right]_x$$

13

Although the successful syntheses of poly-*p*-xylylidene has been reported, none of these syntheses have yielded polymer of high molecular weight. Campbell and McDonald (79) and Vansheidt and Krakovyah (80) using the Wittig reaction have reported the preparation of *p*-xylylidene decamers. By the reaction of bis(triphenyl phosphonium)*p*-xylylidene dichloride with terephthalaldehyde, in the presence of lithium ethylate, these workers obtained a poly-*p*-xylylidene melting above 400°C.

$$\left[(C_6H_5)_3-P-CH_2-\bigcirc-CH_2-P-(C_6H_5)_3 \right]^{2+} 2Cl^-$$

$$+ \; OHC-\bigcirc-CHO \xrightarrow{LiOC_2H_5} \quad (15)$$

$$OHC-\bigcirc\left[CH=CH-\bigcirc \right]_9 CHO$$

Hoeg and co-workers (66) have prepared a polymer consisting predominantly of *p*-xylylidene structure via the reaction of α,α'-dichloro-*p*-xylene with sodium amide (sodamide) and ammonia at low temperatures (see reaction 16). The poly-*p*-xylylidene prepared by both routes was a

$$ClCH_2-\bigcirc-CH_2Cl \xrightarrow[NH_3]{NaNH_2} \left[ClCH_2-\bigcirc-CHCl \right] Na^+ \longrightarrow$$

$$\left[\bigcirc-CHCl-CH_2 \right] \xrightarrow[NH_2]{NaNH_2} \left[\bigcirc-CH=CH \right] \quad (16)$$

yellow solid, melting over 400°C which was insoluble in all solvents, even those found best for linear poly-*p*-xylylene such as boiling Aroclor 1248

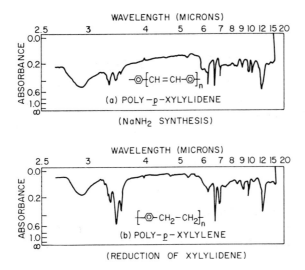

Fig. 27. Infrared absorption spectra (in KBr pellets) of (a) poly-p-xylylidene (oligomer) prepared by the sodamide route and (b) poly-p-xylylene prepared by the sodium/NH$_3$-alcohol reduction of (a) (66).

and benzyl benzoate. The infrared spectrum exhibited as its main distinguishing feature absorption bands at 825 cm^{-1} (p-disubstituted benzene rings) and 960 cm^{-1} (*trans*-ethylenic unsaturation) (Fig. 27). Based on elemental analysis, bromine addition, and endgroups analysis, products by both synthetic methods were shown to be decamers.

As further proof of the poly-p-xylylidene structure, reduction with sodium and alcohol in liquid ammonia (sodium naphthalene-alcohol also reduced the polymer, but some residual unsaturation was found) yielded a white powder which was shown to be identical with poly-p-xylylene (Fig. 27).

$$\left[\bigcirc\!\!\!-\!\!\text{CH}\!\!=\!\!\text{CH} \right]_n \xrightarrow[\text{2. ROH}]{\text{1. Na/NH}_3} \left[\bigcirc\!\!\!-\!\!\text{CH}_2\!\!-\!\!\text{CH}_2 \right]_n \quad (17)$$

The curve from the thermogravimetric analysis in nitrogen of poly-p-xylylidene exhibited a substantial break slightly above 400°C with 55% of material still present at 900°C. In air the TGA curve exhibited a sharp break at 400°C with rapid and continuing loss of weight up to 100% loss at approximately 525°C. In addition this aromatic polymer containing conjugated unsaturation in the main chain exhibited only a low level of photoconduction (66).

5. Polycyanoterephthalylidine

Another polymer class which might be considered as derivatives of poly-*p*-xylylene containing functional groups comprises the polycyanoterephthalylidenes, **14**,

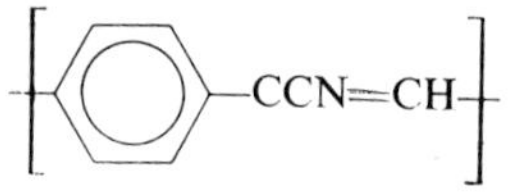

$$\left[-\!\!\left\langle\bigcirc\right\rangle\!\!-CCN\!\!=\!\!CH- \right]$$

14

Those interesting polymers have been prepared (81) via the condensation polymerization of α,α′-dicyano-*p*-xylylene and terephthaladehyde (see reaction 18). Polycyanoterephthalylidene was an insoluble, intractable

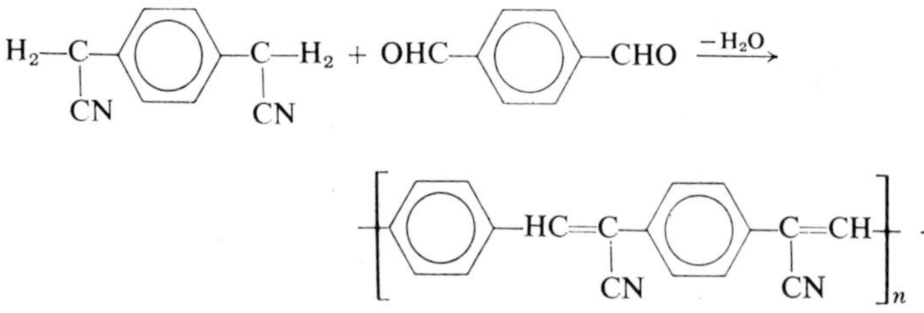

$$H_2\text{-}C\!\!-\!\!\left\langle\bigcirc\right\rangle\!\!-C\!\!-\!\!H_2 + OHC\!\!-\!\!\left\langle\bigcirc\right\rangle\!\!-CHO \xrightarrow{-H_2O}$$
$$\quad\ \ |\qquad\qquad\ \ |$$
$$\ \ CN\qquad\quad\ \ CN$$

$$\left[-\!\!\left\langle\bigcirc\right\rangle\!\!-HC\!\!=\!\!C\!\!-\!\!\left\langle\bigcirc\right\rangle\!\!-C\!\!=\!\!CH- \right]_n \longrightarrow \qquad (18)$$
$$\qquad\qquad\qquad |\qquad\qquad\quad |$$
$$\qquad\qquad\quad CN\qquad\qquad\ CN$$

material exhibiting solubility in a number of potential solvents and exhibiting no signs of softening or melting. When heated to 500°C for 50 min, the polymer lost 15% by weight, and when reheated at 500°C for 100 min, it lost 25% by weight. However, the infrared spectra (Fig. 28) of the

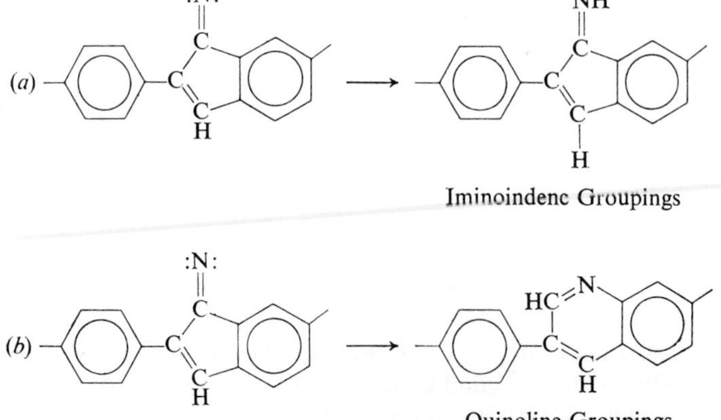

(a) [structure with :N: imine] → [structure with NH]

Iminoindene Groupings

(19)

(b) [structure with :N:] → [structure with quinoline N]

Quinoline Groupings

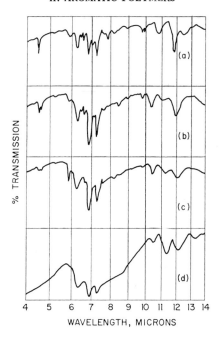

Fig. 28. Infrared spectra of polycyanoterephthalylidenes: (*a*) polymer C of Table I, (*b*) same pyrolyzed at 450°, (*c*) same pyrolyzed at 500°, (*d*) same pyrolyzed at 550°, (Nujol mulls) (81).

heated and unheated polymers were quite different. On heating, absorption bands due to the olefin and nitrile groups gradually disappeared, suggesting that cyclization was taking place with the formation of iminoindene and quinoline groups as shown schematically in reactions 19.

The heated polymers all showed improved stability. In fact, these materials showed weight losses of only 2% after 30 min at 500°C. Unfortunately, the intractability of the samples precluded any statement as to molecular weight of the polymer or assessment of the significance of the weight loss data.

E. OTHER AROMATIC POLYMERS

Although numerous aromatic polymers have been prepared which contain aromatic rings and alkylene linkages, only a few have sufficient thermal stability to be of interest in this discussion. In the migration polymerization of diisopropylenediphenylmethane and diisopropenyldiphenylethane in the presence of Lewis acids, Mitin and Glukhov (82) obtained saturated linear polymers with the following chain structure, **15**:

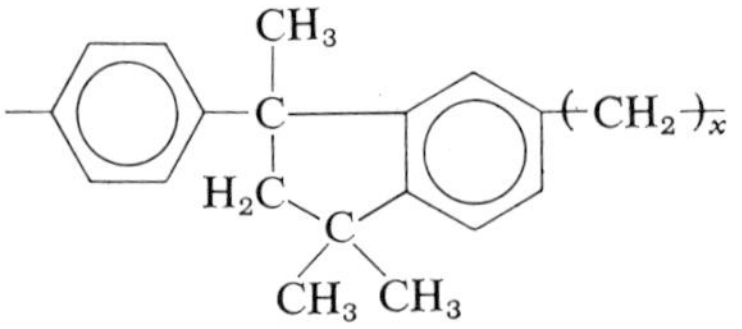

$$\left[\text{structure 15} \right]$$

15

The structures of the polymers were confirmed by a study of infrared spectra. Although the polymers have high resistance to oxidative and thermal attack, due to the low melting point (where $x = 1$, PMT = 170°C; where $x = 2$, PMT = 130°C), these polymers were of little interest as thermally stable polymers (82). As already discussed, polymers containing methylene groups alternating with phenylene groups can be prepared by the polycondensation of various aromatic hydrocarbons with dihaloalkanes in the presence of aluminum chloride and/or other aluminum halides (83, 84). Those polymers, with the general formula

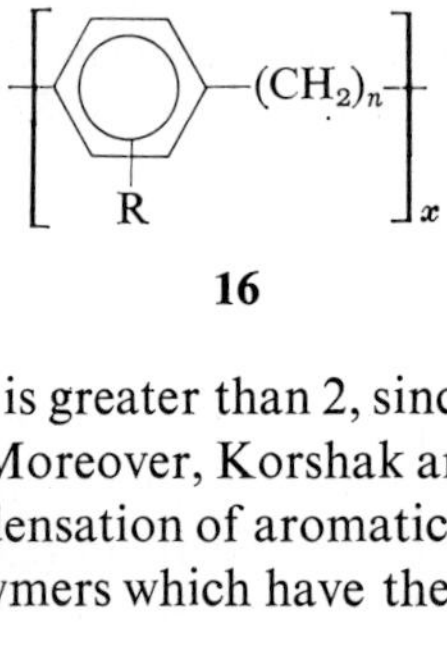

$$\left[\underset{R}{}\!\!\!\!\!\!\bigcirc\!\!-(CH_2)_n \right]_x$$

16

are of little interest when n is greater than 2, since the melting point of such polymers is below 400°C. Moreover, Korshak and Kolesnikov (85–87) have reported that the polycondensation of aromatic hydrocarbons with methylene dichloride yielded polymers which have the repeating units, **17** and **18**, and melt at 380–400°C.

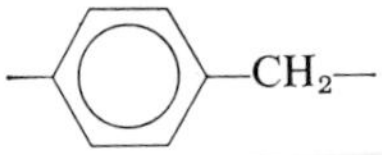

17 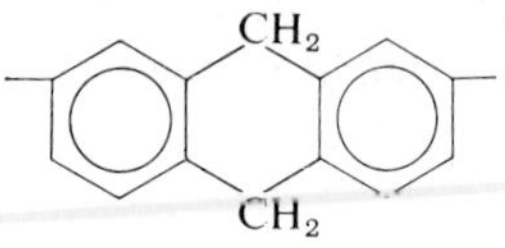   **18**

Unfortunately, the methylene linkages provide a point for oxidative attack and these polymers exhibited thermal stability similar to or slightly lower than that of poly-*p*-xylylene. A related aromatic polymer structure was that obtained from the copolymerization of anthracene and styrene (88). On heating at 300°C for 4 hr, weight losses of 10% for **19** were reported. However, at higher temperatures this polymer was badly degraded and susceptible to oxidative attack. Similarly, the polymer of acenaphthalene (89)

19

which has been prepared in molecular weights in excess of 200,000 was found to be very heat resistant to temperatures of 300°C but at higher temperatures decomposed rapidly.

References

1. J. W. Dale, J. B. John, E. A. McElhill, and J. O. Smith, WADC Tech. Rept. 59-55, Aeronautical Systems Division, Wright-Patterson AFB, Ohio, 1959.
2. E. Clar, *Aromatische Kohlenwasserstoffe*, 2nd ed., Springer, Berlin, 1952.
3. M. Bennett, N. B. Sunshine, and G. F. Woods, *J. Am. Chem. Soc.*, **28**, 2514 (1963).
4. N. B. Sunshine and G. F. Woods, *J. Am. Chem. Soc.*, **28**, 2517 (1963).
5. G. Goldfinger, *J. Polymer Sci.*, **4**, 93 (1949).
6. G. A. Edwards and G. Goldfinger, *J. Polymer Sci.*, **6**, 125 (1951).
7. G. A. Edwards and G. Goldfinger, *J. Polymer Sci.*, **16**, 589 (1955).
8. M. Hellmann, A. J. Bilbo, and W. J. Pummer, *J. Am. Chem. Soc.*, **77**, 3650 (1955).
9. *Chemical Age (London)*, **81**, 659 (1959).
10. V. P. Parini and A. A. Berlin, *Izv. Akad. Nauk SSSR, Otd. Khim. Nauk*, **1958**, 1499.
11. A. A. Berlin and V. P. Parini, *Izv. Akad. Nauk SSSR, Otd. Khim. Nauk*, **1958**, 1674.
12. A. A. Berlin, *J. Polymer Sci.*, **55**, 621 (1961).
13. M. Busch and W. Weber, *J. Prakt. Chem.*, **146**, 1 (1936).
14. W. H. Williams, U.S. Pat. 1,976,468 (1934).
15. G. Wittig and F. Bickelhaupt, *Ber.*, **91**, 883 (1958).
16. V. Y. Bogomalnyy and B. A. Dolgoplosk, *Izv. Akad. Nauk SSSR Otd. Khim. Nauk*, **1961**, 1912.
17. W. V. Evans, R. Pearson, and D. Braithwaite, *J. Am. Chem. Soc.*, **63**, 2574 (1941).
18. F. Hoffman and P. Damm, *Mitt. Schles. Kohlenforsch. Inst. Kaiser-Wilhelm Ges.*, **2**, 97 (1925).
19. H. Stuchlen, H. Thayer, and P. Willis, *J. Am. Chem. Soc.*, **62**, 171 (1940).
20. P. S. Santarovic and I. A. Slyapnikova, *Vysokomolekul. Soedin.*, **2**, 1171 (1960).
21. P. S. Santarovic and I. A. Slyapnikova, *Vysokomolekul. Soedin.*, **3**, 1364 (1961).
22. S. F. Naumova and L. G. Tsykalo, *Vysokomolekul. Soedin.*, **3**, 1031 (1961).
23. B. V. Erofeev, S. F. Naumova, and L. G. Tsykalo, *Sb. Nauchn. Rabot, Akad. Nauk Belorussk. SSR Inst. Fiz.-Organ. Khim.*, **9**, 71 (1961).

24. R. C. Houtz and H. Adkins, *J. Am. Chem. Soc.*, **55**, 1609 (1933).
25. R. Truffault, *Bull. Soc. Chim. France*, **1**, 391 (1934).
26. N. D. Zelinskii, Ya. I Denisenko, M. S. Eventova, and S. I. Kromov, *Sintetich. Kauchuka*, **4**, 11 (1933).
27. K. Ziegler, *Brennstoff-Chem.*, **30**, 181 (1949).
28a. C. S. Marvel and G. E. Hartzell, *J. Am. Chem. Soc.*, **81**, 448 (1959).
28b. C. S. Marvel, P. E. Cassidy, and S. Ray, *J. Polymer Sci. A*, **1**, 1553 (1965).
29. B. V. Erofrev, S. F. Naumova, and L. G. Tsylako, *Dokl. Akad. Nauk Belorussk. SSR*, **3**, 95 (1959).
30. B. V. Erofeev, S. F. Naumova, and T. P. Maksimova, *Sb. Nauchn. Rabot, Akad. Nauk Belorussk. SSR Inst. Fiz.-Organ. Khim.*, **9**, 80 (1961).
31. A. V. Topchiev, Ya. M. Paushkin, M. V. Kurashev, L. S. Polak, and L. S. Tversakaya, *Izv. Akad. Nauk SSSR, Otd. Khim. Nauk*, **1960**, No. 6, 1065.
32. Ger. Pat. 1,097,982 (1961).
33. J. F. Brown and D. M. White, *J. Am. Chem. Soc.*, **82**, 5671 (1960).
34. F. Dawans and G. Lefebvre, *J. Polymer Sci. A*, **2**, 3277 (1964).
35. D. A. Frey, M. Hasegawa, and C. S. Marvel, *J. Polymer Sci. A*, **1**, 2057 (1963).
36. P. Kovacic and A. Kyriakis, *Tetrahedron Letters*, **1962**, 467.
37. P. Kovacic and A. Kyriakis, *J. Am. Chem. Soc.*, **85**, 454 (1963).
38. P. Kovacic and F. W. Koch, *J. Org. Chem.*, **28**, 1864 (1963).
39a. P. Kovacic and R. M. Lange, *J. Org. Chem.*, **28**, 968 (1963).
39b. P. Kovacic, V. J. Marchionna, and J. P. Kovacic, *J. Polymer Sci. A*, **3**, 4297 (1965).
39c. P. Kovacic and L. C. Hsu, *J. Polymer Sci. A-1*, **4**, 5 (1966).
39d. P. Kovacic and R. J. Hopper, *J. Polymer Sci. A-1*, **4**, 1445 (1966).
40a. A. E. Gillam and D. H. Hey, *J. Chem. Soc.*, **1939**, 1170.
40b. I. M. Sarasohn, private communication.
41. C. K. Ingold and E. H. Ingold, *J. Chem. Soc.*, **1928**, 2249.
42. R. L. Shriner and L. Berger, *J. Org. Chem.*, **6**, 305 (1941).
43a. H. C. Haas, D. I. Livingston, and M. Saunders, *J. Polymer Sci.*, **15**, 503 (1955).
43b. W. C. Overhults and A. D. Ketley, *Makromol. Chem.*, **95**, 143 (1966).
44. A. A. Vansheidt, E. P. Mellnikova, and G. A. Gladkovskii, *Vysokomolekul. Soedin.*, **4**, 1178 (1962).
45. A. A. Vansheidt, E. P. Mellnikova, and G. A. Gladkovskii, *Vysokomolekul. Soedin.*, **4**, 1303 (1962).
46. C. S. Marvel and J. K. Stille, *J. Am. Chem. Soc.*, **80**, 1740 (1958).
47. L. A. Errede, *J. Polymer Sci.*, **69**, 253 (1961).
48. F. R. Dammont, *J. Polymer Sci. B*, **1**, 339 (1963).
49. M. Szwarc, *Discussions Faraday Soc.*, **2**, 46 (1947).
50. M. Szwarc, *J. Chem. Phys.*, **16**, 128 (1948).
51. M. Szwarc, *J. Polymer Sci.*, **6**, 319 (1951).
52. M. Szwarc and A. Shaw, *J. Am. Chem. Soc.*, **73**, 1379 (1951).
53. M. H. Kaufman, H. F. Mark, and R. B. Mesrobian, *J. Polymer Sci.*, **13**, 3 (1954).
54. R. S. Corley, H. C. Haas, M. W. Kane, and D. I. Livingstone, *J. Polymer Sci.*, **13**, 137 (1954).
55. L. A. Auspos, L. A. R. Hall, J. K. Hubbard, W. Kirk, J. R. Schaefgen, and S. B. Speck, *J. Polymer Sci.*, **15**, 19 (1955).
56. L. A. Auspos, C. W. Burnham, L. A. R. Hall, J. K. Hubbard, W. Kirk, J. R. Schaefgen, and S. B. Speck, *J. Polymer Sci.*, **15**, 19 (1955).
57. L. A. Errede and B. F. Landrum, *J. Am. Chem. Soc.*, **79**, 4952 (1957).

58. L. A. Errede and S. L. Hopwood, *J. Am. Chem. Soc.*, **79**, 6507 (1957).
59. L. A. Errede and J. M. Hout, *J. Am. Chem. Soc.*, **82**, 436 (1960).
60. L. A. Errede, R. S. Gregorian, and J. M. Hout, *J. Am. Chem. Soc.*, **82**, 5218 (1960).
61. A. C. Farthing and C. J. Brown, *Nature*, **164**, 915 (1949).
62. D. J. Cram and H. Steinberg, *J. Am. Chem. Soc.*, **73**, 5691 (1951).
63. A. C. Farthing and C. J. Brown, *J. Chem. Soc.*, **1953**, 3270.
64. A. A. Vansheidt, E. P. Melnikova, L. V. Kukhareva, and M. G. Kraskovyak, *Zh. Prikl. Khim.*, **31**, 1898 (1958).
65. Brit. Pat. 807,196 (1959).
66. D. F. Hoeg, D. I. Lusk, and E. P. Goldberg, *J. Polymer Sci. B*, **2**, 697 (1964).
67a. L. A. R. Hall, U.S. Pat. 2,863,571 (May, 1958).
67b. H. G. Gilch, *J. Polymer Sci. A-1*, **4**, 438 (1966); **4**, 1351 (1966).
67c. H. G. Gilch and W. L. Wheelwright, *J. Polymer Sci. A-1*, **4**, 1337 (1966).
67d. W. F. Gorham, *J. Polymer Sci. A-1*, **4**, 3027 (1966).
68. C. J. Brown, *J. Chem. Soc.*, **1953**, 3265.
69. J. R. Schaefgen, *J. Polymer Sci.*, **15**, 230 (1955).
70. C. A. Coulson, D. P. Craig, A. Maccoli, and A. Pullman, *Discussions Faraday Soc.*, **2**, 36 (1947).
71. S. Namiot, R. Diatkina, and Y. A. Syrkin, *Compt. Rend. URSS*, **48**, 233 (1945).
72. M. Koton, *J. Polymer Sci.*, **52**, 97 (1961).
73. L. A. Errede and M. Szwarc, *Quart. Rev. (London)*, **12**, 320 (1958).
74. E. Daudel, *Discussions Faraday Soc.*, **2**, 69 (1947).
75. J. L. Madorsky and S. Straus, *J. Res. Natl. Bur. Std.*, **55**, 223 (1955).
76. J. R. Schaefgen, *J Polymer Sci.*, **41**, 133 (1959).
77. V. V. Korshak, S. L. Sosin, and M. V. Chistyakova, *Vysokomolekul. Soedin.*, **1**, 937 (1959).
78. G. F. L. Ehlers, Paper presented at Twelfth Canadian High Polymer Forum, May, 1964.
79. T. W. Campbell and R. N. McDonald, *J. Am. Chem. Soc.*, **82**, 4669 (1960).
80. A. A. Vansheidt and M. G. Krakovyah, *Vysokomolekul. Soedin.*, **5**, 805 (1963).
81. R. W. Lenz and C. E. Handlovits, *J. Org. Chem.*, **25**, 813 (1960).
82. Yu. V. Mitin and N. A. Glukhov, *Dokl. Akad. Nauk SSSR*, **115**, 97 (1957).
83. G. S. Kalinsnikov and V. V. Korshak, *Izv. Akad. Nauk SSSR, Otd. Khim. Nauk*, **1955**, 1100.
84. L. Nicolesen and M. Ioba, *Ind. Plastiques Mod. (Paris)*, **10**, 46 (1958).
85. G S. Kolesnikov and V. V. Korshak, *Izv. Akad. Nauk SSSR, Otd. Khim. Nauk*, **1955**, 172, 359, 1090, 1095.
86. G S. Kolesnikov, V. V. Korshak, T. V. Smirnova, and L. S. Fedorova, *Izv. Akad. Nauk SSSR, Otd. Khim. Nauk*, **1957**, 375, 1478.
87. G. S. Kolesnikov, V. V. Korshak, and T. V. Smirnova, *Izv. Akad. Nauk SSSR, Otd. Khim. Nauk*, **1958**, 353.
88. M. Imoto, *Yuki Gosei Kagaku Kyokai Shi*, **17**, 579 (1959).
89. M. Imoto, *Khim. i Tekhnol., Polim.*, **1960**, 67.

III. AROMATIC POLYMERS CONTAINING FUNCTIONAL GROUPS IN THE CHAIN

Although the limited thermal data discussed in Chapter II clearly indicate that poly-p-phenylene is indeed an extremely stable polymer, a true measure of its thermal stability in use or application tests has not been possible due to its infusibility and intractability. Attempts to increase tractability with retention of high thermal stability by the insertion of methylene groups between phenylene groups have resulted in more tractable polymer systems but, unfortunately, with an accompanying decrease in thermal stability. In order to obtain processible products with thermal stabilities comparable to that of poly-p-phenylene, research has been directed toward the synthesis of aromatic polymers in which groups other than hydrocarbon are inserted in the polymer chain between phenylene groupings. This area of research, though quite broad, has been extensively studied, and a large number of thermally stable polymers which can be

$$\left[\!\!\bigcirc\!\!-R \right]_x$$

classified as phenylene-R polymers have been prepared. These polymer systems will be considered in alphabetical order.

A. AROMATIC POLYAMIDES

The aromatic polyamides, especially those derived from short-chain aliphatic primary and secondary diamines, cycloaliphatic secondary diamines, aromatic polyhydrazides, and aromatic diamines have been among the most difficult polyamides to prepare and have been quite intractable once obtained. Numerous attempts (1–9) have been made to prepare these polymers through conventional melt polymerization methods.

Since 1959, however, great strides have been made in the preparation of these high-melting polymers with the elegant work of Wittbecker, Morgan, and co-workers (10–28) on low temperature polymerization techniques. The methods for the preparation of these polymers will not be discussed in great detail since they are thoroughly covered in the literature.

Attention in this section, for the most part, will be directed to the properties of the polymers.

1. Aromatic Polyamides from Short-Chain Aliphatic Primary and Secondary Diamines and Cycloaliphatic Secondary Diamines

Numerous attempts to prepare aromatic polyamides from short-chain aliphatic primary and secondary diamines and cycloaliphatic secondary diamines through conventional melt polymerization methods are described in the literature. For example, Akiyoshi, Hashimoto, and Takami (1) attempted the reaction of aliphatic primary diamines with dimethyl terephthalate but were unable to obtain any high molecular weight products except with long-chain diamines such as octamethylene- and decamethylenediamine. Similar results were also noted by Thomas (2) and by Edgar and Hill (3). In each case the melt method gave low molecular weight degraded materials.

Polyterephthalamides have been prepared with molecular weights as high as 30,000 both by interfacial (13,15) and solution polymerization (24,27). Shashoua and Eareckson (15), using short-chain primary and secondary aliphatic diamines and terephthaloyl chloride, were able to prepare high melting polyamides in the manner outlined in reaction 1,

$$H_2N-(CH_2)_n-NH_2 + Cl-\overset{O}{\overset{\|}{C}}-\underset{}{\bigcirc}-\overset{O}{\overset{\|}{C}}-Cl \longrightarrow$$

$$\left[\overset{O}{\overset{\|}{C}}-\underset{}{\bigcirc}-\overset{O}{\overset{\|}{C}}-NH-(CH_2)_n-NH\right]_x \quad (1)$$

where $n = 2-7$

As shown in Table I these workers were able to obtain polymers with melting points (as indicated by differential thermal analysis) of 340°C and above with short-chain diamines up to heptamethylenediamine. Even the N-alkylated diamine, N,N'-dimethylethylenediamine, yielded a polymer with a crystalline melting point of 379°C. The polyamides of the corresponding N-substituted secondary diamines were also prepared and their melting points compared with those of the unsubstituted polyamides (see Fig. 1). The odd-even alternation of melting points commonly noted in the aliphatic polyamide series (28,29) is quite apparent in this aliphatic–aromatic series and in the N-substituted aliphatic–aromatic series. In addition to having extremely high polymer melt temperatures, these polyterephthalamides were soluble only in such strong acids as sulfuric and trifluoracetic acids. Thus, their properties were determined by the

TABLE I
Polyterephthalamides from Short-Chain Diamines (15)

Diamine	$\eta_{inh}{}^a$ (dl/g)	T_m (°C)^b	PMT (°C)^c
1,2-Ethylenediamine	1.00	455	>400
N,N'-Dimethyl-1,2-ethylenediamine	0.33	379	~370
Trimethylenediamine	1.70	399	>400
Tetramethylenediamine	1.30	436	>400
Pentamethylenediamine	2.00	353	~360
Hexamethylenediamine	0.9	371	~370
Heptamethylenediamine	0.6	341	~350

^{a} η_{inh}: inherent viscosity determined in 0.5 wt % solution of conc. sulfuric acid at 30°C.

^{b} T_m: crystalline melting point determined by DTA (heating rate 3°C/min under N_2).

^{c} PMT: polymer melt temperature. See R. G. Beaman and F. B. Cramer, *J. Polymer Sci.*, **21**, 223 (1956).

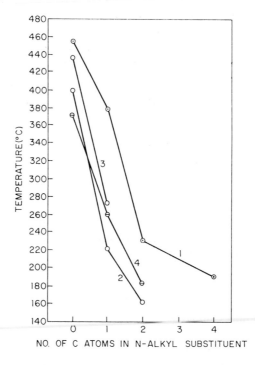

Fig. 1. Influence of *n*-alkyl chain length on the polymer melt temperature of polyterephthalamides (15). (*1*) poly(ethylene terephthalamide) series; (*2*) poly(propylene terephthalamide) series; (*3*) poly(tetramethylene terephthalamide) series; (*4*) poly(hexamethylene terephthalamide) series.

high concentration of strong polar bonds and the rigidity of the *p*-phenylene unit. Complete *N*-alkylation not only reduced the melting point markedly but also changed the solubility characteristics since hydrogen bonding was no longer possible.

Using interfacial (14) and solution polymerization (25–27) techniques, numerous workers have prepared aromatic polyamides from cycloaliphatic secondary diamines such as piperazine and its C-methylated derivatives. These polyamides as can be seen from Table II were extremely

TABLE II
Polyphthalamides (14,25–27)

Diacid	Diamine	η_{inh}^a (dl/g)	T_m (°C)[b]	PMT (°C)[c]
Phthalic	Piperazine	1.5	325	330
	2-Methylpiperazine	1.07	350	350
	trans-2,5-Dimethylpiperazine	1.06	>350	360
Isophthalic	Piperazine	0.62	340	340
	2-Methylpiperazine	0.53	280	285
	trans-2,5-Dimethylpiperazine	0.72	315	315
Terephthalic	Piperazine	1.35	350	>400
	2-Methylpiperazine	1.92	350	>375
	trans-2,5-Dimethylpiperazine	2.20	350	>400
	cis-2,5-Dimethylpiperazine	0.80	—	335
	cis-2,6-Dimethylpiperazine	1.09	—	>400
	2,2,5,5-Tetramethylpiperazine	0.98	—	>400
	α-2,3,5,6-Tetramethylpiperazine	0.33	—	>400
	β-2,3,5,6-Tetramethylpiperazine	0.18	—	345
2-Chloroterephthalic	*trans*-2,5-Dimethylpiperazine	0.74	—	>400

[a] η_{inh}: inherent viscosity determined in 0.5 wt % solution of conc. sulfuric acid at 30°C.

[b] T_m: crystalline melting point determined by DTA (heating rate 3°C/min under N_2.

[c] PMT: polymer melt temperature. See R. G. Beaman and F. B. Cramer, *J. Polymer Sci.*, **21**, 223 (1956).

high melting. Unlike the primary diamines, piperazine and its derivatives yielded polyamides in high molecular weight even with phthalic acid. Here again the reaction of a diacid chloride with the diamine at room temperature yielded high molecular weight film- and fiber-forming polymers (see reaction 2). One interesting property of the polyphthalamides was that unlike the polyamides from isophthalic and terephthalic acids, the *trans*-2,5-dimethylpiperazine polyamide melted higher than the piperazine polyamide. A possible explanation is that the restriction of free

$$\text{HN} \underset{R_1}{\overset{R_2}{\bigcirc}} \text{NH} + \text{Cl}-\overset{O}{\underset{\|}{C}}-\underset{}{\bigcirc}-\overset{O}{\underset{\|}{C}}-\text{Cl} \longrightarrow \left[-N \underset{R_1}{\overset{R_2}{\bigcirc}} N-\overset{O}{\underset{\|}{C}}-\underset{}{\bigcirc}-\overset{O}{\underset{\|}{C}}- \right]_x \quad (2)$$

where $R_1 = R_2 = H$ $R_1 = CH_3$; $R_2 = H$ $R_1 = R_2 = CH_3$

rotation in this laterally methyl-substituted piperazine polymer is much greater than in the piperazine polymer.

Another interesting property of these phthalamides was that the polymer melt temperatures of the 2-methyl- and 2,5-dimethylpiperazine polymers were higher than those of the isophthalamides. A comparison of the atomic models of the poly-1,4-phthaloyl-*trans*-2,5-dimethylpiperazine, **1**, and poly-1,4-isophthaloyl-*trans*-2,5-dimethylpiperazine, **2**, showed that the

1   **2**

chain of **1** could actually be better aligned than that of **2** and then pack more efficiently. This, coupled with the restricted free rotation of the chain due to the lateral methyl groups on the piperazine ring, probably explains why these phthalamides melt higher than the corresponding isophthalamides.

Another class of aromatic polyamides is that derived from 4,4'-sulfonyldibenzoic acid and short-chain aliphatic diamines and cycloaliphatic secondary diamines. Stephens (16) has prepared high melting polyamides of this type by interfacial polymerization of 4,4'-sulfonyldibenzoyl dichloride (see reaction 3).

As indicated in Table III, these polymers, although high melting, were

$$\text{Cl}-\overset{O}{\underset{\|}{C}}-\bigcirc-SO_2-\bigcirc-\overset{O}{\underset{\|}{C}}-\text{Cl} + \text{HN}-\underset{R_1}{|}-R-\underset{R_1}{|}-\text{NH} \longrightarrow$$

$$\left[-\overset{O}{\underset{\|}{C}}-\bigcirc-SO_2-\bigcirc-\overset{O}{\underset{\|}{C}}-\underset{R_1}{\overset{|}{N}}-R-\underset{R_1}{\overset{|}{N}} \right]_x \quad (3)$$

III. AROMATIC POLYMERS: FUNCTIONAL GROUPS IN CHAIN

TABLE III
Polyamides Derived from 4,4′-Sulfonyl Dibenzoic Acid (16)

Diamine	$\eta_{inh}{}^a$ (dl/g)	PMT (°C)[b]
1,2-Ethylenediamine	0.45	380
1,2-Propylenediamine	0.38	335
Tetramethylenediamine	0.63	358
Piperazine	1.28	> 380
2-Methylpiperazine	1.43	> 350
trans-2,5-Dimethylpiperazine	2.62	> 350
Hexamethylenediamine	1.28	310

[a] η_{inh}: inherent viscosity determined in 0.5 wt % solution of conc. sulfuric acid at 30°C.
[b] PMT: polymer melt temperature. See R. G. Beaman and F. B. Cramer, *J. Polymer Sci.*, **21**. 223 (1956).

somewhat lower melting than the corresponding terephthalamides, probably owing to the greater rigidity of the *p*-phenylene grouping over that of the sulfonylbiphenyl grouping. The melting points of these polymers especially those based on short-chain aliphatic diamines exhibited the usual odd-even alternation as observed previously in the all-aliphatic polyamides and in the aliphatic–aromatic and *N*-alkylated aliphatic–aromatic polyamides (Fig. 2). Inspection of Table III also shows that the ring-containing diamines gave higher melting polymers than the straight chain diamines containing a similar number of carbon atoms. This may be the case also in the corresponding terephthalamide series, but such an effect was not observable since the decomposition of the terephthalamides occurs long before the true melting point is reached. At least in the case of these

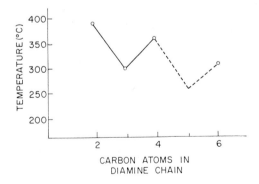

Fig. 2. Melting point vs. chain length for polyamides of 4,4′-sulfonyldibenzoic acid (16).

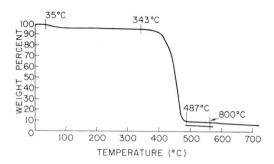

Fig. 3. TGA in nitrogen of polyamide from tetramethylene diamine and terephthalic acid (heating rate, 10°C/min).

sulfone-containing polymers, it appears that the entropy contribution to the melting point imparted by a ring structure is greater than the effect of the loss of hydrogen bonding due to the use of secondary diamines.

The differential thermal analysis and thermogravimetric analysis of the polyamide derived from tetramethylenediamine and terephthalic acid are representative of the previously discussed polymers (see Figs. 3–6). Both in air and in nitrogen, the curves from the thermogravimetric analysis of this polymer exhibit an initial break between 300 and 350°C, which appears to be characteristic of the aliphatic C—C bond, with less than 10% of original material remaining at 600°C. In the differential thermal analysis of this polymer in air and nitrogen, the curves exhibit sharp endotherms in the 419–431°C range, corresponding to those regions in the thermogravimetry curves where the most rapid weight losses occur, with additional endothermic activity up to 500°C.

Thus, in spite of the high melting point, the upper temperature use limit of these phenylene-R polymers is governed by the aliphatic portion of the polymer chain and precludes their utility under conditions involving long-term exposure at elevated temperature, i.e., above 350°C.

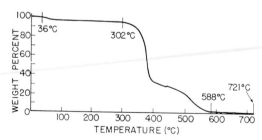

Fig. 4. TGA in air of polyamide from tetramethylene diamine and terephthalic acid (heating rate, 10°C/min).

III. AROMATIC POLYMERS: FUNCTIONAL GROUPS IN CHAIN 85

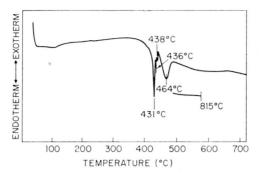

Fig. 5. DTA in nitrogen of polyamide from tetramethylene diamine and terephthalic acid (heating rate, 10°C/min).

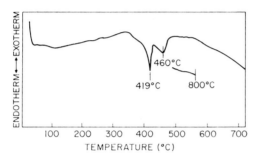

Fig. 6. DTA in air of polyamide from tetramethylene diamine and terephthalic acid (heating rate, 10°C/min).

2. Aromatic Polyhydrazides

The use of hydrazine as a diamine in the preparation of polyamides has been given considerable attention in the past (4). As might be expected, the difference in reactivity of the adjacent amino groups has made the preparation of polyhydrazides by orthodox polymerization techniques not wholly successful.

It has been reported (4) that the dihydrazide of sebacic acid on heating yielded a polymeric product which was too intractable to be spun successfully, and monohydrazine sebacate on melt polymerization yielded a brittle solid, melting at 295–300°C (5). Also the melt polymerization of dihydrazides in the presence of dicarboxylic acids has been reported to yield polymers with unidentified structures (4). On the other hand, the reaction product from the thermal polymerization of adipic acid and hydrazine was a fiber-forming polymer (32–34).

A somewhat more successful route to polyhydrazides by a high temperature solution process has been reported (35). This procedure involved the reaction of stoichiometric amounts of a dihydrazide with carboxylic acid esters or dicarbonyl chlorides in such solvents as nitrobenzene and mixed xylenols at temperatures of 170–200°C. It has been suggested (36) that the attainment of relatively low molecular weights (see Table IV) was due to the insolubility of the products in the reaction medium.

TABLE IV
Polyhydrazide Preparations in Nitrobenzene (36)

Polyhydrazide	Isophthalic dihydrazide plus	η_{inh} (DMSO)	PMT (°C)
Isophthalic	Isophthalic acid	0.08	250
Iosphthalic	Isophthaloyl chloride	0.15	280
Isophthalic	Dimethyl isophthalate	0.14	260
Isophthalic–terephthalic	Terephthalic acid	0.06	290
Isophthalic–terephthalic	Dimethyl terephthalate	0.21	255

By far the most successful route to high molecular weight polyhydrazides has been a low temperature solution technique involving the reaction of one or more acid chlorides with hydrazine or one or more dihydrazides in a basic medium which functions both as an acid acceptor and as a solvent for the produced polyhydrazide or copolyhydrazide (37).

$$NH_2-NH-\underset{\underset{O}{\|}}{C}-R_1-\underset{\underset{O}{\|}}{C}-NH-NH_2 + Cl-\underset{\underset{O}{\|}}{C}-R_2-\underset{\underset{O}{\|}}{C}-Cl \xrightarrow{\text{Basic medium}}$$

$$\left[NH-NH-\underset{\underset{O}{\|}}{C}-R_1-\underset{\underset{O}{\|}}{C}-NH-NH-\underset{\underset{O}{\|}}{C}-R_2-\underset{\underset{O}{\|}}{C}\right]_x + \text{Basic medium HCl} \quad (4)$$

Solvents which have been successfully used as reaction media are hexamethylphosphoramide and N-methylpyrrolidone. As in the case of most low temperature condensation solution polymerization methods, optimum molecular weights and yields of polyhydrazides were obtained when intermediates and solvents were of the highest purity and moisture rigorously excluded (36).

The aromatic and heterocyclic aromatic polyhydrazides (36) which have been prepared in high molecular weight have polymer melt temperatures ranging from 205 to over 400°C (see Table V). Surprisingly, all of these polymers with the exception of poly(terephthalic hydrazide) were soluble in cold dimethyl sulfoxide and hot tetramethylene sulfone. It has been suggested that this solubility is a function of the basicity of the hydrazide group and the acidity of these solvents (37).

Similar solubilities and, in some cases, high polymer melt temperatures

TABLE V
Aromatic and Heterocyclic–Aromatic Polyhydrazides (36)

Dihydrazide	Dicarbonyl chloride	η_{inh}[a] (dl/g)	PMT[b] (°C)
Isophthalic	Isophthaloyl	1.00	350
Isophthalic	Terephthaloyl	1.03	390
Isophthalic	Naphthalene-2,6	0.90	375
Isophthalic	5-Chloroisophthaloyl	0.90	350
Isophthalic	2,5-Dichloroterephthaloyl	0.60	390
Isophthalic	5-*tert*-Butylisophthaloyl	0.45	320
Isophthalic	Phenylether-4,4'	0.80	305
Isophthalic	1,10-Bis(3,formylphenoxy)-decane	0.30	205
Isophthalic	Pyrazine-2,5	0.70	380
Isophthalic	Isocinchomeronyl	0.46	345
Terephthalic	Terephthaloyl	ins.	>400
Isocinchomeronic	Isocinchomeronyl	0.81	370
Isocinchomeronic	Thiophene-2,5	0.77	360
Isocinchomeronic	Dinicotinoyl	0.62	350
Isocinchomeronic	Dipicolinoyl	0.41	350
Dipincolinic	Dipicolinoyl	0.72	360
Pyrazine-2,5	Pyrazine-2,5	0.72	350

[a] η_{inh}: inherent viscosity determined in 0.5 wt % solutions of dimethyl sulfoxide at 30°C.
[b] PMT: polymer melt temperature. See R. G. Beaman and F. B. Cramer, *J. Polymer Sci.*, **21**, 223 (1956).

were reported for certain copolyhydrazides, namely, those consisting of aliphatic–aromatic structures and oxalic–aromatic structures (37) (Table VI). The wholly aliphatic polyhydrazides not only melted at lower temperatures but were insoluble in dimethyl sulfoxide even at the boil and were found to be only slightly soluble in *m*-cresol (38).

The aromatic polyhydrazides gave polymeric chelates with mono- and divalent metal salts (39), as will be discussed in greater detail in Chapter V. Similarly, it has also been reported that these polyhydrazides yield interesting high temperature fibers (36). These fibers will be discussed in greater detail in Chapter VII.

Although the all-aromatic polyhydrazide and the aliphatic–aromatic and oxalic–aromatic polyhydrazides have polymer melt temperatures in excess of 300°C, these temperatures are neither true melting points nor true decomposition points (40). At 250°C and above the polyhydrazides undergo intramolecular cyclodehydration to poly-1,3,4-oxadiazoles which in themselves are quite heat resistant and thermally stable (40). The differential thermal analysis and thermogravimetric analysis of aliphatic,

TABLE VI
Aliphatic and Mixed Copolyhydrazide (37)

Dihydrazide	Dicarbonyl chloride	$\eta_{inh}{}^a$ (dl/g)	PMT[b] (°C)	DMSO[c] sol.
Oxalic	Isophthalic	1.43	342	+
Oxalic	Dipicolinic	1.01	350	+
Oxalic	Isocinchomeric	0.33	380	+
Adipic	Isophthalic	0.60	290	+
Sebacic	Isophthalic	1.00	260	+
Hexahydroterephthalic	Isophthalic	0.60	320	+
Hexahydroterephthalic	Isophthalic	0.40	310	+
Oxalic	Adipic	0.30	295	−
Oxalic	Sebacic	0.28	300	−
Adipic	Adipic	0.30	328	−
Adipic	Azelaic	0.27	318	−
Adipic	Sebacic	0.25	308	−
Adipic	Azelaic	0.23	280	−
Sebacic	Sebacic	0.20	280	−

[a] η_{inh}: inherent viscosity determined in 0.5 wt % solutions of dimethyl sulfoxide at 30°C.

[b] PMT: polymer melt temperature. See R. G. Beaman and F. B. Cramer, *J. Polymer Sci.*, **21**, 223 (1956).

[c] DMSO: dimethyl sulfoxide.

aliphatic–aromatic, aromatic, and oxalic–aromatic polyhydrazides (see Table VII) have been the subject of detailed investigation (41). The results are summarized in Tables VIII and IX. In this work unusually high rates of heating, 30°C/min, were required to obtain weight losses and exotherms which corresponded to decomposition of the polyhydrazide structure. At

TABLE VII
Polyhydrazides (41)

$$\left(R-\underset{\substack{\| \\ O}}{C}-\underset{NH-NH}{}-\underset{\substack{\| \\ O}}{C}-R'-\underset{\substack{\| \\ O}}{C}-\underset{NH-NH}{}-\underset{\substack{\| \\ O}}{C}- \right)_x$$

Polymer	R	R'
A	1,3-Phenylene	1,4-Phenylene
B	1,3-Phenylene	1,3-Phenylene
C	1,3-Phenylene	
D	1,3-Phenylene	1,4-Tetramethylene
E	1,3-Phenylene	1,8-Octamethylene
F	1,4-Tetramethylene	1,4-Tetramethylene
G	1,8-Octamethylene	1,8-Octamethylene

TABLE VIII
Thermogravimetry of Polyhydrazides (41)

Polymer[a]	$(T_i/T_f)_1$[b]	Δ_1[c]	$(T_i/T_f)_2$[d]	Δ_2[e]	$(w/w_0)_{700}$[f]	$(\Delta_1)_{calc}$
A	275/390	10.6	466/550	32.8	40.6	11.0
B	280/376	10.4	428/611	32.5	55.0	11.0
C	275/367	34.3	367/541	30.5	22.8	14.5
D	275/352	11.8	352/495	35.5	38.9	11.8
E	251/357	9.7	357/513	63.9	21.2	10.0
F	285/362	17.2	362/499	55.8	23.5	12.7
G	266/381	10.7	381/513	79.0	8.6	9.8

[a] Polymer: see Table VII.
[b] $(T_i/T_f)_1$ = (onset temp., °C)/(terminus temp., °C), step 1.
[c] $\Delta_1 = (\Delta w/w_0)100$, step 1.
[d] $(T_i/T_f)_2$ = (onset temp., °C)/(terminus temp., °C), step 2.
[e] $\Delta_2 = (\Delta w/w_0)100$, step 2.
[f] $(w/w_0)_{700}$ = % original weight remaining at 700°C; w = sample weight at temperature T; w_0 = original sample weight (based on dry polymer).

lower heating rates the polyhydrazides quantitatively converted to poly-1,3,4-oxadiazoles.

One concludes that the aromatic polyhydrazides, like the aliphatic short chain polyamides and the polyamides derived from cyclic secondary

TABLE IX
Differential Thermometry of Polyhydrazides (41)[a]

Polymer	
A	−132(s), −285(sh)* → −307(s)*, +390(w), +502(s)**
B	−122(s), −264(ww), −324(s)*, +483(s)**
C	−135(s), +285(m)*, −321(s), +338(s)**
D	−134(sh) → −151(s), −209(s) ← −236(sh), −288(m)*, +367(s)**, +398(s)**
E	−122(sh) → −138(s), −182(s), −238(s), −260(s)* → −269(m)*, +343(m)**, +430(w)**, +500(ww)**
F	−122(s), −316(s)*, +353(s)**, +397(s)**
G	−110(s), −269(sh)* → −278(s)*, +333(s), +381(m)**

[a] Polymer: see Table VII.
Endotherm peak: −.
Exotherm peak: +.
* Peak corresponding to first break in TG curve (excluding adsorbed water).
** Peak corresponding to second break in TG curve (excluding adsorbed water).
Relative peak size: ww = very weak; w = weak; m = medium; s = strong; sh = shoulder; x(sh) → y or y ← z(sh) denotes a shoulder at temperature x (or z) which precedes (i.e., x) or follows (i.e., z) the main peak. The main peak itself reaches a maximum temperature at y.

diamines, would have a ceiling temperature in use, as indicated by differential thermal analysis and thermogravimetric analysis, in a 300–400°C range. Although they are superior to the poly-*p*-xylyenes, they are still inferior in stability to poly-*p*-phenylene.

3. Polyamides from Primary Aromatic Diamines and Aromatic Dicarboxylic Acids

The polyamides derived from aromatic diamines and aromatic dicarboxylic acids have been extremely difficult to prepare via melt polymerization. The major difficulty introduced by the use of these intermediates, where the functional amide forming groups are attached directly to aromatic nuclei, lies in the high melting point of the resulting polymers, so high that they cannot be fused without decomposition. This high melting character can be attributed to the combined effects of chain stiffening phenylene group and the intramolecular forces of hydrogen bonding. Aromatic diamines such as *meta*- and *para*-phenylenediamines are also difficult to employ in melt polycondensation because they do not form well-defined salts with organic acids, are slower in amidation (poorer nucleophiles than aliphatic diamines) and undergo side reactions before and during polymerization (7).

Some success has been reported in the preparation of this type of polymer by way of melt polycondensations of (*1*) aromatic diamines with dimethyl esters of aromatic dicarboxylic acids (8), and (*2*) *N*-acetylamino-aromatic acids (9). Although high molecular weight products were claimed for both routes, polymer characterization and properties were so scanty that a definitive statement as to the efficacy of these polymerization methods is impossible.

High molecular weight all-aromatic polyamides have been prepared by both solution and interfacial polymerization methods (42–51). Utilizing the reaction of aromatic diacid chlorides and aromatic diamines in these low temperatures polycondensation processes, numerous workers have

$$Cl-\overset{O}{\overset{\|}{C}}-Ar_1-\overset{O}{\overset{\|}{C}}-Cl + H_2N-Ar_2-NH_2 \longrightarrow \left[\overset{O}{\overset{\|}{C}}-Ar_1-\overset{O}{\overset{\|}{C}}-NH-Ar_2-NH\right]_x \quad (5)$$

reported the preparation of all-aromatic polyamides and copolyamides. Representative examples of these polyamides are listed in Table X.

It is apparent from these data that the polyamides derived from phthaloyl chloride and/or *o*-phenylenediamine have polymer melt temperatures too low to be of interest within the framework of this discussion. The polyamides from isophthaloyl chloride and/or *m*-phenylenediamines,

TABLE X
Aromatic Polyamides and Copolyamides

Acid chlorides	Diamine	$\eta_{inh}{}^a$ (dl/g)	PMT[b] (°C)	Ref.
	A. Polymers			
Isophthaloyl (I)	2,4-Diaminotoluene	2.30	330	43
	2,5-Diaminotoluene	0.67	350	42
	2,4-Diaminoanisole	0.84	300	43
	3,3'-Dimethylbenzidine	1.70	365	42
	Bis(4-aminophenyl)methane	1.86	350	42
	2,2'-Bis(4-aminophenyl)propane	1.2	375	42
	4,6-*m*-Xylenediamine	0.81	352	42
	4,4'-Sulfonyldiphenyl diamine	1.66	>400	43
	o-Phenylenediamine	—	~240	45
	m-Phenylenediamine	0.72	>400	42, 44, 45
	p-Phenylenediamine	0.39	>400	44, 45
Terephthaloyl (T)	4,6-*m*-Xylenediamine	0.72	365	43
	2,4-Diaminotoluene	1.37	365	43
	o-Phenylenediamine	—	~295	45
	m-Phenylenediamine	1.04	>400	43, 44, 45
	p-Phenylenediamine	—	>400	44, 45
	2,2'-Bis(4-aminophenyl)propane	1.8	>400	43
	Benzidine	—	>360	45
	3,3'-Dimethoxybenzidine	—	>200 d	45
	Bis(4-aminophenyl)methane	—	>360	45
	4,4'-Diaminostilbene	—	>360	45
	4,4'-Diaminostilbene 2,2'-Disulfonic acid	—	>360	45
	4,4-Sulfonyldiphenyldiamine	—	~275	45
	4,4-Sulfoxyldiphenyldiamine	—	~260	45
	2,4-Diaminoazobenzene	—	~320 d	45
	1,5-Diaminonaphthalene	—	~320 d	45
	3,6-Diaminoacridine	—	—	45
4-Chloroisophthalyl	*m*-Phenylenediamine	0.84	305	43
Phthaloyl chloride (P)	*o*-Phenylenediamine	—	~185	45
	m-Phenylenediamine	—	~200	45
	p-Phenylenediamine	—	~185	45
	B. Copolymers			
T/I (20/80)	*m*-Phenylenediamine	0.57	340	44
(40/60)		0.73	340	44
(60/40)		0.46	>450	44
(80/20)		0.38	>480	44

(continued)

TABLE X (continued)

Acid chlorides	Diamine	η_{inh}^a (dl/g)	PMT[b] (°C)	Ref.
T/I (20/80)	p-Phenylenediamine	0.33	470	44
(40/60)		0.85	455	44
(60/40)		0.20	485	44
(80/20)		0.14	>500	44

[a] η_{inh}: inherent viscosity determined in 0.5 wt % solution of conc. sulfuric acid at 30°C.

[b] PMT: polymer melt temperature. See R. G. Beaman and F. B. Cramer, *J. Polymer Sci.*, **21**, 223 (1956).

as might be expected, have much lower polymer melt temperatures and are more tractable than those from terephthaloyl chloride and/or p-substituted aromatic diamines.

Using these low temperature polycondensation methods, polymers from AB monomers of the type **3**, where R may be sulfonyl or carbonyl chloride

$$\bar{X}\ H_3N^+ \underset{R'}{\underset{|}{\bigcirc}}\!\!-R$$

3

and R' may be hydrogen, halogen, aryl, nitro, or alkyl up to 6 carbons have been prepared (46). These polyamides were reported to have polymer melt temperatures in excess of 400°C and, in general, to be soluble in dialkylamide solvents such as *N,N*-dimethylacetamide and *N,N*-dimethylformamide. Similarly, copolyamides of these AB type of monomers with aromatic diamines and aromatic diacid chlorides with comparable properties have been reported.

In order to increase the tractability of aromatic polyamides, yet retain the higher polymer melt temperatures, recent work has been directed toward the preparation of ordered alternating copolyamides containing *m*- or *p*-aminobenzoyl moieties (47–51). Copolyamides with structures such as **4** or **5**, where Ar_1, Ar_2, Ar_3, Ar_4 are divalent aromatic groups

$$-RN-Ar_1-NR-\overset{O}{\overset{\|}{C}}-Ar_1-\overset{O}{\overset{\|}{C}}-NR-Ar_1-NR-\overset{O}{\overset{\|}{C}}-Ar_2-\overset{O}{\overset{\|}{C}}-$$

4

$$-RN-Ar_1-\overset{\overset{O}{\|}}{C}-NR-Ar_2-NR-\overset{\overset{O}{\|}}{C}-Ar_3-NR-\overset{\overset{O}{\|}}{C}-Ar_4-\overset{\overset{O}{\|}}{C}-$$
5

containing one or more rings or a condensed ring system, at least one having multicyclic constitution and R is hydrogen, 1 to 3 carbon alkyl, or phenyl, have been prepared by the reaction of aromatic diacid chlorides with diamines of structures **6** and **7** (47–51).

6

$$R_2N-Ar_1-\overset{\overset{O}{\|}}{C}-NR-Ar_2-NR-\overset{\overset{O}{\|}}{C}-Ar_3-NR_2$$
7

In the case of the class of diamines, **7**, five such aromatic diamines have been synthesized and used as polymer intermediates. These were:

N,N'-m-phenylene bis(m-aminobenzamide)	**7A**
N,N'-m-phenylene bis(p-aminobenzamide)	**7B**
N,N'-p-phenylene bis(m-aminobenzamide)	**7C**
N,N'-p-phenylene bis(p-aminobenzamide)	**7D**
2,7-naphthalene bis(m-aminobenzamide)	**7E**

The alternating copolyamides prepared from these diamines are listed in Table XI.

The alternating ordered copolymers containing a high percentage of p-substituted rings exhibited higher polymer melt temperatures than the m-substituted aromatic polyamides and approach that of the p-substituted aromatic polyamides. This effect was obtained with increased solubility, as all of these alternating copolyamides were reported to be soluble in N,N-dimethylacetamide containing 5% lithium chloride. The increased solubility was demonstrated by the preparation of tough films and fibers from these N,N-dimethylacetamide solutions.

A large amount of data is available on the thermal stability of fibers, films, papers, etc., prepared from aromatic polyamides, copolyamides and alternating ordered copolyamides, and these data, along with the physical and chemical properties of the shaped articles, will be discussed in greater detail in Chapter VII.

TABLE XI
Ordered Aromatic Copolyamides (47–51)

Acid chloride	Diamine	$\eta_{inh}{}^a$ (dl/g)	PMT[b] (°C)	Ref.
Isophthaloyl	6	1.86	>400	47
	7A	2.24	430	49, 51
	7B	1.36	475	49, 51
	7C	—	460	49, 51
	7D	—	>500	49, 51
Terephthaloyl	7A	0.65	>430	49, 51
	7B	0.68	490	49, 51
	7C	—	467	49, 51
	7D	—	>500	49, 51
2,6-Naphthalenedicarbonyl chloride	7A	2.24	495	48, 51
	7B	1.38	>500	48
	7E	1.15	455	48
4,4′-Bibenzoyl chloride	7A	0.86	432	48, 51

[a] η_{inh}: inherent viscosity determined in 0.5 wt % solution of conc. sulfuric acid at 30°C.
[b] PMT: polymer melt temperature. See R. G. Beaman and F. B. Cramer, *J. Polymer Sci.*, **21**, 223 (1956).

The most comprehensive study of the thermal stability of the aromatic polyamides was carried out by Wright and co-workers (45). They studied the relative thermal stabilities in vacuum of polyamides based on phenylenediamines with phthalic, isophthalic, and terephthalic acid using a quartz spring balance and a rate of temperature rise of approximately 3°C/min. The results are summarized in Figure 7.

The polymers could be divided into two groups, the stabilities of which differed by almost 200°C. The group with poorer stability consisted of polymers containing at least one *ortho*-linked component. In the second group, the stability increased as the structure changed from *meta–meta* to *meta–para* to *para–para* linkages. The lower stability of the polymers containing *ortho*-links may be attributed to the increased steric hindrance to rotation and the possibility of the intramolecular hydrogen bonding, which may facilitate cleavage of the chain. Although the lower molecular weight of *ortho*-linked polymers may influence the stability, the increased steric hindrance is a real effect as indicated by molecular models.

Since the *para–para* linkages gave the greatest thermal stability by this test, other aromatic polyamides from terephthaloyl chloride were prepared. The diamines used were *para*-substituted wherever possible and were

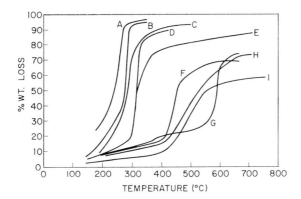

Fig. 7. Relative thermal stability of polyamides derived from phthaloyl, isophthaloyl, and terephthaloyl chlorides and o-, m-, and p-phenylene diamines (45). (A) Polymer from phthaloyl chloride and m-phenylene diamine; (B) polymer from phthaloyl chloride and o-phenylene diamine; (C) polymer from phthaloyl chloride and p-phenylene diamine; (D) polymer from isophthaloyl chloride and o-phenylene diamine; (E) polymer from terephthaloyl chloride and o-phenylene diamine; (F) polymer from isophthaloyl chloride and m-phenylene diamine; (G) polymer from terephthaloyl chloride and p-phenylene diamine; (H) polymer from isophthaloyl chloride and p-phenylene diamine; (I) polymer from terephthaloyl chloride and m-phenylene diamine.

selected to give only slight variations in overall structure of the resultant polymer. The oxidative stabilities of these polymers were compared by heating a 100-mg sample in air at 300°C for periods up to 74 hr. The results are given in Table XII.

It was found that the most stable polymers were those derived from diaminostilbenedisulfonic acid, benzidene, diaminonaphthalene, diaminodiphenyl sulfone, and diaminobenzene. The polymer from diaminostilbenedisulfonic acid after initial loss of 4% by weight showed a rate of loss of approximately 0.01%/hr. The polymers from the other four diamines named had rates of weight loss between 0.04 and 0.05%/hr. The initial weight gain of the polymer derived from diaminodiphenyl methane was indicated by infrared spectral data to be due to the oxidation of the methylene group to carbonyl.

Preston (50) has compared the thermal stability of aromatic ordered copolyamides and aromatic random copolyamides by thermogravimetric analysis and differential thermal analysis. He found that the ordered copolymers were much more stable and that thermal stability increased as the p-phenylene content increased.

Although not all the polyamides and copolyamides listed have been the subject of such comprehensive study, in practically all cases, melting was

TABLE XII
Oxidative Stabilities of Polyamides Based on Terephthaloyl Chloride (45)

Diamine	Wt loss % in air at 300°C after						
	2 hr	2.5 hr	7 hr	8 hr	40 hr	72 hr	74 hr
m-Phenylenediamine	1.8			2.4	5.3	6.8	
p-Phenylenediamine	2.7			3.0	7.4	10.6	
Benzidine	1.2			1.9	3.2	4.4	
4,4'-Diaminodiphenylmethane		+1.4		0.9	7.7	14.2	
4,4'-Diaminostilbene	2.2			8.2		21.7	
4,4'-Diaminostilbene-2,2'-disulfonic acid		3.3		4.2	4.8		5.1
4,4'-Diaminodiphenyl sulfoxide	4.2			4.5	7.4	8.7	
4,4'-Diaminodiphenyl sulfone	2.8			2.2	4.3	5.5	
2,4-Diaminoazobenzene	2.3		3.0		4.7	6.2	
1,5-Diaminophthalene	1.0		1.9		3.3	4.7	
3,6-Diaminoacridine		4.2	5.1		12.3		19.7

accompanied by decomposition. Thus, the polymer melt temperatures which were determined by differential thermal analysis were indicative of thermal stability.

By this measure of thermal stability, stability increased in going from *meta–meta* to *meta–para* to *para–para* linkages. Symmetrical substitution in the benzene ring increased thermal stability whereas unsymmetrical substitution decreased thermal stability. In those polymers where the benzene rings were connected by a functional group other than an amide, the following order of decreasing stability was observed:

When resistance to oxidative attack was taken as the criterion of stability, the substituents which were readily oxidizable such as alkyl groups or which activated the ring such as methoxyl groups decreased the stability of the polymer. Similarly for those polymers where the benzene rings were connected by a functional group other than amide the order of stability changed in the following manner:

—Ph—Ph— > —Ph—O—Ph— >

—Ph—SO$_2$—Ph— > —Ph(CH$_3$)—Ph(CH$_3$)— >

—Ph—CH$_2$—Ph— > —Ph—C(CH$_3$)(CH$_2$)—Ph— >

—Ph—CH$_2$—CH$_2$—Ph—

B. AROMATIC POLYANHYDRIDES

The aromatic polyanhydrides unlike the aromatic polyamides have been prepared in high molecular weight by melt condensation methods.

The first successful preparation of aromatic polyanhydrides in high molecular weight was reported by Conix (52) utilizing a dicarboxylic acid and acetic anhydride (see reaction 6).

$$\text{HO}-\overset{\overset{\text{O}}{\|}}{\text{C}}-\text{Ar}-\overset{\overset{\text{O}}{\|}}{\text{C}}-\text{OH} + \text{CH}_3-\text{CO}-\text{O}-\text{CO}-\text{CH}_3 \xrightarrow{-\text{CH}_3\text{CO}_2\text{H}}$$

$$\text{CH}_3-\overset{\overset{\text{O}}{\|}}{\text{C}}-\text{O}-\overset{\overset{\text{O}}{\|}}{\text{C}}-\text{Ar}-\overset{\overset{\text{O}}{\|}}{\text{C}}-\text{O}-\overset{\overset{\text{O}}{\|}}{\text{C}}-\text{CH}_3 \xrightarrow{-\text{CH}_3\text{CO}-\text{O}-\text{COCH}_3} \quad (6)$$

$$\left[-\text{O}-\overset{\overset{\text{O}}{\|}}{\text{C}}-\text{O}-\text{Ar}-\overset{\overset{\text{O}}{\|}}{\text{C}}-\right]$$

Using this general procedure, Conix (52,53) and Yoda (54–60) have prepared a large number of aromatic polyanhydrides, aromatic copolyanhydrides, heterocyclic polyanhydrides, and aromatic heterocyclic copolyanhydrides. Some of these polymers are listed in Tables XIII and XIV.

TABLE XIII
Aromatic Polyanhydrides (52,53,54–60)

Polyanhydride from	T_m^a (°C)	T_g^b (°C)
Terephthalic acid	400	—
Isophthalic acid	259	—
1,1-Bis(4-carboxyphenyl)methane	332	122
1,2-Bis(4-carboxyphenyl)ethane	340 d	—
1,4-Bis(4-carboxyphenyl)butane	263	46
Bis(4-carboxyphenyl) ether	296	—
1,2-Bis(4-carboxyphenoxy)ethane	340	122

^{a} T_m: crystalline melting point determined by DTA (heating rate 3°C/min under N_2).
^{b} T_g: glass transition temperature.

TABLE XIV
Aromatic–Heterocyclic Copolyanhydrides (54–60)

Heterocyclic acid^a	Mole % heterocyclic acid	PMT^b (°C)
Thiophene-2,5-dicarboxylic acid	100	80
	40	280
	30	
	20	350
	10	295
Furan-2,5-dicarboxylic acid	100	80
	40	310
	30	330
	20	380
	10	400
Tetrahydrofuran-2,5-dicarboxylic acid	100	90
	40	280
	30	320
	20	360
	10	400
N-Methyl-pyrrole-2,5-dicarboxylic acid	100	120
	40	320
	30	240
	20	390
	10	405
N-Methyl-tetrahydropyrrole-2,5-dicarboxylic acid	100	120
	40	280
	30	310
	20	370
	10	400

^{a} Copolymer with terephthalic acid.
^{b} PMT: polymer melt temperature. See R. G. Beaman and F. B. Cramer, *J. Polymer Sci.*, **21**, 223 (1956).

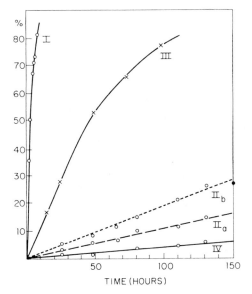

Fig. 8. Stability against alkaline hydrolysis of different polyanhydrides (weight percent hydrolysis vs. time of exposure to N NaOH) (53). (I) Poly(sebacic anhydride); (IIa) crystalline polyanhydride from di-p-carboxyphenoxy-1,3-propane; (IIb) amorphous polyanhydride from di-p-carboxyphenoxy-1,3-propane; (III) polyanhydride from di-p-carboxyphenoxy methane (amorphous, difficult to crystallize); (IV) crystalline poly(ethylene terephthalate).

A large number of these polymers have been fabricated into fibers and films (53,54) with interesting properties. The aromatic crystalline polyanhydrides, unlike the aliphatic polyanhydrides, were hydrolytically stable (53) (see Fig. 8). In fact, the poly(terephthalic anhydride) has been shown to be as stable as poly(ethylene terephthalate) toward alkaline hydrolysis. In addition, this particular aromatic polyanhydride was found (61) to be resistant to attack by hydrochloric acid, nitric acid, and even aqua regia.

From an inspection of Tables XIII and XIV, it is apparent that only a limited number of these polyanhydrides have polymer melt temperatures in excess of 300°C. By the standards applied in the previous section, only the poly(terephthalic anhydride) is of interest. Although this polymer has a crystalline melting point in excess of 400°C, no pertinent thermal data such as differential thermal analysis or thermogravimetric analysis are available. However, it has been shown (30,31) for lower melting polyanhydrides that, based on thermogravimetry data, the anhydride linkage and carbonate linkage are equivalent, exhibiting thermal breakdown in the 320–380°C region. This relatively low level of thermal stability was to be

C. AROMATIC POLYCARBONATES

Aromatic polycarbonates have been prepared in high molecular weight by numerous methods. The methods which have been most widely applied are: (*1*) phosgenation in the presence of pyridine; (*2*) interfacial polycondensation; and (*3*) transesterification.

Phosgenation of aromatic dihydroxy compounds in pyridine or pyridine–solvent mixtures is carried out at room temperature in the absence of water (62–64) (see reaction 7). In this procedure the reaction medium

$$HO-Ar-OH + COCl_2 \xrightarrow[\text{Reaction medium}]{\text{Pyridine}} \left[-OAr-O-\overset{\overset{O}{\|}}{C}- \right] + \langle\bigcirc\rangle N \cdot HCl \quad (7)$$

is a solvent for the polymer produced. The molecular weight of the polymer is influenced by reaction temperature, quantity of pyridine, rate of addition of phosgene, and purity of reactants and solvents.

In the interfacial polycondensation process, aqueous alkaline solutions of aromatic dihydroxy compounds were made to react with phosgene or with bischlorocarbonic acid esters of aromatic dihydroxy compounds in the presence of inert solvent (65–68). In this process, reaction temperatures

$$NaO-Ar-ONa + \begin{array}{c} COCl_2 \\ \text{or} \\ \overset{O}{\underset{\|}{Cl-C}}-O-Ar-O-\overset{O}{\underset{\|}{C}}-Cl \end{array} \longrightarrow \left[-O-Ar-O-\overset{\overset{O}{\|}}{C} \right]_x \quad (8)$$

of 0–40°C, pH in excess of 10, and the exclusion of oxygen were required for high molecular weight products. Catalytic amounts of tertiary amines, quaternary ammonium bases, quaternary phosphonium compounds, quaternary arsenium compounds, and tertiary sulfonium compounds accelerated this reaction (68–70).

In the two previous methods, high molecular weight polymers have been prepared only when the product polycarbonate was soluble in the inert organic reaction mixture. The transesterification route does not suffer from this limitation and involves the reaction of (*1*) an aromatic

III. AROMATIC POLYMERS: FUNCTIONAL GROUPS IN CHAIN

dihydroxy compound with diphenyl carbonate or bis-aryl carbonate or (2) aryl-bis carbonates with themselves (71–73) (see reactions 9a and 9b).

(1) $HO-Ar_1-OH + Ar_2-O-\overset{O}{\underset{\|}{C}}-OAr_2 \rightleftharpoons$

$\left[-O-Ar_1-O-\overset{O}{\underset{\|}{C}}-\right] + Ar_2OH$ (9a)

(2) $Ar_2-O-\overset{O}{\underset{\|}{C}}-O-Ar_1-O-\overset{O}{\underset{\|}{C}}-O-Ar_2 \rightleftharpoons$

$\left[-O-Ar_1-O-\overset{O}{\underset{\|}{C}}-\right] + Ar_2-O-\overset{O}{\underset{\|}{C}}-O-Ar_2$ (9b)

In these equilibrium reactions, high yields and high molecular weight required the removal of the by-product diphenyl carbonate or phenol. Basic catalysts such as alkali metals and alkaline earth metals, oxides, hydrides, etc., were found to accelerate transesterification to a significant degree (74,75).

TABLE XV
Aromatic Polycarbonates

Structure	PMT (°C)[a]	T_m (°C)[b]	Ref.
(bisphenol diphenyl carbonate structure)	> 350	225–235	74
(triphenyl ether carbonate structure)	250	—	74
(dichloro biphenyl carbonate structure)	~350	—	74
(bisphenol-CH-phenyl carbonate with chlorinated bicyclic R group)	> 350	297	75

(continued)

TABLE XV (*continued*)

Structure	PMT (°C)[a]	T_m (°C)[b]	Ref.
(structure with CH-bridged bisphenol ester and hexachloro bicyclic R group)	> 350	285	75
—O—⟨⟩—O—C(=O)—	> 350	—	65
—O—⟨⟩—O—C(=O)— (meta)	~300	—	65
[naphthalene-1,5-diyl with O and O—C(=O)]	> 350	—	65
[naphthalene-1,6-diyl with O and O—C(=O)]	> 350	—	65
—O—⟨⟩—⟨⟩—O—C(=O)— (biphenyl)	> 350	—	65

[a] PMT: polymer melt temperature. See R. G. Beaman and F. B. Cramer, *J. Polymer Sci.*, **21**, 223 (1956).

[b] T_m: crystalline melting point determined by DTA (heating rate 3°C/min N_2).

III. AROMATIC POLYMERS: FUNCTIONAL GROUPS IN CHAIN 103

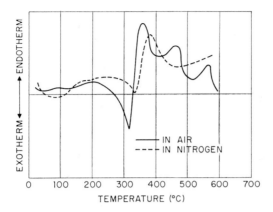

Fig. 9. Effect of oxygen upon thermal stability of poly-2,2-propane-bis-4-phenyl carbonate (differential thermogram) (76).

Although since 1956 over 300 patent applications, issued patents, and technical articles on polycarbonates have appeared, only a few polycarbonates have been described which melt over 300°C and these are inadequately characterized structures. In Table XV are listed some of the polycarbonate structures which fall within the framework of this discussion.

The reduced interchain forces in the polycarbonates, as in the polyanhydrides, as evidenced by the low cohesion energy per polymer unit, result in lower melting polymers for a given aromatic structure; i.e., a p,p-substituted polyamide of a given structure has a polymer melt temperature much greater than a p,p-substituted polycarbonate of the same

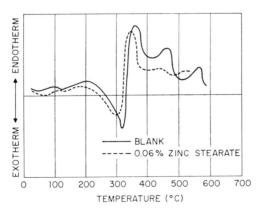

Fig. 10. Effect of zinc stearate upon thermal stability of poly-2,2-propane-bis-4-phenyl carbonate (differential thermogram) (76).

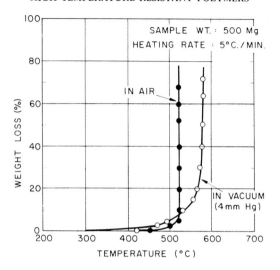

Fig. 11. TGA curves for the degradation of poly-2,2-propane-bis-4-phenyl carbonate (76).

structure. Although no thermal data are available on these higher melting polycarbonates, the thermal stability of lower melting polycarbonates has been studied.

Lee (76) found by differential thermal analysis that poly-2,2-propane-bis-4-phenyl carbonate degraded at temperatures above 310°C (Fig. 9). This first stage of degradation of the polymer appeared to be induced by oxygen. Zinc stearate acted as an oxidation catalyst (see Fig. 10) with the

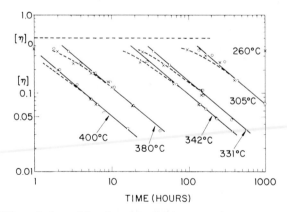

Fig. 12. Effect of time of heating, in vacuum, at various temperatures, on the intrinsic viscosity, $[\eta]$, of poly-2,2-propane-bis-4-phenyl carbonate (77). Experimental points, (t) -----×-----; corrected points, $(t + t_0)$ ——○——.

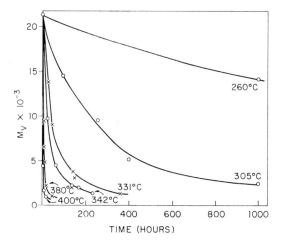

Fig. 13. Effect of time of heating, in vacuum, at various temperatures, on the molecular weight, $[M_v]$, of poly-2,2-propane-bis-4-phenyl carbonate (77).

initial site of oxygen attack being at the isopropylidene linkage in the polymer chain. The second stage of degradation showed a characteristic endothermic peak between 340 and 380°C associated with depolymerization or thermolysis of the carbonate linkage itself. Similarly, the thermogravimetric analysis of these polymers show that there was little weight loss up to 380°C, but above 400°C, the rate rapidly increased and became uncontrollable at 500°C or above (Fig. 11).

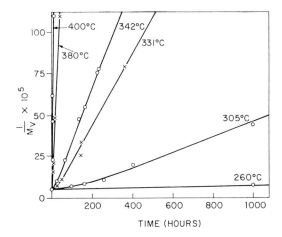

Fig. 14. Relationship between the reciprocal molecular weight, $[M_v^{-1}]$, and time of heating of poly-2,2-propane-bis-4-phenyl carbonate at different temperatures (77).

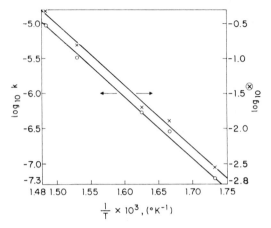

Fig. 15. Effect of temperature on rate constant, k (sec^{-1}), ──O──. Effect of temperature on viscosity, expressed by $\otimes$ (hr^{-1}) ($\otimes = (t[\eta]' - t[\eta]'')^{-1}$), ──×── for poly-2,2-propane-bis-4-phenyl carbonate (77).

Golden and co-workers (77) studied the thermal degradation of this same polycarbonate in vacuum (see Figs. 12–15) and found that the degradation was a random scission process with an activation energy of 39.5 kcal and a frequency factor of 4.8×10^7.

Thus, it would appear that the aromatic polycarbonates like the aromatic polyanhydrides have a much lower order of thermal stability than the previously discussed polyamides and probably will not find utility in high temperature applications.

D. AROMATIC POLYESTERS

The aromatic polyesters, like the preceding aromatic polycarbonates, have been prepared by several methods. Among the more widely used methods were melt polycondensation, low temperature interfacial polymerization, and high temperature solution polymerization.

A typical example of the melt polycondensation route to these polymers is the polymerization of *m*-hydroxybenzoic acid and copolymerization of this acid with *p*-hydroxybenzoic acid (78–80). In this procedure the acetoxy derivative of the hydroxy acid was made to react in the manner outlined in reaction 10. The acetic acid formed was displaced by the aromatic carboxy group, and elimination of acetic acid resulted in the formation of high molecular weight polymer. This reaction proceeded without a catalyst, but higher molecular weight polymers were formed and faster reaction resulted when a catalyst such as magnesium, sodium acetate, or zinc acetate was used. This general procedure has also been applied to the reaction of bisphenols (diacyloxy derivatives) with dicarboxylic acids (82–100).

$$\text{HO}_2\text{C}-\!\!\bigcirc\!\!-\text{O}-\!\!\overset{\overset{\text{O}}{\|}}{\text{C}}\!\!-\text{CH}_3 \longrightarrow$$

$$\text{HO}_2\text{C}-\!\!\bigcirc\!\!-\text{O}\left[\overset{\overset{\text{O}}{\|}}{\text{C}}\!\!-\!\!\bigcirc\!\!-\text{O}\right]_{n-2}\!\!\overset{\overset{\text{O}}{\|}}{\text{C}}\!\!-\!\!\bigcirc\!\!-\text{O}-\!\!\overset{\overset{\text{O}}{\|}}{\text{C}}\!\!-\text{CH}_3 \;+$$

$$(n-1)\text{CH}_3-\!\!\overset{\overset{\text{O}}{\|}}{\text{C}}\!\!-\text{OH} \quad (10)$$

The low temperature interfacial route was exemplified by the preparation of the aromatic polyester from 2,2-bis(hydroxyphenyl)propane and isophthaloyl chloride (101,102). In this procedure, an equivalent of alkali diphenate in water was rapidly stirred at room temperature with an equivalent of the diacid chloride in a suitable organic solvent, the solvent being a swelling agent or a solvent for the polymer formed (see reaction 11).

$$\text{Na}-\text{O}-\!\!\bigcirc\!\!-\!\!\underset{\underset{\text{CH}_3}{|}}{\overset{\overset{\text{CH}_3}{|}}{\text{C}}}\!\!-\!\!\bigcirc\!\!-\text{O}-\text{Na} + \text{Cl}-\!\!\overset{\overset{\text{O}}{\|}}{\text{C}}\!\!-\!\!\bigcirc\!\!-\!\!\overset{\overset{\text{O}}{\|}}{\text{C}}\!\!-\text{Cl} \longrightarrow$$

$$\left[\overset{\overset{\text{O}}{\|}}{\text{C}}\!\!-\!\!\bigcirc\!\!-\!\!\overset{\overset{\text{O}}{\|}}{\text{C}}\!\!-\text{O}-\!\!\bigcirc\!\!-\!\!\underset{\underset{\text{CH}_3}{|}}{\overset{\overset{\text{CH}_3}{|}}{\text{C}}}\!\!-\!\!\bigcirc\!\!-\text{O}\right]_x \quad (11)$$

These reactions were markedly catalyzed by small amounts of quaternary ammonium or sulfonium compounds. This polymerization method has been broadly applied, especially in the preparation of aromic polyesters derived from polynuclear bisphenols and bisphenols of the general formula **8** (84,85,88,103–105).

$$\text{HO}-\!\!\bigcirc\!\!-\text{X}-\!\!\bigcirc\!\!-\text{OH}$$
$$\mathbf{8}$$

The high temperature solution polymerization of aromatic polyesters has been utilized for the preparation of aromatic polyesters from bisphenols and aromatic diacid chlorides (106–117). In this general procedure, the bisphenols and the diacid chlorides were made to react in a suitable high boiling reaction medium at elevated temperatures in the manner outlined in reaction 12.

$$HO-Ar_1-OH + HO-Ar_2-OH + Cl-\overset{O}{\underset{\|}{C}}-\!\!\!\bigcirc\!\!\!-\overset{O}{\underset{\|}{C}}-Cl \xrightarrow[\text{reaction medium}]{\text{High boiling}}$$

$$\left[-O-Ar_1-O-\overset{O}{\underset{\|}{C}}-\!\!\!\bigcirc\!\!\!-\overset{O}{\underset{\|}{C}}-O-Ar_2-O-\overset{O}{\underset{\|}{C}}-\!\!\!\bigcirc\!\!\!-\overset{O}{\underset{\|}{C}}-\right]_x \quad (12)$$

where Ar_1 and Ar_2 are aromatic

In the case of the reaction of hydroquinone and 4,4'-dihydroxydiphenyl with isophthaloyl chloride (106), suitable reaction media were benzophenone, *m*-terphenyl, chlorinated biphenyls, brominated biphenyls, chlorinated diphenyl oxides, brominated diphenyl oxides, chlorinated naphthalenes, and brominated naphthalenes. The unique feature of this method of polymerization was that the reaction medium was a compound which was a nonsolvent at room temperature but an excellent solvent at elevated temperatures. The polymer so prepared was amorphous and could be fabricated into suitable shapes and subsequently crystallized.

In Tables XVI and XVII are listed some of the aromatic polyesters and aromatic copolyesters which have been prepared. It is obvious from an inspection of these tables that many of these polyesters have polymer melt temperatures which place them within the framework of this discussion.

The relationship between chemical structure and physical properties of polyesters has been studied by numerous investigators (118–122), including a summary of this matter in terms of cohesion energies, molecular flexibility, and shape factors (123). The polymer melt temperatures of polyesters have been shown to be related to interchain forces and molecular flexibility (82). High polymer melt temperatures have been attributed to (*1*) regularity of structure, i.e., molecular symmetry, (*2*) stiff intrachain bonds, (*3*) linear chains capable of close packing, and (*4*) the presence of dipole interaction and polarization (82).

Wilfong (81) has reviewed the effect of resonance and strong interchain polarization forces on melting points and microcrystallinity. Thus, as shown in Table XVIII, the strongly coupled resonance polyester from *p*-hydroxybenzoic acid did not melt but decomposed at 350°C. The true melting point was obscured by decomposition. Uncoupling the benzene nuclei by use of *p*-hydroxyethylbenzoic acid resulted in a polyester of much lower melting point, 185°C (129). Resonance between benzene rings and the ester group led to a high melting point, 265°C for poly(ethylene

III. AROMATIC POLYMERS: FUNCTIONAL GROUPS IN CHAIN

TABLE XVI
Aromatic Polyesters

Bisphenol	Diacid	PMT[a] (°C)	Ref.
Hydroquinone	Isophthalic	~370	81, 82, 101
Hydroquinone	Terephthalic	>500	82, 101, 109
Hydroquinone	2,5-Dichloroisophthalic	~360	82
Hydroquinone	2,5-Dichloroterephthalic	>360	82
Hydroquinone	4,4'-Methylenedibenzoic	360	82
Hydroquinone	2,5-Dimethylterephthalic	305	91, 92
Hydroquinone	Bibenzoic	395	84, 92
Hydroquinone	4,4'-Oxydibenzoic	355	107, 108
Hydroquinone	4,4'-Isopropylidinedibenzoic	335	83, 91, 92
Resorcinol	Isophthalic	245	101, 105
Resorcinol	Terephthalic	>400	101, 105
4,4'-Dihydroxybiphenyl	4,4'-Isopropylidinedibenzoic	400	114, 115
4,4'-Dihydroxybiphenyl	Terephthalic	350	84, 92
2,2'-Dihydroxybiphenyl	Isophthalic	>300	84
Bis(2-methyl-4-hydroxyphenyl)	Isophthalic	>370	101, 102
Bis(3-methyl-4-hydroxyphenyl)	Isophthalic	>300	83, 84
Bis(4-hydroxyphenyl)methane	Isophthalic	~300	101
Bis(4-hydroxyphenyl)methane	Terephthalic	>400	101
2,2-Bis(4-hydroxyphenyl)propane	Isophthalic	280	101
2,2-Bis(4-hydroxyphenyl)propane	Terephthalic	~350	101
2,2-Bis(4-hydroxyphenyl)propane	4,4'-Isopropylidinedibenzoic	315	91, 92
2,2-Bis(4-hydroxyphenyl)propane	2,6-Naphthalenedicarboxylic	340	107, 110
2,2-Bis(4-hydroxyphenyl)propane	5-Methylisophthalic	325	108, 110
2,2-Bis(4-hydroxyphenyl)propane	2,5-Dimethylterephthalic	300	92
2,2-Bis(4-hydroxyphenyl)propane	4,4'-Bibenzoic	380	114
2,2-Bis(4-hydroxyphenyl)propane	4,4'-Sulfonyldibenzoic	335	101
1,3-Dihydroxynaphthalene	Isophthalic	~350	105
1,3-Dihydroxynaphthalene	Terephthalic	>500	105
1,5-Dihydroxynaphthalene	Isophthalic	>500	109
1,5-Dihydroxynaphthalene	Terephthalic	>500	109
1,6-Dihydroxynaphthalene	Isophthalic	>500	109
1,7-Dihydroxynaphthalene	Isophthalic	320	109
2,6-Dihydroxynaphthalene	Isophthalic	>500	105
2,6-Dihydroxynaphthalene	Terephthalic	>500	105
4,4'-Dihydroxydiphenyl ketone	Isophthalic	>300	84
4,4'-Dihydroxydiphenyl sulfone	Bibenzoic	325	81, 110
4,4'-Dihydroxydiphenyl sulfone	Isophthalic	331	101
4,4'-Dihydroxydiphenyl sulfone	4,4'-Sulfonyldibenzoic	>400	101
4,4'-Dihydroxydiphenyl sulfone	Terephthalic	>400	101

[a] PMT: polymer melt temperature. See R.G. Beaman and F. B. Cramer, *J. Polymer Sci.*, **21**, 223 (1956).

TABLE XVII
Aromatic Copolyesters

Bisphenols[a]			Dicarboxylic acids[b]			PMT	
A	B	Composition	A	B	Composition[d]	(°C)[c]	Ref.
1,5NG	—		I	T	(50/50)	~290	101
PSO$_2$PG	—				(50/50)	~325	101
H	PPG	(80/20)			(80/20)	~305	106, 117
		(90/10)			(90/10)	~340	
		(70/30)			(50/50)	~315	
		(80/20)[d]			(60/40)	~311	
		(70/30)[d]			(50/50)	~302	
		(70/30)[d]			(90/10)	~291	
H	R	(80/20)			(80/20)	~323	
		(50/50)			(50/50)	~285	
		(80/20)			(60/40)	~320	
		(50/50)[d]			(40/60)	~340	
		(60/40)[d]			(40/60)	~350	
		(60/40)[d]			(40/60)	~330	
		(50/50)[d]			(30/70)	~340	
		(50/50)[d]			(20/80)	~500	
		(80/20)[d]			(80/20)	~340	
H	C	(70/30)[d]			(70/30)	~297	111, 112
		(70/30)[d]			(80/20)	~328	
		(80/20)[d]			(50/50)	~500	
		(70/30)[d]			(60/40)	~360	
		(70/30)[d]			(40/60)	~500	
					(95/5)	~395	111, 112, 113
					(90/10)	~392	
					(85/15)	~384	
H	—				(80/20)	~364	
					(75/25)	~390	
					(70/30)	~437	
					(65/35)	~500	
					(90/10)[e]	~405	
					(85/15)[e]	~400	
					(80/20)[e]	~400	
					(72/25)[e]	~400	
					(70/30)[e]	~390	
					(65/35)[e]	~440	
					(60/40)[e]	~470	

(continued)

III. AROMATIC POLYMERS: FUNCTIONAL GROUPS IN CHAIN

TABLE XVII (continued)

Structures						PMT	
Bisphenols[a]			Dicarboxylic acids[b]				
A	B	Composition	A	B	Composition[d]	(°C)[c]	Ref.
H	PPG	(90/10)	I	—		~394	130–132
		(80/20)				~380	
		(70/30)				~370	
		(60/40)				~370	
		(50/50)				~380	
		(40/60)				~405	
		(30/70)				~440	

[a] Codes for bisphenols: 1,5NG, 1,5-dihydroxynaphthalene; PSO$_2$PG, bis-4-hydroxyphenyl sulfone; H, 1,4-hydroquinone; PPG, 4,4'-dihydroxydiphenyl; R, resorcinol; C, catechol.
[b] Codes for dicarboxylic acids: I, isophthalic acid; T, terephthalic acid.
[c] PMT: polymer melt temperature. See R. G. Beaman and F. B. Cramer, *J. Polymer Sci.*, **21**, 223 (1956).
[d] Bisphenol A prereacted with dicarboxylic acids.
[e] Dicarboxylic acid A prereacted with bisphenol.

TABLE XVIII
Effect of Resonance and Polarization Forces on Polymer Melting Points (81)

Structure of monomer unit	PMT[a] (°C)
1. —C(O)—C$_6$H$_4$—OC(O)—C$_6$H$_4$—O—	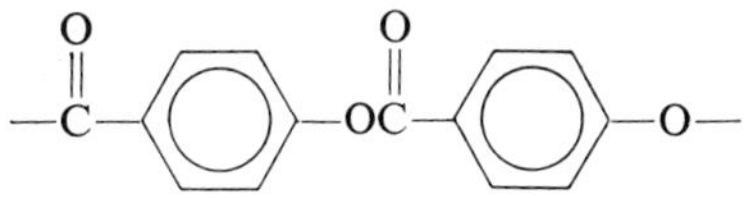 350
2. —C(O)—C$_6$H$_4$—CH$_2$CH$_2$OC(O)—C$_6$H$_4$—CH$_2$CH$_2$O—	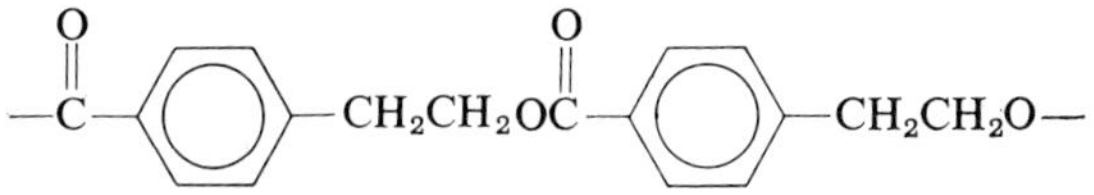 185
3. —OCH$_2$CH$_2$OC(O)—C$_6$H$_4$—C(O)—	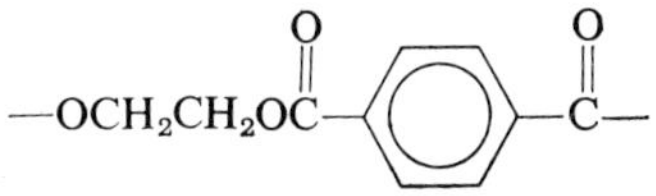 265
4. —OCH$_2$CH$_2$OC(O)—C$_{10}$H$_6$—C(O)—	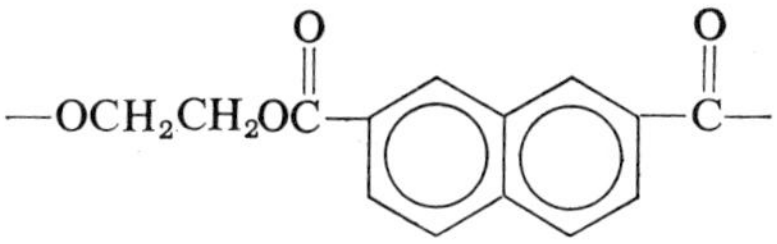 270

(continued)

TABLE XVIII (continued)

	Structure of monomer unit	PMT[a] (°C)
5.	—OCH$_2$CH$_2$OC(O)—C$_6$H$_4$—C$_6$H$_4$—C(O)—	355
6.	—OCH$_2$CH$_2$OC(O)—C$_6$H$_4$—CH=CH—C$_6$H$_4$—C(O)—	420
7.	—OCH$_2$CH$_2$OC(O)—C$_6$H$_4$—S(O)$_2$—C$_6$H$_4$—C(O)—	Very high
8.	—OCH$_2$CH$_2$OCCH$_2$—C$_6$H$_4$—CH$_2$C(O)—	137
9.	—OCH$_2$CH$_2$OCCH$_2$—C$_6$H$_4$—C$_6$H$_4$—CH$_2$C(O)—	150
10.	—OCH$_2$CH$_2$OC(O)—C$_6$H$_4$—CH$_2$CH$_2$—C$_6$H$_4$—C(O)—	220
11.	—OCH$_2$CH$_2$OC(O)—C$_6$H$_{10}$—C(O)—	120 (trans) < 30 (cis)
12.	—OCH$_2$CH$_2$OC(O)—C$_6$H$_4$—(CH$_2$)$_4$—C$_6$H$_4$—C(O)—	170
13.	—OCH$_2$CH$_2$OC(O)—C$_6$H$_4$—S(CH$_2$)$_2$S—C$_6$H$_4$—C(O)—	200
14.	—OCH$_2$CH$_2$OC(O)—C$_6$H$_4$—O(CH$_2$)$_2$O—C$_6$H$_4$—C(O)—	240
15.	—OCH$_2$CH$_2$OC(O)—C$_6$H$_4$—NH(CH$_2$)$_2$NH—C$_6$H$_4$—C(O)—	273

[a] PMT: polymer melt temperature. See R. G. Beaman and F. B. Cramer, *J. Polymer Sci.*, **21**, 223 (1956).

terephthalate). Poly(ethylene-2,7-naphthalate) melted even higher, since the resonance energy of the naphthalene nucleus has been estimated to be about 90% greater than that of a single benzene nucleus (126). The resonance of the biphenyl nucleus has been estimated to exceed the resonance of two benzene rings by 7–8 kcal/mole; the melting point of poly(ethylene-p,p'-diphenylate) was very high at 355°C. A conjugated double bond was roughly equivalent to adding another benzene ring in increasing resonance energy (126). Therefore, the polyester from ethylene glycol and p,p'-stilbene dicarboxylic acid melted higher, at 420°C. The sulfone group can also couple strongly with benzene nuclei, and, as expected, the corresponding polyester was very high melting. Uncoupling the benzene or diphenylene groups from the ester groups reduced resonance and dropped the melting point sharply. Uncoupling the benzene ring in diphenylene polyesters had a similar, but less profound effect. Destroying ring resonance by hydrogenation resulted in a major drop in melting point as illustrated by the low melting points of poly(ethylene glycol) esters of cyclohexane-1,4-dicarboxylic acid. Finally, the last four polyesters listed in Table XVIII form an interesting demonstration of the increasing melting point associated with increasing resonance and interchain polarization forces, as the methylene groups, alpha to the benzene nuclei, were progressively replaced with sulfur, oxygen, and NH. In the case of the latter, hydrogen bonding may contribute to the high melting point. Usually replacing a methylene group with a sulfur atom has less effect on the melting point than shown by the above example.

Unfortunately, the relationship between chemical structure and thermal stability of polyesters has been the subject of little investigation. Only limited thermal data such as differential thermal analysis and thermogravimetric analysis are available.

Some indication of the thermal stability of aromatic polyesters can be obtained from polymer melt temperatures. From an inspection of Tables XVI, XVII, and XVIII, the polymers melting at 350°C, or below, melt without decomposition. For those melting above this temperature, there is no quantitative indication of a level of thermal stability, but for the most part, these polymers (qualitatively) decompose above this temperature.

The differential thermal analysis, along with the other thermal behavior, of poly (m-hydroxybenzoic acid), poly(p-hydroxybenzoic acid) and copolymers derived from these hydroxy acids has been studied (80). The differential thermal analysis of poly(m-hydroxybenzoic acid) gave a value of 176°C for the crystalline melting point. The copolymers were heterogeneous and did not have well-defined melting points. However, when temperatures in excess of 350°C were used to press films from these

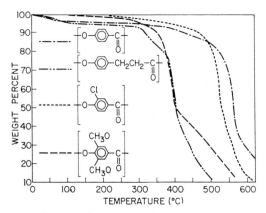

Fig. 16. TGA of aromatic polyesters under nitrogen (heating rate, 10°C/min) (30,31).

copolymers, they decomposed. Similarly, the differential thermal analysis of poly(p-hydroxybenzoic acid) indicated decomposition above 350°C but no melting even up to temperatures of 450°C.

The thermogravimetric analyses of some aromatic polyesters in nitrogen have been reported (30,31) and are reproduced in Figure 16. Inspection of these thermograms indicate that poly(p-hydroxybenzoic acid) was the most stable of these polyesters. As might be expected, the least stable of these polyesters was the one containing aliphatic carbon–carbon linkages. Like the aromatic polyamides, substitution of the aromatic rings reduced stability, the chlorinated ring being more stable than the methoxylated ring.

Thus, based on these limited data it would appear that the ester linkage has the same order of thermal stability as the anhydride and carbonate linkages and is inferior to that of the amide linkage.

E. AROMATIC POLYETHERS (POLYPHENYLENE OXIDES)

Aromatic ethers possess considerable thermal stability and good chemical resistance which had led to their use as high temperature hydraulic fluids and heat exchange media (133,134). Since it is known (135,136) that the incorporation of a preponderance of rigid aromatic p-phenylene linkages into the main chain of a polymer generally imparts a relatively high softening point to the product, it is apparent that aromatic polyethers or more specifically poly-p-phenylene oxide should have high thermal stability, good chemical resistance and high polymer melt temperature. Polymers of the phenylene oxide series have been obtained through oxidative and free-radical reactions and through displacement reactions such as the Ullman ether synthesis.

The polyphenylene oxides were first described in detail by Hunter (137–143) who studied the decomposition of metal halogenophenoxides (see reaction 13). Subsequently, Golden (144) extended this synthesis to a number of other aromatic polyethers such as poly(2,6-dibromo-

$$MO-\underset{X}{\overset{X}{\bigcirc}}-X \xrightarrow[\phi-\underset{O}{\overset{I_2}{C}}-\phi]{} \left[O-\underset{X}{\overset{X}{\bigcirc}} \right]_n \quad (13)$$

where X = I, Br; M = metal ion

phenylene oxide), poly(2,6-dichlorophenyl oxide), poly(2,5-dichlorophenylene oxide), poly(2,3,5,6-tetrachlorophenylene oxide), and poly-(4,4′-diphenylene oxide), and obtained these in useful molecular weights.

Wall (145) has also reported the preparation of the polyperfluorophenylene oxides of low molecular weight by the decomposition of sodium, silver, and ammonium pentafluorophenoxides.

Price and Staffin (146,147) prepared polyphenylene oxides by the catalytic oxidation of metal-2,4,6-trisubstituted phenoxides. Price and

$$MO-\underset{R}{\overset{R}{\bigcirc}}-X \xrightarrow{\text{Oxidizing agent}} \left[O-\underset{R}{\overset{R}{\bigcirc}} \right]_n \quad (14)$$

where R may be alkyl or aryl

co-workers (148–151) subsequently explored the mechanism of this reaction, studied various oxidizing agents, and prepared a number of polyphenylene oxides.

Recently, Dewar and James (152) studied the polyphenylene oxides formed by the decomposition and self-condensation of benzene-1,4-diazooxides (see reaction 15). In this synthesis, molecular weight was

$$\bar{O}-\underset{Br}{\overset{Br}{\bigcirc}}-N_2^+ \longrightarrow \cdot O-\underset{Br}{\overset{Br}{\bigcirc}}\cdot + N_2 \longrightarrow \left[O-\underset{Br}{\overset{Br}{\bigcirc}} \right]_n \quad (15)$$

severely limited by side reactions, e.g., incorporation of solvent molecules into the polymer.

More recently Hay and co-workers (153–158) have described the catalytic oxidation of certain 2,6-disubstituted phenols which yielded linear polyphenylene oxides of high molecular weight, e.g., poly(2,6-dimethylene oxide) with molecular weights of 300,000–700,000 (see reaction 16). In this method 2,6-disubstituted phenols were oxidized in the presence

$$\text{2,6-(CH}_3\text{)}_2\text{C}_6\text{H}_3\text{OH} \xrightarrow[\text{O}_2]{\text{CuCl, Pyridine}} {+}\text{O-C}_6\text{H}_2\text{(CH}_3\text{)}_2{+} \quad (16)$$

of an aliphatic tertiary amine and a soluble Cu^+ salt at temperatures up to 70°C. These workers (159) also prepared polyethers with the following repeating units:

           9   10

where R_1R_2 = methyl, ethyl; ethyl, ethyl; methyl, isopropyl; chloro, phenyl; chloro, methyl; methyl, methoxyl.

Brown (160–162) and Wright (163) have reported the preparation of polyphenylene oxides by the Ullman reaction (see reaction 17). Using

$$\begin{array}{c}\text{MO—Ar—Br}\\\text{or}\\\text{MO—Ar—OM + Br—Ar—Br}\end{array} \longrightarrow {+}\text{O—Ar}{+}_n + \text{MBr} \quad (17)$$

copper powder as a catalyst, anhydrous conditions, hydrocarbon reaction media, and temperatures between 180 and 200°C, both the m- and p-phenylene oxide polymers and derivatives were prepared.

Summarized in Table XIX are some of the polyphenylene oxides prepared by the aforementioned methods. From an inspection of this table it is apparent that very few of these polymers have polymer melt temperatures in excess of 300°C, but, as in the case of the aromatic polycarbonates, some idea as to the stability of higher melting polyphenylene oxides can be obtained from studies of these polyphenylene oxides.

TABLE XIX
Polyphenylene Oxides

Polymer	PMT^a (°C)	Ref.
Poly-2,6-dichloro-1,4-phenylene oxide	>300	144, 154, 159
Poly-2-methyl-6-*t*-butyl-1,4-phenylene oxide	217	154, 159
Poly-2,6-diisopropylidine-1,4-phenylene oxide	225	154, 159
Poly-2,6-di-*t*-butyl-1,4-phenylene oxide	246	154, 159
Poly-2,6-dimethoxy-1,4-phenylene oxide	>300	154, 159
Poly-2,6-diphenyl-1,4-phenylene oxide	>300	154
Poly-2,6-diethyl-1,4-phenylene oxide	~180	159
Poly-2,6-diallyl-1,4-phenylene oxide	~160	148
Poly-2-isopropylidine-6-phenyl-1,4-phenylene oxide	~116	148
Poly-2-allyl-6-phenyl-1,4-phenylene oxide	~117	148
Poly-2-isopropylidine-6-methyl-1,4-phenylene oxide	~117	148
Poly-2-allyl-6-methyl-1,4-phenylene oxide	~117	148
Poly-1,4-phenylene oxide	~260	144
Poly-2-chloro-1,4-phenylene oxide	~270	144
Poly-2-bromo-1,4-phenylene oxide	~240	144
Poly-2,5-dichloro-1,4-phenylene oxide	~265	144
Poly-2,6-dichloro-1,4-phenylene oxide	~188	144
Poly-2,6-dibromo-1,4-phenylene oxide	~245	144
Poly-2,3,6-trichloro-1,4-phenylene oxide	~222	144
Poly-2,3,5,6-tetrachloro-1,4-phenylene oxide	~192	144
Poly-2,3,5,6-tetrabromo-1,4-phenylene oxide	>300	144
Poly-4,4'-biphenylene oxide	~290	144
Poly-2,6-dibromo-1,4-phenylphenylene oxide	~240	144
Poly-2,6-dibromo-1,4-phenylene oxide	~220	144

^{a} PMT: polymer melt temperature. See R. G. Beaman and F. B. Cramer, *J. Polymer Sci.*, **21**, 223 (1956).

Golden (144) has studied the thermal stability in vacuum of some representative polyphenylene oxides using a quartz spring balance to determine weight loss of samples at a series of elevated temperatures. A fresh sample was used for each temperature and the time of heating fixed at 2 hr. The results are shown in Figure 17. The temperatures of initial loss in weight and subsequent losses in weight on heating polymer samples continuously from ambient temperatures to 800°C at approximately 5°/min are in Table XX.

The order of stability of polymers examined was polyphenylene oxide > poly-2,6-dichlorophenylene oxide > poly-2,6-dibromophenylene oxide > polytetrachlorophenylene oxide. Increase in substitution of the benzene nucleus led to a decrease in thermal stability due probably to the fact that the C—Cl and C—Br bonds were the weakest in the system and provided sites for initiation of degradation. A further contributory cause for the

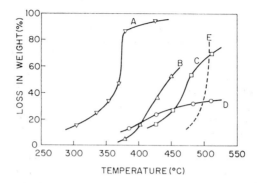

Fig. 17. Loss in weight of polyphenylene oxides after heating in a vacuum (each point represents a sample heated for 2 hr) (144). (*A*) Polytetrachlorophenylene oxide; (*B*) poly-2,6-dibromophenylene oxide; (*C*) poly-2,6-dichlorophenylene oxide; (*D*) polyphenylene oxide; (*E*) polydiphenylene oxide.

lower stability of the substituted polymers could be the restriction of rotation in the more hindered molecules which could reduce the ease of dissipation of thermal energy. In the case of the dibromo and tetrachloro derivatives, this may be of particular importance since molecular models indicated that these were rigid molecules incapable of rotation about the oxygen atoms.

Similar studies were carried out by Wright (163). In this work, the rates of volatilization of polyphenylene oxides *in vacuo* at elevated temperatures were studied along with thermogravimetric analysis (Table XXI). Using a quartz spring balance, he found that polyphenylene oxide showed a rapid weight loss during the initial heating period, followed by little further change. If samples of these polymers were heated at comparatively low temperatures until stabilization occurred, and the residue then heated to a higher temperature, the same pattern was observed, i.e., there was an

TABLE XX
Weight Loss of Polyphenylene Oxides with Increasing Temperature (144)

Polyphenylene oxides	Temp. of initial wt loss (°C)	Total loss in wt % at temp. (°C)					
		300	400	500	600	700	800
Phenylene	285	3	10	18	24	31	13
2,6-Dichlorophenylene	425	—	—	20	69	76	77
2,6-Dibromophenylene	220	2	5	52	76	85	88
Tetrachlorophenylene	275	3	25	88	93	98	98
Diphenylene	155	13	27	41	52	55	55

TABLE XXI
Thermogravimetric Analysis Results on Poly(phenylene Oxides) (163)

Polymer	Temp. at which loss commences (°C)	Total loss in wt. at various temp. (%)							
		200°C	300°C	400°C	500°C	600°C	700°C	800°C
Poly-1,3-phenylene oxide	300	—	—	4	10	23	33	38
Poly-1,4-phenylene oxide (1)	290	—	3	8	18	24	31	31
Poly-4-bromo-1,2-phenylene oxide-1,4-phenylene oxide	260	—	4	14	22	50	61	68
Poly-2,6-dichloro-1,4-phenylene oxide	430	—	—	—	18	69	76	77
Poly-1,4-phenylene oxide (2)	260	—	2	7	18	46	57	61
Poly-6-bromo-2-chloro-1,4-phenylphenylene oxide	360	—	—	4	29	77	89	93
Poly-4,4'-biphenylene oxide	160	3	13	26	41	52	55	55
Poly-2-bromo-1,4-phenylene oxide	280	—	3	28	53	64	70	76
Poly-2,6-dibromo-1,4-phenylene oxide-4,6-dibromo-1,4-phenylene oxide	200	—	5	24	51	66	78	81
Poly-2,3,5,6-tetrachloro-1,4-phenylene oxide	280	—	2	22	88	92	98	98
Poly-4-bromo-1,2-phenylene oxide-1,4-phenylene oxide	150	3	9	20	33	42	49	55
Poly-5-bromo-1,2-phenylene oxide-1,4-phenylene oxide	170	2	4	14	25	46	60	70
Poly-4-fluoro-1,2-phenylene oxide-1,4-phenylene oxide	160	3	6	20	40	43	50	59
Poly-6-fluoro-1,2-phenylene oxide-1,4-phenylene oxide	180	2	12	25	37	55	63	67
Poly-2,3,6-trifluoro-1,4-phenylene oxide	150	3	13	30	38	45	54	76
Poly-2,6-difluoro-1,4-phenylene oxide-2,6-dibromo-1,4-phenylene oxide	170	1	6	14	41	61	73	77
Poly-2,6-difluoro-1,4-phenylene oxide-2,6-diiodo-1,4-phenylene oxide	150	2	4	32	51	58	62	65
Poly-2-bromo-6-fluoro-1,4-phenylene oxide-2,6-dibromo-1,4-phenylene oxide	180	1	11	29	49	59	66	70
Poly-2-bromo-6-fluoro-1,4-phenylene oxide-2,6-diiodo-1,4-phenylene oxide	150	6	16	34	48	57	63	66
Poly-4,6-pyrimidylene oxide-2,3,5,6-tetrachloro-1,4-phenylene oxide	270	—	5	14	30	43	49	—
Poly-2,4-quinazolylene oxide-1,4-phenylene oxide	240	—	22	36	42	52	58	—
Poly-4,6-pyrimidylene oxide-1,4-phenylene oxide	270	—	6	23	61	77	83	—
Poly-2,5-pyrazylene oxide-1,4-phenylene oxide	210	—	34	60	74	80	85	—

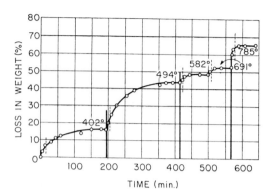

Fig. 18. Cumulative weight loss on heating poly-*p*-phenylene oxide to successively higher temperatures. Dotted lines indicate points at which temperature was actually attained (163).

initial rapid loss in weight followed by restabilization. This could be repeated for a series of temperatures as shown in Figures 18–20 and Table XXII. Elemental analysis of the residues indicated little change in basic structure for temperatures up to 400°C. Above this temperature the proportion of carbon increased and of hydrogen decreased. Similar behavior of stepwise pyrolysis has been observed for polyvinyl and polyvinylidine chlorides (164). It was considered in those cases that

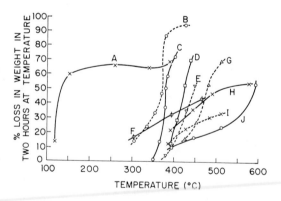

Fig. 19. Relative thermal stabilities of polyphenylene oxides prepared by decomposition of metal halogenophenoxides (163). (*A*) Poly-*p*-2,5-dichlorophenylene oxide; (*B*) poly-*p*-2,3,5,6-tetrachlorophenylene oxide; (*C*) poly-*p*-2,3,5,6-tetrabromophenylene oxide; (*D*) poly-*p*-2,3,6-trichlorophenylene oxide; (*E*) poly-*p*-2,6-dibromophenylene oxide; (*F*) poly-*p*,-*p*-diphenylene oxide; (*G*) poly-*p*-2,6-dichlorophenylene oxide; (*H*) poly-*p*-2-bromophenylene oxide; (*I*) poly-*p*-phenylene oxide; (*J*) poly-*p*-2-chlorophenylene oxide.

III. AROMATIC POLYMERS: FUNCTIONAL GROUPS IN CHAIN

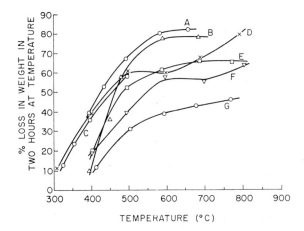

Fig. 20. Relative thermal stabilities of polyphenylene oxides prepared by oxidation of halogenophenols or Ullman condensations (163). (*A*) Poly-*p*-2,6-dibromophenylene oxide-*o*-2,6-dibromophenylene oxide; (*B*) poly-*o*-6-bromo-4-*p*-chlorophenylphenylene oxide; (*C*) poly-*o*-4-fluorophenylene oxide-*p*-phenylene oxide; (*D*) poly-*o*-6-bromo-4-phenylphenylene oxide; (*E*) poly-*p*-phenylene oxide; (*F*) poly-*o*-4-bromophenylene oxide-*p*-phenylene oxide; (*G*) poly-*m*-phenylene oxide.

TABLE XXII
Order of Thermal Stability of Polyphenylene Oxides Examined Quantitatively (163)

Polymer	Temp. at which 50% weight loss occurs in 2 hr (°C)
Poly-1,3-phenylene oxide	40% loss at 610°C
Poly-2-chloro-1,4-phenylene oxide	590
Poly-1,4-phenylene oxide (1)	33% loss at 515°C
Poly-1,2-bromophenylene oxide-1,4-phenylene oxide	550
Poly-2,6-dichloro-1,4-phenylene oxide	480
Poly-1,4-phenylene oxide (2)	480
Poly-2-bromo-1,4-phenylene oxide	510
Poly-6-bromo-4-chloro-1,4-phenylphenylene oxide	470
Poly-2,6-dibromo-1,4-phenylene oxide	445
Poly-4,4-diphenylene oxide	43% loss at 470°C
Poly-4-fluoro-1,2-phenylene oxide-1,4-phenylene oxide	445
Poly-2,3,6-trichloro-1,4-phenylene oxide	425
Poly-6-bromo-1,4-phenylene oxide	445
Poly-2,6-dibromo-1,4-phenyle oxide-4,6-dibromo-1,2-phenylene oxide	425
Poly-2,3,5,6-tetrabromo-1,4-phenylene oxide	380
Poly-2,3,5,6-tetrachloro-1,4-phenylene oxide	370
Poly-2,5-dichloro-1,4-phenylene oxide	135

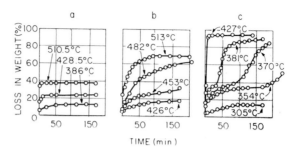

Fig. 21. Loss in weight of various poly-*p*-phenylene oxides when heated in vacuum (163). (*a*) Poly-*p*-phenylene oxide; (*b*) poly-*p*-2,6-dichlorophenylene oxide; (*c*) poly-*p*-tetrachlorophenylene oxide.

hydrogen chloride was first lost from the molecule, followed by the formation of aromatic groups which rearranged to larger fused ring systems, giving a coke- or graphite-like structure. For these aromatic polyethers, the initial step might be hydrogen abstraction, followed by cyclization to larger fused ring systems.

As in the work of Golden, substitution of the benzene nucleus had an adverse effect on the thermal stability of the polyphenylene oxides (Fig. 21). The unsubstituted polyphenylene oxide showed rapid stabilization and lost only approximately 37% of weight at 511°C. The poly-2,6-dichlorophenylene oxide also lost weight rapidly at first but this was followed, except at the highest temperature used, by a further slow but steady loss in weight. At 513°C, the total loss in weight was 69%. The tetrachloro compound, at temperatures at which the total weight loss did not exceed

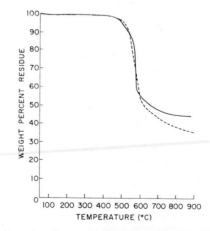

Fig. 22. TGA of poly-*p*-phenylene oxide (heating rate, 3°C/min) (30,31); ———, dry nitrogen; -----, dry air.

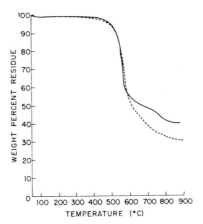

Fig. 23. TGA of poly-*m*-phenylene oxide (heating rate, 3°C/min) (30,31); ———, dry nitrogen; - - - - -, dry air.

36–37% during the course of the experiment, showed finally either stabilization or a very slow rate of breakdown. If, however, the total weight loss exceeded 36–37%, a second rapid loss in weight commenced and at the highest temperature used (427°C), volatilization of 94% of the material occurred. Without supplementary analytical data, it is difficult to understand exactly what these results imply and impossible to suggest the mechanism for this degradative process.

The findings of Golden and Wright were in substantial agreement with the thermogravimetric analysis of *m*- and *p*-phenylene oxide polymers in air and nitrogen by Ehlers (30,31). It was found in these studies that in nitrogen for both polymers the initial break in the thermograms occurred at 400–450°C with major weight losses observed at 450–580°C and 40–45% of the original weight remaining at 900°C. In air the initial break in the TGA curves was still in the 400–450°C range, with major weight losses at 450–680°C, and 30 to 40% of original weight remaining at 900°C (Figs. 22 and 23).

Thus, it would appear that polyphenylene oxides or aromatic polyethers even if prepared with higher polymer melt temperatures would exhibit thermal behavior similar to that of the aromatic polycarbonates, aromatic polyesters, and aromatic polyanhydrides and would be inferior to the aromatic polyamides.

F. AROMATIC POLYSULFIDES (POLYPHENYLENE SULFIDES)

As in the case of the aromatic ethers it would be expected that aromatic polysulfides or polyphenylene sulfides would have good thermal stability and good chemical resistance. Incorporation of aromatic *p*-phenylene linkages in the main chain of the polymer would be expected to yield a relatively high-melting, aromatic polysulfide.

Hilditch (165) reported that the self-condensation of thiophenol in cold concentrated sulfuric acid gave a cream-colored insoluble amorphous powder in yields of about 60%, softening and decomposing at 300°C or above, and having an empirical formula of C_6H_4S. Also Tasker and Jones (166) obtained an amorphous, brown material having the same empirical formula by the reaction of thiophenol with thionyl chloride.

The first reported preparation of polyphenylene sulfide was that by Macallum (167–169) via the reaction of *p*-dichlorobenzene with sulfur and dry sodium carbonate at 300–340°C (see reaction 18). The properties of

$$Cl-\underset{}{\bigcirc}-Cl + S + Na_2CO_3 \longrightarrow$$

$$\left[\underset{}{\bigcirc}-S\right]_n + NaCl + CO_2 + Na_2SO_4 \quad (18)$$

the polymer varied with the ratio of sulfur and *p*-dichlorobenzene. It had an empirical formula of $C_6H_4S_{1.2}$ to $C_6H_4S_{2.3}$, melting points from 285 to over 350°C and molecular weight from 9,000 to 17,000.

Lenz and co-workers (170–172) have reinvestigated the Macallum polymerization and suggested a mechanism for the same. In addition, they have also prepared the linear polyphenylene sulfide via the self-condensation of alkali metal salts of *p*-halothiophenol both in the presence and

$$Cl-\underset{}{\bigcirc}-SM \longrightarrow \left[\underset{}{\bigcirc}-S\right]_x + MCl \quad (19)$$

absence of solvents (see reaction 19). They have reported x-ray diffraction patterns, infrared spectra and thermal stability.

Tsunawaki and Price (173) have also studied the preparation of polyarylene sulfides via the polymerization of metal salts of halothiophenols. These workers have suggested that the mechanism of this reaction involves the nucleophilic attack on the aryl halides by thiophenoxide anions.

Lenz and co-workers studied the thermal stability of linear polyphenylene sulfide by thermogravimetric analysis in air and nitrogen and differential thermal analysis in air. The thermogravimetric analyses (172) in air and nitrogen (Figs. 24 and 25) indicated that a linear polyphenylene sulfide was kinetically stable to a temperature above 400°C. These analyses

III. AROMATIC POLYMERS: FUNCTIONAL GROUPS IN CHAIN

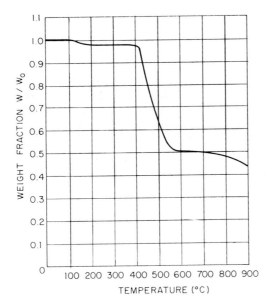

Fig. 24. TGA of linear polyphenylene sulfide under an atmosphere of nitrogen (172).

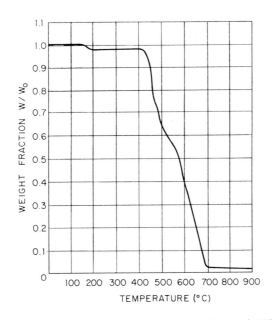

Fig. 25. TGA of linear polyphenylene sulfide in air (172).

which were conducted at a heating rate of 150°C/hr showed an initial volatilization of approximately 2% occurring both in air or in nitrogen at 125–175°C. Up to 500°C the thermogravimetry curves for the polymer in nitrogen and in air were virtually identical. The stabilities differ, however, in that the curve for the experiment in nitrogen atmosphere leveled off to 50% weight loss at approximately 600°C and remained essentially unchanged up to 900°C while the curve for the experiment in air continued to fall sharply until approximately 96% weight loss occurred at 700°C.

Tsunawaki and Price (173) also studied the thermal stability of the polyphenylene sulfides. Weight losses after heating for 1 hr at various temperatures up to 600°C, under reduced pressure were determined.

TABLE XXIII
Thermal Stability Evaluation of Polyphenylene Sulfides (173)

MW	Temp. (°C)	Time (hr)	Yield (%)	MW	mp (°C)	Sublimate on condenser surface (S)		
						Yield (%)^a	MW	MW^b
740	250	½	90.3	1420	150–230	9.7	310	800^c
740	250	½	89.6	1440	160–170	10.4	320	770^c
1050	250	½	88.2	1290		11.8	450	1050
1490	250	1	99.5		250–280	0.5		
1490	300	1	97.3		250–280	2.7		
1490	400	1	92.0		250–280	8.0		
1490	500	1	71.7		260–280	28.3		
1490	500	2	60.8	6000^d	260–280	39.7	770^d	1640
1490	600	1	38.6		300	61.4		
2810	300	1	97.3		260–280	2.7		
2810	400	1	95.2		270–280	4.8		
2810	500	1	76.3		270–280	23.7		
2810	600	1	46.2		300	53.8		
980	250	1	86.9	1400		13.1	330	980
1040	250	1	82.6	2100	150–160	17.4	400	1190
2580	250	1	95.5	1890	150–160	4.5	520	1710
1280	250	1	72.0	1240	60–100	28.0	310	610
1500	250	½	81.0	2020	90–110	8.6	350	1560

^{a} Calculated as $\dfrac{\text{(weight of polymer)} - \text{[weight of residue (R)]}}{\text{(weight of polymer)}} \times 100$.

^{b} Calculated as $100 \bigg/ \left(\dfrac{\text{yield (\%) of R}}{\text{MW of R}} + \dfrac{\text{yield (\%) of S}}{\text{MW of S}} \right)$.

^{c} Calculated as $100 \bigg/ \left(\dfrac{\text{yield (\%) of R}}{\text{MW of R}} + \dfrac{\text{yield (\%) of S}}{\text{(MW of S)}(\frac{1}{2})} \right)$ (assuming R to be polyphenylene sulfides containing one disulfide bond).

^{d} Calculated from elemental analysis.

III. AROMATIC POLYMERS: FUNCTIONAL GROUPS IN CHAIN

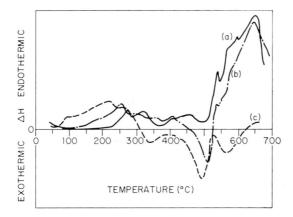

Fig. 26. Differential thermograms of phenylene sulfide polymers (171). (*a*) Linear polymer; (*b*) Macallumhomopolymer; (*c*) Macallum copolymer. Rate of heating, approximately 10°C/min; run in air.

These data which are summarized in Table XXIII clearly show that significant weight losses did not occur under these experimental conditions until temperatures in excess of 400°C were reached. This is in agreement with the thermogravimetric analysis data reported by Lenz.

The differential thermal analysis of linear and the Macallum polyphenylene sulfides in air (170) which are collected in Figure 26 showed sharp endothermic peaks at 275 and 255°C, respectively, which undoubtedly were associated with some second-order transition. All polymers showed endothermic peaks of different sharpness at approximately 400°C. These peaks which are suggestive of a phase transition also coincide with the initial weight loss in the thermogravimetric analysis curves. All curves showed large exothermic peaks with a maximum at 500°C and this coincided with the highest rate of weight loss in the thermogravimetry curve.

This thermal behavior is quite reminiscent of that of the polyphenylene oxides and suggests that the polyphenylene sulfides and phenylene oxides are of equivalent thermal stability but are inferior to the aromatic polyamides.

G. AROMATIC POLYSULFONATES AND AROMATIC SULFONAMIDES

Since aromatic polymers containing sulfur groups such as polyphenylene sulfide show a certain level of thermal stability, it was of interest to prepare and study the thermal stability of aromatic polysulfonates and aromatic polysulfonamides.

These polymers were readily available in high molecular weight via interfacial polymerization or low temperature solution polymerization (27,174–179). Some of the aromatic polysulfonates and aromatic sulfonamides which have been prepared are listed in Table XXIV. From an

TABLE XXIV
Polysulfonates and Polysulfonamides

Diols or diamines	Disulfonic acid	PMT[a] (°C)	Ref.
Resorcinol	1,3-Phenylene	~160	179
4,4′-Dihydroxydiphenyl	4,4′-Biphenylene	~260	179
4,4′-Dihydroxydiphenyl sulfone	Sulfonyl-bis(4-phenylene)	~280	179
4,4′-Dihydroxydiphenyl methane	Methylene-bis(4-phenylene)	~200	179
4,4′-Dihydroxydiphenyl ether	Oxy-bis(4-phenylene)	~200	179
2,2-Bis(2,6-dibromo-4-hydroxyphenyl)propane	4,4′-Biphenylene	~210	175
Ethylenediamine	1,3-Phenylene	~250	19
Ethylenediamine	1,4-Phenylene	~290	19
Ethylenediamine	1,5-Naphthalene	~300	19
Ethylenediamine	4,4′-Biphenylene	~302	19
Ethylenediamine	Oxy-bis(4-phenylene)	~380	19
Hexamethylenediamine	1,5-Naphthalene	265	19
Hexamethylenediamine	4,4′-Biphenylene	288	19

[a] PMT: polymer melt temperature. See R. G. Beaman and F. B. Cramer, *J. Polymer Sci.*, **21**, 223 (1956).

inspection of this table, it is quite apparent that practically none of these polymers are of interest within the framework of this discussion. The aromatic polysulfonates and aromatic polysulfonamides melt 150–200°C below that of the comparable structures in the carboxylic acid series.

In spite of these lower melting temperatures, Ehlers and Thomson (179) have studied the thermal stability of the aromatic sulfonates. Thermogravimetric analysis in a nitrogen atmosphere of the polysulfonates listed in Figure 27 showed a sharp onset of weight loss around 300°C, which leveled off at 500°C leaving approximately 50% residue up to 900°C. Since all these polymers showed this loss of weight in the 300–340°C region, the upper limits of stability appeared to be governed by the stability of the sulfonate linkage. The polysulfonates decomposed rapidly with the evolution of sulfur dioxide, indicating that the sulfonate linkage is the weak spot. Thus, even if these structures could be prepared with

III. AROMATIC POLYMERS: FUNCTIONAL GROUPS IN CHAIN

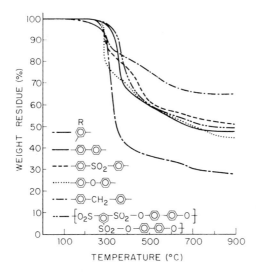

Fig. 27. TGA curves of polysulfonates $-(O_2S-R-SO_2-O-R-O)_{\overline{n}}$ (179).

higher polymer melt temperatures, their upper temperature limit of usefulness would be in the 300–340°C region.

Surprisingly, the polysulfonates were extremely hydrolytically stable, showing no change in inherent viscosity on exposure to 10% sodium hydroxide solution and 10% sulfuric acid solution at room temperature and only slight decreases in molecular weight when these solutions were heated at temperatures of 80–85°C for 24 hr (179).

It is apparent that the aromatic polysulfonates and sulfonamides like the other phenylene-R structures discussed previously are inferior in thermal stability to the aromatic polyamides.

H. AROMATIC POLYSULFONES (POLYPHENYLENE SULFONES)

Another class of polymers containing a sulfur atom between phenylene linkages comprises the aromatic sulfones or the polyphenylene sulfones. As in the case of the aromatic ethers, aromatic sulfones possess considerable thermal stability and good chemical resistance. Here again, incorporation of a rigid aromatic *p*-phenylene linkage in the main chain should yield a relatively high melting aromatic polysulfone. In addition, thermal stability measurements (30,31,45,180–184) on other aromatic polymers such as the aromatic polyamides, the aromatic imides, etc., have indicated that the sulfone group itself is quite thermally stable in those systems, for the most part not appreciably lowering the stability of

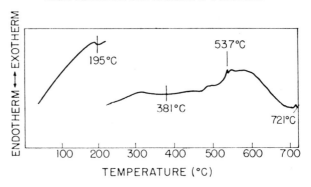

Fig. 28. DTA in nitrogen of aromatic polyether sulfone (heating rate, 10°C/min).

these stable polymers. Thus, the aromatic polysulfones would be expected to have a high degree of thermal stability.

Surprisingly there are only two references to the preparation of aromatic polysulfones in literature (185). In earlier work aromatic polysulfones were prepared via the condensation of dihalosulfones with alkali metal *p*-phenylene dimercaptides, followed by oxidation of the sulfide linkage to the sulfone. It was claimed that these oxidized polymers, the poly-*p*-phenylene sulfones, are useful in applications such as coatings, films, and molded articles requiring stability and high temperature. No indication of the level of stability at high temperatures was given. In the more recent references (186), these polymers were prepared by the ferric chloride-catalyzed polycondensation of disulfonyl chlorides with reactive dinuclear aromatic compound or the self-condensation of certain dinuclear arylsulfonyl chlorides. Although in most cases, high molecular weight products were obtained, no thermal stability data were given.

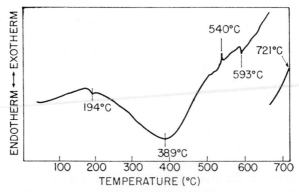

Fig. 29. DTA in air of aromatic polyether sulfone (heating rate, 10°C/min).

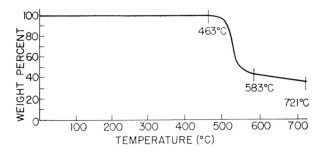

Fig. 30. TGA in nitrogen of aromatic polyether sulfone (heating rate, 10°C/min).

An aromatic polyethersulfone with the structure **9** has recently been reported (187). The differential thermal analysis (Fig. 28) and thermogravimetric analysis (Fig. 30) of this polymer in nitrogen yielded curves

$$\left[\bigcirc\!\!-\!\!\underset{\underset{CH_3}{|}}{\overset{\overset{CH_3}{|}}{C}}\!\!-\!\!\bigcirc\!\!-\!\!O\!\!-\!\!\bigcirc\!\!-\!\!SO_2\!\!-\!\!\bigcirc\!\!-\!\!O \right]_{n=50-80}$$

9

$$\left[\bigcirc\!\!-\!\!O\!\!-\!\!\bigcirc\!\!-\!\!SO_2 \right]$$

10

which were practically identical with those of aromatic polyethers. Curves from similar analyses in air (Figs. 29 and 31) reflected oxidative attack at the isopropylidene group in the polymer chain.

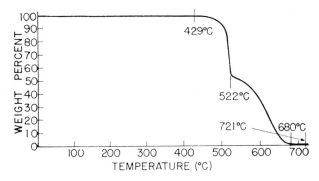

Fig. 31. TGA in air of aromatic polyether sulfone (heating rate, 10°C/min).

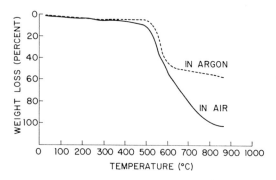

Fig. 32. TGA of polymer 10 (heating rate, 12°C/min).

The polyether sulfone with the structure shown in 10 exhibited similar behavior in thermogravimetric analysis (Fig. 32).

I. AROMATIC POLYUREAS AND AROMATIC POLYURETHANES

Other classes of aromatic polymers which have been studied extensively are the aromatic polyureas and the aromatic polyurethanes. It was early recognized in these classes of polymers that the urea and urethane linkages were not thermally stable (188,189). Extensive work has been carried out on the stabilization of polyurethanes and polyureas from thermal degradation with only limited success (190–198). On the basis of all the available evidence, the urea linkage thermally dissociates at 150–200°C and the urethane linkage at 225–250°C (188,189).

Ehlers (30,31) has studied the thermal stability of these polymer systems using thermogravimetric analysis. He found that polyurethanes started decomposing at about 260°C with complete weight loss at temperatures around 400°C. Similarly, in his studies on the aromatic polyureas, initial breakdown was observed at 150°C, with ureas containing triazine rings showing decomposition at slightly higher temperatures (200°C), and complete weight loss in the 400–500°C region.

Thus the data show that aromatic polyureas and polyurethanes are inherently thermally unstable and are of no utility as high temperature resistant polymers.

References

1. S. Akiyoshi, S. Hashimoto, and K. Takami, *J. Chem. Soc. Japan, Ind. Chem. Sect.*, **57**, 212 (1954); *Chem. Abstr.*, **49**, 2774a (1955).
2. A. M. Thomas, Brit. Pat. 794, 365 (1958).
3. O. B. Edgar and R. Hill, *J. Polymer Sci.*, **8**, 1 (1952).
4. J. W. Fisher, *Chem. Ind.* (London), **1952**, 244.

5. J. W. Fisher and E. W. Whatley, U.S. Pat. 2,512,633 (June 1960).
6. P. J. Flory, U.S. Pat. 2,244,192 (June 1941).
7. R. Hill, *Fibres from Synthetic Polymers*, Elsevier, New York, 1953, Chap. 6, p. 136.
8. O. Ya. Fedotova, I. P. Losev, and Yu. P. Brysin, *Vysokomolekul. Soedin.*, **2** (6), 899 (1960).
9. H. Hasegawa, *Bull. Chem. Soc. Japan*, **27**, 227–230 (1954).
10. E. L. Wittbecker and P. W. Morgan, *J. Polymer Sci.*, **40**, 289 (1959).
11. P. W. Morgan and S. L. Kwolek, *J. Chem. Educ.*, **36**, 182, 530 (1959).
12. P. W. Morgan and S. L. Kwolek, *J. Polymer Sci.*, **40**, 299 (1959).
13. R. G. Beaman, P. W. Morgan, C. R. Koller, E. L. Wittbecker, and E. E. Magat, *J. Polymer Sci.*, **40**, 329 (1959).
14. M. Katz, *J. Polymer Sci.*, **40**, 337 (1959).
15. V. E. Shashoua and W. M. Eareckson, *J. Polymer Sci.*, **40**, 343 (1959).
16. C. W. Stephens, *J. Polymer Sci.*, **40**, 359 (1959).
17. E. L. Wittbecker and M. Katz, *J. Polymer Sci.*, **40**, 367 (1959).
18. J. R. Schaefgen, F. H. Koontz, and R. F. Tietz, *J. Polymer Sci.*, **40**, 377 (1959).
19. S. A. Sundet, W. A. Murphey, and S. B. Speck, *J. Polymer Sci.*, **40**, 389 (1959).
20. W. M. Eareckson, *J. Polymer Sci.*, **40**, 399 (1959).
21. D. J. Lyman and S. L. Jung, *J. Polymer Sci.*, **40**, 407 (1959).
22. P. W. Morgan and S. L. Kwolek, *J. Polymer Sci.*, **62**, 33 (1962).
23. P. W. Morgan and S. L. Kwolek, *J. Polymer Sci. A*, **1**, 1147 (1963).
24. P. W. Morgan, *J. Polymer Sci. A*, **1**, 1075 (1963).
25. P. W. Morgan and S. L. Kwolek, *J. Polymer Sci. A*, **2**, 181 (1964).
26. P. W. Morgan and S. L. Kwolek, *J. Polymer Sci. A*, **2**, 209 (1964).
27. S. L. Kwolek and P. W. Morgan, *J. Polymer. Sci. A*, **2**, 2693 (1964).
28. D. D. Coffman, G. F. Berchet, W. R. Peterson, and E. W. Spanagel, *J. Polymer Sci.*, **2**, 306 (1947).
29. W. O. Statton, *Ann. N.Y. Acad. Sci.*, **83**, 27 (1959).
30. G. F. L. Ehlers, paper presented at 12th Canadian High Polymer Forum, May 1964.
31. G. F. L. Ehlers, WADC TR 61–622, Feb. 1962.
32. P. Alexander and C. S. Whencel, "Some Aspects of Textile Research in Germany," B.I.O.S. Final Report No. 1472, H. M. Stationary Office, London, 1947.
33. Belg. Pat. 443,955 (1945).
34. O. Moldenhauer and H. Bock, U.S. Pat. 2,349,979 (May 1944).
35. S. B. McFarlane, Jr. and A. L. Miller, U.S. Pat. 2,615,862 (Oct. 1952).
36. A. H. Frazer and F. T. Wallenberger, *J. Polymer Sci. A*, **2**, 1147 (1964).
37. E. J. Vandenberg and C. G. Overberger, *Science*, **141**, 176–177 (1963).
38. A. H. Frazer and F. T. Wallenberger, *J. Polymer Sci. A*, **2**, 1137 (1964).
39. A. H. Frazer and F. T. Wallenberger, *J. Polymer Sci. A*, **2**, 1825 (1964).
40. A. H. Frazer, W. Sweeny, and F. T. Wallenberger, *J. Polymer Sci. A.*, **2**, 1157 (1964).
41. A. H. Frazer and I. M. Sarasohn, *Am. Chem. Soc., Div. Polymer Chem., Preprints*, **5**, No. 1, 114 (1964); *J. Polymer Sci. A-1*, **4**, 1649 (1966).
42. H. W. Hill, S. L. Kwolek, and P. W. Morgan, U.S. Pat. 3,006,899 (Oct. 1961).
43. S. L. Kwolek, P. W. Morgan, and W. R. Sorenson, U.S. Pat. 3,063,966 (Oct. 1961).

44. (a) H. F. Mark, S. M. Atlas, and N. Ogata, *J. Polymer Sci.*, **61**, 549 (1962). (b) S. Nishizaki and A. Kukami, *J. Polymer Sci. A-1*, **4**, 2337 (1966). (c) L. Starr, *J. Polymer Sci. A-1*, **4**, 3041 (1966).
45. R. A. Dine-Hart, B. J. C. Moore, and W. W. Wright, *J. Polymer Sci. B*, **2**, 369 (1964).
46. (a) Belg. Pat. 637,260 (1964). (b) French Pat. 1,422,829 (1965). (c) W. A. H. Huffman, R. W. Smith, and W. T. Dye, Jr., U.S. Pat. 3,203,933 (Aug. 1965). (d) Ger. Pat. 1,222,674 (1966).
47. C. W. Stephens, U.S. Pat. 3,049,518 (Aug. 1962).
48. Belg. Pats. 637,257 (1964) and 637,258 (1964).
49. Brit. Pats. 980,906 (1965) and 980,907 (1965).
50. J. Preston, *Am. Chem. Soc., Div. Polymer Chem., Preprints*, **6**, No. 1, 42 (1966).
51. (a) J. Preston and F. Dobinson, *J. Polymer Sci. B*, **2**, 1171 (1964). (b) J. Preston and F. Dobinson, U.S. Pat. 3,205,199 (Aug. 1965). (c) French Pat. 1,402,552 (1965). (d) French Pat. 1,430,480 (1966). (e) J. Preston, U.S. Pat. 3,232,910 (Feb. 1966). (f) J. Preston and F. Dobinson, U.S. Pat. 3,240,760 (March 1966). (g) J. Preston and W. B. Black, *J. Polymer Sci. B*, **3**, 845 (1965). (h) J. Preston and W. B. Black, *J. Polymer Sci. B*, **4**, 267 (1966). (i) J. Preston and R. W. Smith, *J. Polymer Sci. B*, **4**, 1033 (1966). (j) J. Preston, *J. Polymer Sci. A-1*, **4**, 529 (1966). (k) J. Preston and F. Dobinson, *J. Polymer Sci. A-1*, **4**, 2093 (1966).
52. A. Conix, *Makromol. Chem.*, **24**, 76 (1957).
53. A. Conix, *J. Polymer Sci.*, **29**, 343 (1958).
54. N. Yoda, *Makromol. Chem.*, **32**, 1 (1959).
55. N. Yoda and A. Miyake, *Bull. Chem. Soc. Japan*, **32**, 1120 (1959).
56. N. Yoda, *Chem. High Polymer Japan*, **19**, 490, 495, 553 (1962).
57. N. Yoda, *Makromol. Chem.*, **55**, 174 (1962).
58. N. Yoda, *Makromol. Chem.*, **56**, 10, 36 (1962).
59. N. Yoda, *J. Chem. Soc. Japan, Ind. Chem. Sect.*, **65**, 667 676 (1962).
60. N. Yoda, *J. Polymer Sci. A*, **1**, 1323 (1963).
61. F. Henglein and H. Tarrasch, *Kunststoffe*, **6**, 5 (1959).
62. German Patent 971,790 (1959); Belg. Pat. 532,543 (1954).
63. N. I. Shirokova, E. F. Russkova, A. B. Alishoeva, R. M. Gitina, J. J. Levkoey, and P. V. Kozlov, *Vysokomolekul. Soedin.*, **3** (4), 642 (1961).
64. N. P. Chopey, *Chem. Eng.*, **67**, 174 (1960).
65. H. Schnell, *Angew. Chem.*, **68**, 633 (1956).
66. W. W. Moyer, J. Wynstra, and J. S. Fry, U.S. Pat. 2,970,131 (April 1961).
67. Belg. Pat. 579,497 (1959).
68. Ger. Pats. 959,497 (1957) and 1,046,311 (1958).
69. Belg. Pat. 603,106 (1961).
70. H. Schnell, *Angew. Chem.*, **73**, 629 (1961).
71. C. A. Bischoff and A. Hedenstroem, *Ber.*, **35**, 3431 (1902).
72. Brit. Pat. 839,858 (1960).
73. Ger. Pat. 1,031,512 (1958).
74. D. W. Fox, U.S. Pat. 3,148,172 (1964).
75. Belg. Pat. 952,152 (1962).
76. L. Lee, *J. Polymer Sci. A*, **2**, 2859 (1964).
77. (a) A. Davis and J. H. Golden, *Makromol. Chem.*, **78**, 16 (1964). (b) J. H. Golden, B. L. Hammant, and E. A. Hazell, *J. Appl. Polymer Sci.*, **11**, 1571 (1967).
78. J. R. Caldwell, U.S. Pat. 2,600,376 (June 1952).
79. R. Gilkey and J. R. Caldwell, *J. Appl. Polymer Sci.*, **2**, 198 (1959).
80. J. R. Caldwell, U.S. Pat. 2,593,411 (May 1952).

III. AROMATIC POLYMERS: FUNCTIONAL GROUPS IN CHAIN

81. R. E. Wilfong, *J. Polymer Sci.*, **54**, 385 (1961).
82. French Pat. 1,163,702 (1958).
83. Belg. Pat. 563,173 (1957).
84. V. V. Korshak, S. V. Vinogradova, and S. N. Salazkin, *Vyskomolekul. Soedin.*, **4**, 339 (1962).
85. Brit. Pat. 604,073 (1948).
86. Brit. Pat. 604,985 (1948).
87. Belg. Pat. 565,478 (1957).
88. Brit. Pat. 660,883 (1950).
89. Brit. Pat. 716,877 (1954).
90. Brit. Pat. 719,706 (1954).
91. Can. Pat. 603,582 (1960).
92. Can. Pat. 674,714 (1963).
93. D. Aelony and M. M. Renfrew, U.S. Pat. 2,728,747 (Dec. 1955).
94. L. H. Bock, U.S. Pat. 2,755,273 (July 1956).
95. D. D. Smith, U.S. Pat. 2,902,473 (Sept. 1959).
96. J. W. Fisher, U.S. Pat. 2,621,107 (Dec. 1952).
97. C. J. Frosch, U.S. Pat. 2,315,613 (April 1943).
98. P. Kass, U.S. Pat. 2,634,251 (May 1949).
99. J. G. N. Drewitt and J. Lincoln, U.S. Pat. 2,595,343 (May 1952).
100. E. A. Wielichi and R. D. Evans, U.S. Pat. 3,008,929 (Nov. 1961), U.S. Pat. 3,008,930 (Nov. 1961), U.S. Pat. 3,008,931 (Nov. 1961), U.S. Pat. 3,008,932 (Nov. 1961), U.S. Pat. 3,008,933 (Nov. 1961), U.S. Pat. 3,008,934 (Nov. 1961), U.S. Pat. 3,008,935 (Nov. 1961).
101. W. M. Eareckson, *J. Polymer Sci.*, **40**, 399 (1959).
102. A. Conix, *Ind. Eng. Chem.*, **51**, 147 (1959).
103. Yu. V. Egorova, V. V. Korshak, and N. N. Lebedev, *Vysokomolekul. Soedin.*, **2**, 1475 (1960).
104. V. V. Korshak, S. V. Vinogradova, and S. N. Salazkin, *Vysokomolekul. Soedin.*, **4**, 339 (1962).
105. V. V. Korshak, S. V. Vinogradova, and M. A. Iskenderov, *Vysokomolekul. Soedin.*, **4**, 345 (1962).
106. S. W. Kantor and F. F. Holub, U.S. Pat. 3,036,991 (June 1960).
107. V. V. Korshak, S. V. Vinogradova, and R. M. Valetskii, *Vysokomolekul. Soedin.*, **4**, 987 (1962).
108. M. A. Iskenderov, V. V. Korshak, S. V. Vinogradova, and V. V. Kharlamov, *Vysokomolekul. Soedin.*, **5**, 799 (1963).
109. V. V. Korshak, *Usp. Khim.*, **29**, 569 (1960).
110. V. V. Korshak and S. V. Vinogradova, *Vysokomolekul. Soedin.*, **1**, 834 (1959).
111. S. W. Kantor and F. F. Holub, U.S. Pat. 3,160,605 (June 1960).
112. S. W. Kantor and F. F. Holub, U.S. Pat. 3,036,992 (June 1960).
113. S. W. Kantor and F. F. Holub, U.S. Pat. 3,036,990 (June 1960).
114. Brit. Pat. 968,390 (1964).
115. V. V. Korshak and S. V. Vinogradova, *Vysokomolekul. Soedin.*, **1**, 839 (1959).
116. M. A. Iskenderov, V. V. Korshak, S. V. Vinogradova, and V. V. Kharlamov., *Vysokomolekul. Soedin.*, **5**, 799 (1963).
117. S. W. Kantor and F. F. Holub, French Pat. 1,291,251 (1962).
118. R. Hill and E. E. Walker, *J. Polymer Sci.*, **3**, 609 (1948).
119. A. S. Carpenter, *J. Soc. Dyers Colourists*, **65**, 469 (1949).
120. E. F. Izard, *J. Polymer Sci.*, **9**, 35 (1952).
121. E. F. Izard, *J. Polymer Sci.*, **8**, 503 (1952).

122. H. J. Kolb and E. F. Izard, *J. Appl. Phys.*, **20**, 564 (1949).
123. C. W. Bunn, *J. Polymer Sci.*, **16**, 323 (1955).
124. K. W. Doak and H. N. Campbell, *J. Polymer Sci.*, **17**, 215 (1955).
125. C. S. Fuller, *Chem. Rev.*, **26**, 143 (1940).
126. L. Pauling, *Nature of the Chemical Bond*, Cornell Univ. Press, Ithaca, N.Y., 1953, Chap. 6.
127. P. M. Cachia, *Ann. Chim. (Paris)*, **4**, 5 (1959).
128. C. S. Fuller and W. O. Baker, *J. Chem. Educ.*, **20**, 3 (1943).
129. R. Hill, *Fibres from Synthetic Polymers*, Elsevier, New York, 1953, Chap. 6.
130. Brit. Pat. 604,074 (1948).
131. Brit. Pat. 604,075 (1948).
132. Brit. Pat. 609,792 (1948).
133. J. K. Wolfe, U.S. Pat. 2,547,679 (April 1951).
134. E. F. Blake, WADC TR 54-532, Part III, April 1957.
135. H. Biletch, *Interchem. Rev.*, **13**, 63 (1954).
136. C. W. Bunn, *J. Polymer Sci.*, **16**, 323 (1955).
137. W. H. Hunter, A. O. Olson, and E. A. Daniels, *J. Am. Chem. Soc.*, **38**, 1761 (1916).
138. G. H. Wollett, *J. Am. Chem. Soc.*, **38**, 2472 (1916).
139. W. H. Hunter and F. E. Joyce, *J. Am. Chem. Soc.* **39**, 2640 (1917).
140. W. H. Hunter and G. H. Wollett, *J. Am. Chem. Soc.*, **43**, 131, 135 (1921).
141. W. H. Hunter and L. M. Seyfried, *J. Am. Chem. Soc.*, **43**, 151 (1921).
142. W. H. Hunter and R. B. Whitney, *J. Am. Chem. Soc.*, **54**, 1167 (1932).
143. W. H. Hunter and F. H. Rathmann, *Zh. Obsch. Khim.*, **7**, 2202, 2206, 2226, 2230 (1937).
144. J. H. Golden, *Soc. Chem. Ind. (London)*, Monograph No. 13, 231 (1961).
145. L. A. Wall, R. E. Florin, and W. J. Pummer, *Am. Chem. Soc. Abstr. 134th Mtg.*, pg. 19 (1958).
146. C. C. Price and G. Staffin, *Rubber Age*, **84**, 295 (1958).
147. C. C. Price and G. Staffin, *J. Am. Chem. Soc.*, **82**, 3632 (1960).
148. C. C. Price and C. J. Kurian, *J. Polymer Sci.*, **49**, 267 (1961).
149. C. C. Price and W. A. Butte, *J. Am. Chem. Soc.*, **84**, 3567 (1962).
150. C. C. Price and N. S. Chu, *J. Polymer Sci.*, **61**, 135 (1962).
151. C. C. Price, W. A. Butte, and R. E. Hughes, *J. Polymer Sci.*, **61**, 528 (1962).
152. M. J. S. Dewar and A. N. James, *J. Chem. Soc.*, **1958**, 917.
153. A. S. Hay, H. S. Blanchard, G. F. Endres, and J. W. Eustance, *J. Am. Chem. Soc.*, **81**, 6336 (1959).
154. A. S. Hay, *Am. Chem. Soc., Div. Polymer Chem., Preprints*, **2**, No. 2, 319 (1961)
155. H. S. Blanchard, H. L. Finkbeiner, and G. F. Endres, *Am. Chem. Soc., Div. Polymer Chem., Preprints*, **2**, No. 2, 331 (1961).
156. H. S. Blanchard, A. L. Finkbeiner, G. F. Endres, and J. W. Eustance, *Am. Chem. Soc., Div. Polymer Chem., Preprints*, **2**, No. 2, 340 (1961).
157. A. S. Hay, *J. Polymer Sci.*, **58**, 581 (1962).
158. G. F. Endres and J. Kwiatek, *J. Polymer Sci.*, **58**, 593 (1962).
159. (*a*) French Pat. 1,322,152 (1963). (*b*) H. S. Hay and G. F. Endres, *J. Polymer Sci. B*, **3**, 887 (1965).
160. G. P. Brown, S. Aftergut, and R. J. Balckinton, *J. Chem. Eng. Data*, **6**, 125 (1961).
161. G. P. Brown and A. Goldman, *Am. Chem. Soc., Div. Polymer Chem., Preprints*, **4**, No. 2, 39 (1963).

162. G. P. Brown and A. Goldman, *Am. Chem. Soc., Div. Polymer Chem., Preprints*, **5**, No. 1, 195 (1964).
163. (*a*) J. M. Cox, B. A. Wright, and W. W. Wright, *J. Appl. Polymer Sci.*, **9**, 513 (1965). (*b*) J. M. Lancaster, B. H. Wright, and W. W. Wright, *J. Appl. Polymer Sci.*, **9**, 1955 (1965).
164. F. H. Winslow, W. O. Baker, and W. A. Yager, *Proc. Conf. Carbon, 2nd, Buffalo, 1959*, p. 93.
165. T. P. Hilditch, *J. Chem. Soc.*, **97**, 2579 (1910).
166. H. S. Tasker and H. O. Jones, *J. Chem. Soc.*, **95**, 1910 (1909).
167. A. D. Macallum, *J. Org. Chem.*, **13**, 154 (1948).
168. A. D. Macallum, U.S. Pat. 2,513,188 (Oct. 1950).
169. A. D. Macallum, U.S. Pat. 2,538,941 (Jan. 1951).
170. R. W. Lenz and W. K. Carrington, *J. Polymer Sci.*, **41**, 333 (1959).
171. R. W. Lenz and C. E. Handlovits, *J. Polymer Sci.*, **43**, 167 (1960).
172. R. W. Lenz, C. E. Handlovits, and H. A. Smith, *J. Polymer Sci.*, **58**, 351 (1962).
173. S. Tsunawaki and C. C. Price, *J. Polymer Sci. A*, **2**, 1511 (1964).
174. F. R. Killay and E. R. Wallsgrove, U.S. Pat. 2,680,728 (Aug. 1950).
175. A. Conix and U. Lasidon, *Abstr.*, Symposium über Makromolekule, Weisbaden, Oct. 12–17, 1959, Vol. IV, B. 9.
176. Ger. Pat. 1,171,618 (1964).
177. Brit. Pat. 916,660 (1964).
178. Can. Pat. 685,311 (1964).
179. D. W. Thomson and G. F. L. Ehlers, *J. Polymer Sci. A*, **2**, 1051 (1964).
180. L. E. Amborski, *Am. Chem. Soc., Div. Polymer Chem., Preprints*, **4**, No. 1, 175 (1963).
181. J. T. Jones, F. W. Ochynski, and F. A. Rackley, *Chem. Ind. (London)*, **1962**, 1686.
182. G. M. Bower and L. W. Frost, *J. Polymer Sci. A*, **1**, 3135 (1963).
183. S. I. Nishizaki and A. Fukami, *J. Chem. Soc. Japan, Ind. Chem. Sect.*, **66**, 382 (1963).
184. C. E. Sroog, S. V. Abramo, C. E. Berr, W. M. Edwards, A. L. Endrey, and K. L. Oliver, *Am. Chem. Soc., Div. Polymer Chem., Preprints*, **5**, No. 1, 132 (1964).
185. A. Kreuchunas, U.S. Pat. 2,822,351 (Feb. 1958).
186. (*a*) M. E. A. Cudby, R. G. Feasey, B. E. Jennings, M. E. B. Jones, and J. B. Rose, *Polymer*, **6**, 589 (1965); *J. Polymer Sci. C.*, **16**, 715 (1967). (*b*) H. A. Vogel, U.S. Pat. 3,321,449 (May 1967).
187. (*a*) Dutch Appl. 6,408,130 (1965). (*b*) R. N. Johnson, A. G. Farnham, R. A. Clendinning, W. F. Hale, and C. N. Merriam, *J. Polymer Sci. A-1*, **5**, 5407, 5408, 5409 (1967).
188. R. G. Arnold and J. J. Verbanc, *J. Chem. Educ.*, **34**, 158 (1957).
189. R. G. Arnold, J. A. Nelson, and J. J. Verbanc, *Chem. Rev.*, **57**, 47 (1957).
190. French Pat. 845,917 (1939).
191. French Pat. 951,496 (1949).
192. W. D. Jones and S. B. McFarlane, U.S. Pat. 2,660,574 (Aug. 1950).
193. Brit. Pat. 856,318 (1960).
194. Brit. Pat. 846,175 (1960).
195. C. L. Wilson, U.S. Pat. 2,921,866 (Jan. 1960).
196. Brit. Pat. 915,504 (1963).
197. Brit. Pat. 954,201 (1954).
198. H. C. Beachell and C. P. N. Son, *J. Appl. Polymer Sci.*, **8**, 1089 (1964).

IV. AROMATIC POLYMERS CONTAINING HETEROCYCLIC RINGS

Another approach to the preparation of processible polymers with thermal stabilities comparable to that of poly-*p*-phenylene has been the synthesis of phenylene-R polymers

$$\{-\!\!\langle\bigcirc\rangle\!-\!R\}$$

in which the R grouping is a heterocyclic ring. With the development of the chemistry and technique for intramolecular cyclization of polymers, a large number of thermally stable polymers in which the polymer chain consists of alternating phenylene and heterocyclic rings or condensed aromatic heterocyclic rings have been prepared. Since the literature is so large, these aromatic heterocyclic systems will be considered in alphabetical order.

A. POLYBENZIMIDAZOLES

Benzimidazole derivatives which are synthesized from *o*-phenylenediamine and carboxylic acids or their derivatives (1–3) (see reaction 1),

$$\text{o-C}_6\text{H}_4(\text{NH}_2)_2 + \text{R--COX} \longrightarrow \text{benzimidazole} + \text{HX} + \text{H}_2\text{O} \quad (1)$$

where X = OH, Cl, OR, OAr

are generally high melting (430–645°C), stable crystalline solids possessing both acid and basic characteristics. They exhibit unusual chemical resistance to acid, basic, and oxidizing reagents (3). Chemical inertness of the benzimidazoles is explicable in terms of the aromatic nature of the systems which derives delocalization energy from overlapping of the sextet of π electrons. These properties suggest that a polymer consisting of recurring benzimidazole rings would be high melting and thermally stable with unusual resistance to chemical attack.

The extension of the general synthetic route to benzimidazoles to include

IV. AROMATIC POLYMERS CONTAINING HETEROCYCLIC RINGS

bifunctional materials was first accomplished by Brinker and Robinson with the finding that bis-*o*-diamines and aliphatic dioic acids or their derivatives react to form linear condensation polymers (4) (see reaction 2).

$$\text{diamine structure} + X-\overset{O}{\underset{\|}{C}}-R-\overset{O}{\underset{\|}{C}}-X \longrightarrow$$

$$\left[\text{polybenzimidazole repeat unit}\right]_x \qquad (2)$$

where X = OH, OR′

Though the polymers were not crystalline these polybenzimidazoles had high softening temperatures and showed outstanding retention of properties at elevated testing temperatures. In addition, they had excellent resistance to weatherability, hydrolytic attack, and exposure to oxygen at elevated temperatures.

A modification of the Brinker-Robinson procedure resulting in the preparation of polybenzimidazoles containing recurring aromatic units was developed by Marvel and Vogel (5) by way of the condensation of aromatic tetraamines with the phenyl esters of aromatic dicarboxylic acids. Originally, these workers believed that the superior results arising from the use of phenyl esters were due to the advantageous plasticizing effect of the by-product phenol. However, the preparation of high molecular weight aromatic polybenzimidazoles from 3,3′-diaminobenzidine and phthalic

$$\text{tetraamine} + \phi-O-\overset{O}{\underset{\|}{C}}-\text{Ar}-\overset{O}{\underset{\|}{C}}-O-\phi \xrightarrow[N_2 \text{ atm}]{200-300°C}$$

$$\left(\text{intermediate polymer}\right)_x + H_2O + \phi-OH \xrightarrow[1 \text{ mm}]{300-400°C} \qquad (3)$$

$\eta_{inh} = 0.30$

$$\left(\text{final polybenzimidazole}\right)_x$$

$\eta_{inh} = 0.6-1.0$

anhydride clearly showed that the action of phenol as a plasticizer was not necessary for the achievement of high molecular weight. These results suggest that the successful preparation of high molecular weight aromatic polybenzimidazoles with these reactants might be due to the higher rates of reaction of the phenyl ester and acid anhydrides over those of the free acid and alkyl esters.

(4)

High molecular weight polybenzimidazoles were obtained by heating the aromatic tetraamino compounds and various diphenyl esters of aromatic dibasic acids at high temperatures. This condensation differs from that of the aliphatic polybenzimidazoles in that a vacuum cycle was required for the preparation of polymer in film- or fiber-forming molecular weights. As outlined in the equations above of the reaction of 3,3'-diaminobenzidine with diphenyl isophthalate, the condensation was initiated at temperatures around 250°C under nitrogen atmosphere. The temperature was gradually raised to 300°C with evolution of phenol and water and a vacuum applied. Under high vacuum, the polymerization mixture was then heated for several hours at temperatures in excess of 300°C up to a maximum of 400°C. This final stage of polymerization was a solid state reaction (5) (see reaction 3).

Based on kinetic studies (6a,b), a mechanism for the preparation of aromatic polybenzimidazoles has been proposed. It suggests that the polymeric material at the prepolymer stage (260°C) has both benzimidazole and hydroxybenzimidazoline structure, while heating *in vacuo* to 400°C results in the formation of the polybenzimidazole (see reaction 4).

Using the phenyl ester–tetramine route, Marvel and co-workers have prepared numerous polybenzimidazoles (7–10).

The preparation of high molecular weight polybenzimidazoles has been reported via the reaction of aromatic tetraamine tetrahydrochlorides with dicarboxylic acids or their derivatives in polyphosphoric acid (11) (see reaction 5).

$$\text{(5)}$$

where X is CO_2H, CO_2CH_3, $CONH_2$, or CN.

Using this general procedure Iwakura and co-workers (11–14) have prepared aromatic polybenzimidazoles, polyalkylene benzimidazoles, and polybenzimidazole amides.

The polybenzimidazoles which have been prepared are summarized in Table I. All of the polymers listed in Table I were colored, ranging from yellow to dark brown, and were characterized by superior hydrolytic stability and a high degree of thermal stability under certain conditions. They were soluble in concentrated sulfuric acid or formic acid yielding stable solutions, and surprisingly, many were soluble in dimethyl sulfoxide and *N,N*-dimethylacetamide.

The interrelationship between solubility, crystallinity, and structure has been investigated by Levine and co-workers (16) as summarized in Table II. He found that a comparison of polymer **5** with polymer **6** and polymer **10** with polymer **11** showed that the more symmetrically oriented diphenyl acid (*p* greater than *o*) resulted in crystallinity and decreased solubility in

TABLE I
Aromatic Polybenzimidazoles

Polymer	Tetraamine	Acid	PMT^a (°C)	$\eta_{\text{inh}}^{b,c}$ (dl/g)	Ref.
1	3,3'-Diaminobenzidine	3,4-Diaminobenzoic	>600	1.27	5
2	3,3'-Diaminobenzidine	Adipic	450 d	3.19	5
3	3,3'-Diaminobenzidine	Pinic	400 d	1.03	5
4	3,3'-Diaminobenzidine	Homopinic	400 d	1.38	5
5	3,3'-Diaminobenzidine	Terephthalic	>600	1.00	5, 10
6	3,3'-Diaminobenzidine	Isophthalic	>600	3.34	5, 10
7	3,3'-Diaminobenzidine	Pyridine-3,5-dicarboxylic	>600	1.48	5
8	3,3'-Diaminobenzidine	Furan-2,5-dicarboxylic	480 d	0.74	5
9	3,3'-Diaminobenzidine	Naphthalene-1,6-dicarboxylic	>600	2.70	5
10	3,3'-Diaminobenzidine	Biphenyl-4,4'-dicarboxylic	>600	0.86^c	5
11	3,3'-Diaminobenzidine	Biphenyl-2,2'-dicarboxylic	430	2.99	5
12	1,2,4,5-Tetraaminobenzene	Adipic	490 d	2.51	5
13	1,2,4,5-Tetraaminobenzene	Terephthalic	>600	0.80	5
14	1,2,4,5-Tetraaminobenzene	Isophthalic	>600	1.10	5
15	3,3'-Diaminobenzidine	Phthalic	>500	3.01	7
16	1,3-Dianilino-4,6-diaminobenzene	Isophthalic	>500	0.81^c	7
17	3,3'-Diaminobenzidine	Glutaric	~420	1.19	7

#	Diamine	Diacid	PMT	η_{inh}	Ref
18	3,3′-Diaminobenzidine	Succinic	~470	2.70	7
19	3,3′-Diaminobenzidine	Naphthalene-2,6-dicarboxylic	>500	1.26[c]	8
20	3,3′-Diaminobenzidine	Naphthalene-2,7-dicarboxylic	>400	1.10	8
21	3,3′-Diaminobenzidine	Naphthalene-2,7-dicarboxylic	>500	0.63	8
22	3,3′-Diaminobenzidine	4,5-Imidazoldicarboxylic	~300	0.16	8
23	3,3′-Diaminobenzidine	Maleic	~400	0.10	8
24	3,3′-Diaminobenzidine	Perfluorosuberic	~300	0.20	8
25	3,3′-Diaminobenzidine	Perfluoroglutaric	~300	0.17[c]	8
26	3,3′,4,4′-Tetraaminodiphenylether	Isophthalic	>400	0.53	9
27	3,3′,4,4′-Tetraaminodiphenylether	4,4′-Oxydibenzoic	>400	0.53	9
28	3,3′-Diaminobenzidine	4,4′-Oxydibenzoic	>400	0.72	9
29	3,3′-Diaminobenzidine	Azelaic	>400	0.94[c]	11
30	1,3-Diamino-4,6-di(methylamino)-benzene	Isophthalic	>400	0.20[c]	15
31	3,3′-Diaminobenzidine	Sebacic	>400	2.54[c]	12
32	3,3′-Diaminobenzidine	1,3-Phenylene bis(4-amidobenzoic)	>400	0.50[c]	12
33	3,3′-Diaminobenzidine	1,4-Phenylene bis(4-amidobenzoic)	>400	0.40[c]	12

[a] PMT: Polymer melt temperature. See R. G. Beaman and F. B. Cramer, *J. Polymer Sci.*, **21**, 223 (1956).
[b] η_{Inh}: inherent viscosity determined in 0.3 wt % concentrations in formic acid at 30°C.
[c] η_{Inh}: inherent viscosity determined in 0.3 wt % concentrations in conc. sulfuric acid at 30°C.

TABLE II
Properties of Polybenzimidazoles (16)

Polymer	Reactants		Properties			
	Tetraamine	Acid	Crystallinity[a]	Solubility[b]		
				Formic acid	Dimethyl sulfoxide	N,N-Dimethylacetamide
5	3,3'-Diaminobenzidine	Terephthalic	+	2–3	0.5–1	Insoluble
6	3,3'-Diaminobenzidine	Isophthalic	0	5–6	>20	Partially soluble
10	3,3'-Diaminobenzidine	Biphenyl-4,4'-dicarboxylic	+	Par partially soluble	Insoluble	Insoluble
11	3,3'-Diaminobenzidine	Biphenyl-2,2'-dicarboxylic	0	20	>20	>20
7	3,3'-Diaminobenzidine	Pyridine-3,5-dicarboxylic	+	10–15	>20	Partially soluble
8	3,3'-Diaminobenzidine	Furan-2,5-dicarboxylic	+	3–4	>20	>20
13	1,2,4,5-Tetraaminobenzene	Terephthalic	+	2–3	Insoluble	Insoluble
14	1,2,4,5-Tetraaminobenzene	Isophthalic	+	5–6	Insoluble	Insoluble
12	1,2,4,5-Tetraaminobenzene	Adipic	+	20	Partially soluble	Insoluble

[a] Crystallinity: +, detected; 0, undetected.
[b] Solubility in wt % concentration.

polymers **5** and **10**. The use of 1,2,4,5-tetraaminobenzene in place of 3,3'-diaminobenzidine with the same esters resulted in crystallinity and decreased solubility as shown by the comparison of polymers **6** and **14**; decreased solubility resulted also in crystalline polymer, **5**, when 1,2,4,5-tetraaminobenzene was used, as shown by comparison of polymers **5** and **13**. Apparently these changes were caused by free rotation at the 1,1'-bond in the benzidine molecule and resulted in decreased rigidity. The more rigid 1,2,4,5-tetraaminobenzene moiety even caused crystallization in the aliphatic–aromatic polymer, **12**, but the effect on decreasing solubility was not as great as in the wholly aromatic systems.

Another point of interest was the formation of crystalline polymer using heterocyclic compounds having two esters groups separated by one atom making them somewhat analogous to noncrystalline polymer, **6**, having the same ester group orientation. Although crystalline, the heterocyclic isooriented polymer, **7**, was more soluble in formic acid than was the non-crystalline, carbocyclic isooriented polymer, **6**, perhaps because of the added polyelectrolyte effect resulting from the basic hetero atom.

The outstanding hydrolytic stability of these polybenzimidazole structures has been amply demonstrated (5). Samples of poly-2,2'(m-phenylene)-5,5'-bibenzimidazole, **6**, have been heated to reflux in 70% sulfuric acid

TABLE III
Thermal Stability of Polybenzimidazoles under Nitrogen

Polymer[a] code	Weight lost during 1 hr of heating (wt %)					Total weight loss (wt %)	Ref.
	400°C	450°C	500°C	550°C	600°C		
1	1.1	0.4	0.4	0.4	5.0	7.3	5
5	1.0	1.0	0	1.7	1.0	4.7	5
6	0.6	0	0.4	1.3	2.2	4.5	5
15	0	0.4	0.4	3.7	—	4.5	7
16	0	0.4	1.6	6.0	—	8.0	7
7	0.2	0.8	0.5	1.4	2.7	5.6	5
8	1.4	1.7	2.6	2.3	2.0	10.0	5
9	0.4	0.4	0.8	1.2	3.7	6.5	5
10	0.3	0	0.8	0.3	2.1	3.5	5
11	0	0.5	8.0	8.5	0.5	17.5	5
13	2.8	0.5	1.0	1.9	4.0	10.2	5
11	0	0.5	8.0	8.5	0.5	17.5	5
13	2.8	0.5	1.0	1.9	4.0	10.2	5
14	0.7	1.4	0.3	1.4	1.4	5.2	5

[a] See Table I for polymer code. Each polymer sample was heated for 1 hr at each of the given temperatures, consecutively.

for 10 hr. The precipitated polymer had the same inherent viscosity as the starting polymer. Similarly, a sample of the same polymer was heated in 50 ml of 25% sodium hydroxide for 10 hr and the resulting polymer showed only a slight loss in inherent viscosity.

The outstanding thermal stability of the polybenzimidazoles is indicated by the infrared spectra (5) of heated and unheated samples under nitrogen, which show no change even after exposure to temperatures of 400, 500, and 600°C for 1 hr at each temperature. Marvel and co-workers (5,6) have carried out thermal degradation studies of these polymers by measuring the rate of weight loss in a nitrogen atmosphere at 400, 450, 500, 550, and 600°C consecutively for 1 hr at each temperature. The results are summarized in Table III. The weight loss during heating at 400, 450, and 500°C has been ascribed to endgroup reactions.

The superior thermal behavior of the wholly aromatic polymers was further evidenced by thermogravimetric analysis (5,7) as seen in Figures 1 and 2. These figures showed that aliphatic polybenzimidazoles undergo complete decomposition around 470°C with initial losses in the 350–400°C region because of the presence of the aliphatic C—C linkage, whereas all-aromatic polymers showed weight losses, starting around 600°C, which reached only about 30% at 900°C.

The superior thermal stability in inert atmospheres was not carried over into air, oxygen, or oxidizing atmospheres (7–9). When poly-2,2'(o-phenylene)5,5'-bibenzimidazole, **15**, was heated in a nitrogen atmosphere consecutively for 1 hr each at temperatures of 400, 450, 500, and 550°C, weight losses of 0, 0.4, 0.4, and 3.7%, respectively, were obtained. In air, under the same conditions of heating, the losses were 0, 1.5, 7.0, and 7.6%,

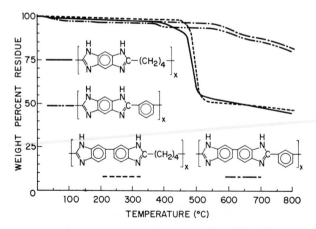

Fig. 1. TGA plot for polybenzimidazoles under nitrogen (7).

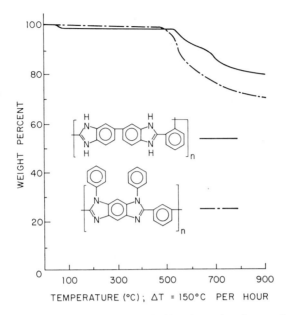

Fig. 2. TGA curves for polybenzimidazoles under nitrogen (7).

respectively. Similarly, poly-2-6′(*m*-phenylene)3,5-diphenyldiimidazolebenzene, **16**, showed weight losses when heated in a nitrogen atmosphere as above of 0, 0.4, 1.6, and 6.0%, respectively, and in air showed losses of 1.0, 4.8, 10.1, and 12.1%, respectively. Even poly-2,2′(*m*-phenylene)5,5′-bibenzimidazole, **6**, which showed weight losses under nitrogen of 0.6, 0, 0.4, and 1.3%, respectively, had losses in air of 0, 1.7, 5.2, and 7.0%, respectively.

Phillips and Wright (17) have studied the thermal stability of poly-2,2′(*m*-phenylene)5,5′-bibenzimidazole, **6**, in air using a Stanton thermobalance and in vacuum, or under 300 mm Hg pressure of oxygen using a quartz spring apparatus. Experiments were carried out both with the temperatures rising at a constant rate and under isothermal conditions.

In vacuum with temperature rising at approximately 3°/min, a slow but steady weight loss was observed between 400 and 700°C, the final total weight loss being 13%. An experiment in which the temperature was increased stepwise, the sample being held for 1 hr at each temperature, gave the results shown in Table IV. At lower temperatures, stabilization occurred within the hour, i.e., the rate of weight loss held to zero, but at 680°C the rate of weight loss was increasing, and in an additional 30 min at that temperature, another 3% weight loss was recorded.

The results in air were quite different, however. Using a programmed

TABLE IV
Percent Weight Loss of Polymer **6** during 1 Hr of Heating at Temperature (17)

Temp. °C	480	530	560	600	665	680
Wt loss, %	6.3	0.5	1.3	1.2	0.9	2.6
		Total wt loss		12.8		

temperature rise on a Stanton thermobalance, a rapid breakdown was observed between 500 and 650°C, the sample having completely volatilized at the latter temperature. A constant rate of 0.5% wt loss/hr was recorded over a 24-hr period at 400°C. A rate of approximately 9%/hr, up to 60% volatilization occurred at 450°C; thereafter the rate decreased steadily, volatilization being substantially complete after 16 hr at that temperature.

In oxygen, a series of isothermal experiments was carried out between 400 and 470°C. A plot of the rate of weight loss vs. the percentage decomposition is given in Figure 3. The curves show a maximum at approximately 30% volatilization. Arrhenius plots were made of the rate of weight loss at fixed percentage decompositions, and the overall activation energies calculated for various stages of the reaction. The results are given in Table V. The rate equation for the major part of the reaction was found to be represented by $K = 10^9 \, e^{-36,000/RT}$.

Thus Wright's results are in agreement with earlier work. The polybenzimidazoles have a high level of thermal stability in inert atmospheres but not in air, oxygen, or oxidizing atmospheres. The stability in the presence of air was found to be comparable to that of the copolymer of tetrafluoroethylene and hexafluoropropylene (18).

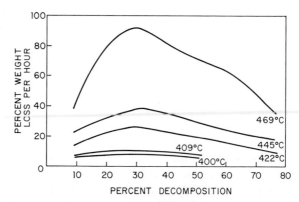

Fig. 3. Rate of weight loss of **6** under 300 mm Hg pressure of oxygen (17).

IV. AROMATIC POLYMERS CONTAINING HETEROCYCLIC RINGS

TABLE V
Overall Activation Energy of Decomposition of Polymer **6** at Various Stages of Reaction (17)

Decomposition (%)	Overall activation energy (kcal/mole)
10	32
20	36
30	36
40	36
50	36
60	40
70	39

Thermogravimetric analysis in air of polybenzimidazoles, Figures 4–6, showed similar behavior (11–13) with sharp losses of weight between 450 and 500°C and complete loss of weight by 600–650°C.

It has been reported (19) that when film strips of polybenzimidazoles were passed over a heated surface at 350–425°C they turned a deep purple color. This color disappeared when the film was removed from the heat source. Electron spin resonance measurements (19) on bulk polymer, although qualitative, did reveal absorption bands for unpaired electrons which increased in intensity on exposure to light or heat and indicated that the process was not reversible. This is reminiscent of the photochromism of triphenylimidazoles and strongly argues for radical formation via NH bond-breaking. Thus the behavior of the polybenzimidazoles in air could

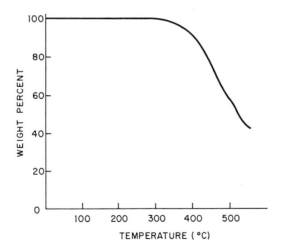

Fig. 4. TGA of poly-2,2'-octamethylene-5,5'-bibenzimidazole (11).

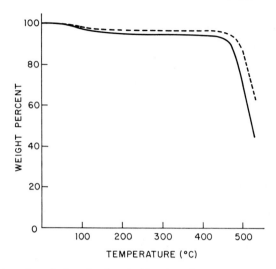

Fig. 5. TGA of polyphenylenebenzimidazoles (12): ———, poly-*m*-phenylene-benzimidazole; -----, poly-*p*-phenylenebenzimidazole.

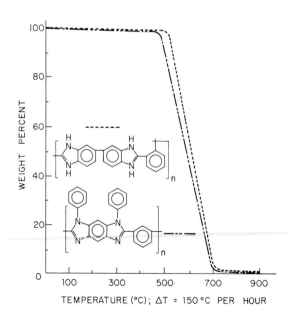

Fig. 6. TGA curves of polybenzimidazoles in air (13).

be readily attributable to the thermolytic dissociation of the NH bond at these elevated temperatures and subsequent oxidative attack on the produced radical.

The thermal stability and superior resistance to acid and basic hydrolytic media, along with toughness and good electrical properties of these aromatic polybenzimidazoles, have made them attractive for adhesive, laminate, sealant, coating, foam, and fiber applications. These applications of the polybenzimidazoles and their thermal stability in such applications will be discussed in Chapter VII.

B. POLYBENZOXAZOLES

Benzoxazole derivatives are synthesized from *o*-hydroxyaminobenzenes and carboxylic acids or their derivatives (20–23) (see reaction 6).

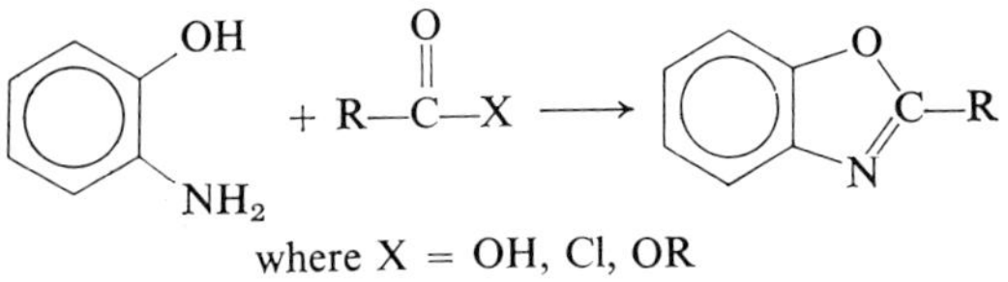

$$\text{where X = OH, Cl, OR} \tag{6}$$

Although the benzoxazoles are not as high melting as the benzimidazoles, they possess both acid and basic characteristics and exhibit chemical resistance to acid, basic, and oxidizing reagents (20,23).

The extension of this general synthetic route to benzoxazoles to include bifunctional materials was first reported by Brinker and co-workers (24) who found that aminohydroxyphenylalkanoic acids or bisaminohydroxyphenyl compounds with dicarboxylic acids or their derivatives yielded high molecular weight linear condensation polymers (see reactions 7a and 7b).

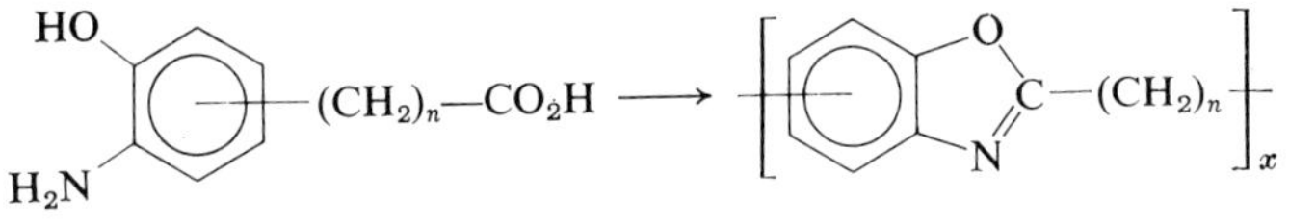

(7a)

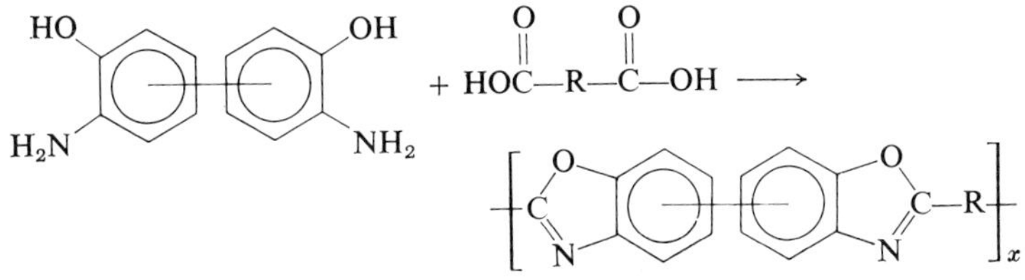

(7b)

These polybenzoxazoles had high softening temperatures and showed outstanding retention of properties at elevated testing temperatures. In addition, they showed excellent resistance to weatherability, hydrolytic attack, and exposure to oxygen at elevated temperatures (22).

Moyer (25) has reported the preparation of polybenzoxazoles via a modification of the Brinker procedure yielding such polymers containing recurring aromatic units. This was accomplished either by the condensation of bisaminohydroxyphenyl compounds with diphenyl esters of aromatic dicarboxylic acids or by the solution polymerization of *o*-aminohydroxybenzoyl chlorides in pyridine to yield the precursor polyamide phenol, followed by thermal cyclodehydration to the benzoxazole structure (see reactions 8a and 8b).

Kubota and Nakanishi (26) have also utilized a similar reaction to prepare the fully aromatic polybenzoxazoles. In this work the reactants were bisaminohydroxyphenyl compounds and aromatic diacid chlorides in a low temperature solution polymerization with such solvents as the

IV. AROMATIC POLYMERS CONTAINING HETEROCYCLIC RINGS

N,N-dialkylamides and pyridine as acid acceptors. The precursor polymers, the polyamide phenols, were subsequently thermally cyclized to the benzoxazole structure (see reaction 9).

Iwakura and co-workers (27,28) have recently prepared polybenzoxazoles by the reaction of bisaminohydroxyphenyls with aromatic

[Reaction scheme 10a showing bisaminohydroxybiphenyl + aromatic dicarboxylic acid → polybenzoxazole via PPA] (10a)

[Reaction scheme 10b showing o-aminohydroxybenzoic acid → polybenzoxazole via PPA] (10b)

dicarboxylic acids or the homopolymerization of o-aminohydroxyaromatic acids in polyphosphoric acid (PPA) (see reactions 10a and 10b).

The aromatic polybenzoxazoles which have been prepared are collected in Table VI. Like the polybenzimidazoles, these polymers were colored and were characterized by superior hydrolytic stability and a high degree of

TABLE VI
Aromatic Polybenzoxazoles

Polymer	Amine	Acid	PMT[a] (°C)	η_{inh}[b] (dl/g)	Ref.
34		3-Amino-4-hydroxybenzoic	>500	1.04	25
35		4-Amino-5-hydroxybenzoic	>500	0.22	25
36	3,3'-Diamino-4,4'-dihydroxybiphenyl	Isophthalic	>600	1.22	25, 26
37	3,3'-Diamino-4,4'-dihydroxybiphenyl	Terephthalic	>600	0.94	25, 26
38	3,3'-Diamino-4,4'-dihydroxybiphenyl	4-Chloroisophthalic	>600	0.25	25

[a] PMT: polymer melt temperature. See R. G. Beaman and F. B. Cramer, *J. Polymer Sci.*, **21**, 223 (1956).
[b] η_{inh}: inherent viscosity determined at 0.5 wt % concentration in conc. sulfuric acid at 30°C.

TABLE VII
Thermal Stability of Polybenzoxazoles under Nitrogen (7,25)

Polymer[a] Code	Weight lost during 1 hr of heating (wt %)				Total wt Loss (wt %)
	400°C	450°C	500°C	550°C	
34	5	3	5	7	20
35	5	4	5	6	20
36	1.2	2.4	3.8	6	13.4
37	1.1	2.3	3.4	5.8	12.6
38	—	3.2	4.8	7	15.1

[a] See Table VI for polymer code; each sample heated for 1 hr at each of the given temperatures, consecutively.

thermal stability. Unlike the polybenzimidazoles, they were soluble only in concentrated sulfuric acid and polyphosphoric acid (25–28).

Moyer (25) has carried out thermal degradation studies of the all-aromatic polybenzoxazoles by measuring the rate of weight loss in nitrogen at 400, 450, 500, and 550°C consecutively for 1 hr at each temperature. These results which are collected in Table VII were remarkably similar to those obtained for the polybenzimidazoles in a similar study.

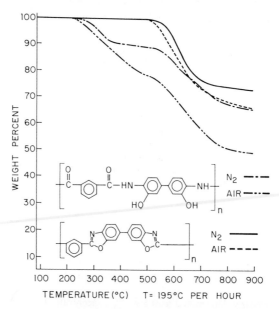

Fig. 7. TGA curves of polyamide and polybenzoxazole in nitrogen and in air (26).

IV. AROMATIC POLYMERS CONTAINING HETEROCYCLIC RINGS 155

The thermal stability of the fully aromatic polybenzoxazole as measured in nitrogen and air by thermogravimetric analysis (26) is illustrated in Figure 7 (also see Fig. 9). As in the case of the polybenzimidazoles, decomposition started at about 500°C in air and nitrogen with loss of 27% of original weight in nitrogen as compared to 35% original weight in air on heating up to 900°C.

Since these polybenzoxazoles were soluble only in such solvents as concentrated sulfuric acid or polyphosphoric acid, the fabrication of the polybenzoxazoles into useful articles has met with only limited success (see Chapter VII). Based on thermogravimetry, however, the polybenzoxazoles appear to be comparable in stability to the polybenzimidazoles in nitrogen and somewhat superior in air.

C. POLYBENZOTHIAZOLES

Benzothiazole derivatives are synthesized from *o*-aminothiophenols and carboxylic acids or their derivatives (29–31) (see reaction 11). Although

$$\text{[benzene ring with SH and NH}_2\text{]} + R\text{—}\overset{O}{\underset{\|}{C}}\text{—}X \longrightarrow \text{[benzothiazole ring]}\text{C—R} \qquad (11)$$

where X = OH, Cl, OR, OAr

the benzothiazoles, like the benzoxazoles, are not as high melting as the benzimidazoles, they possess both acid and basic characteristics and exhibit chemical resistance to acid, basic, and oxidizing reagents (29,30).

Numerous attempts have been made to prepare polybenzothiazoles (32–34). Levine and co-workers (35,36) have reported the synthesis of these polymers in high molecular weight. In this work polybenzothiazoles were prepared by (*1*) the reaction of 3,3'-dimercaptobenzidine with aromatic dicarboxylic acids or benzothiazole-forming dicarboxylic acid derivatives, and (*2*) the self-condensation of 3-mercapto-4-aminobenzoic acid as *N,N*-diethylaniline (see reactions 12a and 12b).

Levine and co-workers (35,36) have suggested that the mechanism for the reaction is as shown in reaction 13. A second possible mechanism similar to that proposed for polybenzimidazole formation (Section A) is unlikely in view of the difference in nucleophilicity between the mercaptan and amino groups.

Iwakura and co-workers (27,28) have also reported the preparation of high molecular weight polybenzothiazoles via the reaction of bis(aminomercaptophenyl) compounds with aromatic dicarboxylic acids or the

$$\text{HS} \underset{H_2N}{\bigcirc}\underset{NH_2}{\bigcirc} \text{SH} + Y-Ar-Y \longrightarrow$$

$$\left[\begin{array}{c} S \\ -C \\ N \end{array} \bigcirc\bigcirc \begin{array}{c} S \\ C-Ar \\ N \end{array} \right]_x \quad (12a)$$

where Y = CO_2H, CO_2R, $CONH_2$, CN, COCl, $C(NH_2)=NH$

Ar = phenyl, pyridyl, diphenyl ether, benzophenone

$$\underset{NH_2}{\underset{HO_2C}{\bigcirc}} \text{SH} \longrightarrow \left[\bigcirc \begin{array}{c} S \\ C \\ N \end{array} \right]_x \quad (12b)$$

homopolymerization of *o*-aminomercaptoaromatic acids in polyphosphoric acid at elevated temperatures.

The aromatic polybenzothiazoles which have been prepared are listed in Table VIII. Like the polybenzimidazoles and the polybenzoxazoles,

$$\text{HS}\underset{H_2N}{\bigcirc}\underset{NH_2}{\bigcirc}\text{SH} + \underset{C-O-\phi}{\underset{\parallel}{\overset{O}{\bigcirc}}}\underset{C-O-\phi}{\overset{\parallel}{O}} \longrightarrow$$

$$\left[\bigcirc \underset{NH_2}{\overset{S-C-\bigcirc}{\underset{OH}{\overset{O-\phi}{|}}}} \right] \underset{+H_2O}{\overset{-H_2O}{\rightleftarrows}} \left[\bigcirc \underset{N}{\overset{S}{\underset{H}{\overset{O-\phi}{C}}}}\bigcirc \right] \xrightarrow{-\phi-OH}$$

$$\left[\bigcirc \begin{array}{c} S \\ C \\ N \end{array} \bigcirc \right] \quad (13)$$

TABLE VIII
Aromatic Polybenzothiazoles (27,28,35,36)

Mercaptan	Acid	Color	PMT[a] (°C)	η_{inh}[b] (dl/g)
3,3'-Diamino-4,4'-dimercaptobiphenyl	Isophthalic	Yellow	600	0.51
3,3'-Diamino-4,4'-dimercaptobiphenyl	Pyridine-3,5-dicarboxylic	Brown	600	0.41
3,3'-Diamino-4,4'-dimercaptobiphenyl	4,4'-Oxydibenzoic	Green	500	0.61
3,3'-Diamino-4,4'-dimercaptobiphenyl	4,4'-Ketodibenzoic	Orange	500	0.42
	3-Amino-4-mercaptobenzoic	Brown	600	1.02

[a] PMT: Polymer melt temperature. See R. G. Beaman and F. B. Cramer, *J. Polymer Sci.*, **21**, 223 (1956).

[b] η_{inh}: Inherent viscosity determined at 0.5 wt % concentration in conc. sulfuric acid at 30°C.

these polymers were colored and were characterized by superior hydrolytic stability and a high degree of thermal stability. Unlike the polybenzimidazoles, but like the polybenzoxazoles, these polymers were soluble only in concentrated sulfuric acid and polyphosphoric acid (27,28,35).

The thermal stability of the fully aromatic polybenzothiazoles prepared by Levine (35) as measured in air by thermogravimetric analysis is illustrated in Figure 8. These have excellent oxidative stability up

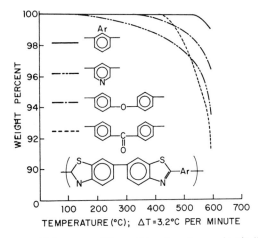

Fig. 8. TGA curves of benzothiazole polymers in air (35).

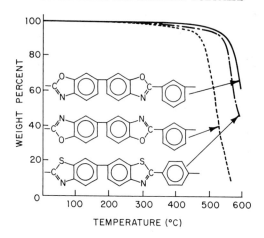

Fig. 9. TGA curves of the aromatic polybenzoxazoles and polybenzothiazoles (27).

to temperatures of 600°C in air as evidenced by their small weight loss.

The fully aromatic polybenzothiazoles prepared by Iwakura show a markedly different thermal stability in air by thermogravimetry (27,28) (Fig. 9). These polybenzothiazoles appear to be equivalent to the polybenzoxazoles showing an upper stability limit below 600°C. In fact, Iwakura (28) has compared identical aromatic polybenzimidazoles, polybenzoxazoles, and polybenzothiazoles and found their behavior in air to be practically identical (Fig. 10).

Until this difference in thermal stability has been reconciled, it can only be concluded that the polybenzimidazoles, the polybenzothiazoles, and the polybenzoxazoles are not markedly different in stability.

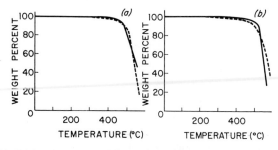

Fig. 10. (a) TGA of poly-2,5(6)-benzimidazole (———) and poly-2,5-benzoxazole (-----). (b) TGA of poly-2,6-benzoxazole (———) and poly-2,6-benzothiazole (-----) (28).

IV. AROMATIC POLYMERS CONTAINING HETEROCYCLIC RINGS

D. AROMATIC POLYIMIDES

Phthalimide is probably the most widely known derivative of isoindole. Due to the fact that the phthalimides are high melting crystalline solids, possessing unusual chemical resistance to acids and bases, they have been of interest as structural elements in high temperature resistant polymers (37).

Probably the first recorded synthesis of a polyimide is that of Bogert and Renshaw (37) who observed that 4-aminophthalic anhydride and dimethyl-4-aminophthalate, on heating, eliminated water and methanol, respectively, yielding a polyimide (see reaction 14). The first reported preparation

$$\text{(14)}$$

of high molecular weight polyimides is found in the work of Edwards and Robinson (38). These workers prepared polyimides via the fusion of the salt from a diamine and tetraacid or a diamine and a diacid/diester (see reaction 15). In these procedures a low molecular weight polymer was prepared by an initial reaction at temperatures below 150°C and this prepolymer subsequently heated at 250–300°C for several hours. This melt–flux method was limited to polyimides with melting points sufficiently low that they remained molten under the conditions of polymerization and was therefore restricted to long chain diamines.

Refinements of this salt–melt method by Edwards and Robinson to yield higher molecular weight polymer involved (*1*) heating the prepolymer at temperatures just below the melting point of the polymer (39), and (*2*) using mixtures of pyromellitic dianhydride and suitable diamines in aqueous alcohol (ethanol) at low temperatures to yield a prepolymer which was subsequently heated (40).

In all cases, these procedures were successful in achieving high molecular weights only in the case of fusible polyimides.

The preparation of infusible and/or all aromatic polyimides has been

$$\text{HO-C(=O)-C}_6\text{H}_2\text{(C(=O)-O-CH}_3\text{)(C(=O)-O-CH}_3\text{)(C(=O)-O-H)} + \text{H}_2\text{N-R-NH}_2 \longleftrightarrow \text{H}_2\text{N-R-NH}_2 + \text{HO-C(=O)-C}_6\text{H}_2\text{(C(=O)-OH)}_3$$

$$\downarrow \text{Salt}$$

$$\left[\text{-N(CO)}_2\text{-C}_6\text{H}_2\text{-(CO)}_2\text{N-R-} \right] + \text{H}_2\text{O} \qquad (15)$$

the subject of many recent publications (41–57). Probably the earliest description of such a preparation is found again in the work of Edwards (42,43). The general polymerization procedure involved reaction of an aromatic dianhydride with an aromatic diamine in a suitable solvent to yield an intermediate soluble precursor polymer, the polyamic acid, which on subsequent dehydration yielded the insoluble polyimide as shown in reaction 16.

Suitable solvents in polyamic acid preparation and operable dianhydrides and diamines are listed in Tables IXA and IXB, respectively (44,45).

$$\text{dianhydride} + \text{H}_2\text{N-Ar-NH}_2 \longrightarrow$$

$$\left[\text{HO-C(=O)-C}_6\text{H}_2\text{(C(=O)-OH)(NH-C(=O))(C(=O)-NH-Ar)} \right] \xrightarrow{-\text{H}_2\text{O}} \left[\text{-H-N(CO)}_2\text{-C}_6\text{H}_2\text{-(CO)}_2\text{N-Ar-} \right] \qquad (16)$$

TABLE IXA
Solvents for Polyamic Acid Synthesis (42–45)

N,N-Dimethylformamide	N-Methyl-2-pyrrolidone
N,N-Diethylformamide	Pyridine
N,N-Diethylacetamide	Dimethyl sulfone
N,N-Dimethylmethoxyacetamide	Hexamethylphosphoramide
N-Methylcaprolactam	Tetramethylene sulfone
Dimethyl sulfoxide	N-Acetyl-2-pyrrolidone

TABLE IXB
Diamines and Dianhydrides for Polyamic Acid Synthesis (42–45)

Diamines

m-Phenylenediamine	Decamethylenediamine
p-Phenylenediamine	3-Methylheptamethylenediamine
4,4′-Diaminodiphenylpropane	4,4-Dimethylheptamethylenediamine
4,4′-Diaminodiphenylmethane	2,11-Diaminododecane
Benzidine	1,2-Bis(3-amino-propoxyethane)
4,4′-Diaminodiphenyl sulfide	2,2-Dimethylpropylenediamine
4,4′-Diaminodiphenyl sulfone	3-Methoxyhexamethylenediamine
4,4′-Diaminodiphenyl ether	2,5-Dimethylhexamethylenediamine
1,5-Diaminonaphthalene	2,5-Dimethylheptamethylenediamine
3,3′-Dimethylbenzidine	3-Methylheptamethylenediamine
3,3′-Dimethoxybenzidine	5-methylnonamethylenediamine
2,4-Bis(β-amino-t-butyl)toluene	2,17-Diaminoeicosadecane
Bis(p-β-amino-t-butylphenyl)ether	1,4-Diaminocyclohexane
Bis(p-β-methyl-δ-aminopentyl)benzene	1,12-Diaminooctadecane
1-Isopropyl-2,4-m-phenylenediamine	$N_2N(CH_2)_3S(CH_2)_3NH_2$
m-Xylylenediamine	di(p-Aminocyclohexyl)methane
p-Xylylenediamine	Heptamethylenediamine
Hexamethylenediamine	$H_2N(CH_2)_3N(CH_3)(CH_2)_3NH_2$
Octamethylenediamine	Piperazine
Nonamethylenediamine	

Dianhydrides

Pyromellitic dianhydride	2,2-Bis(3,4-dicarboxy-phenyl)propane dianhydride
2,3,6,7-Naphthalenetetracarboxylic dianhydride	3,4-Dicarboxyphenyl sulfone dianhydride
3,3′,4,4′-Diphenyltetracarboxylic dianhydride	Perylene-3,4,9,10-tetracarboxylic dianhydride
1,2,5,6-Naphthalenetetracarboxylic dianhydride	Bis(3,4-dicarboxyphenyl)ether dianhydride
2,2′,3,3′-Diphenyltetracarboxylic dianhydride	

As shown by Sroog and co-workers (44,45), the synthesis of high molecular weight aromatic polyimides was for the most part dependent on the preparation of high molecular weight polyamic acid. The molecular weight of the polyamic acid, in turn, has been found to be dependent on the use of extremely pure monomers, rigorous exclusion of moisture, choice of solvent, and maintenance of low to moderate temperatures. Studies by Wallach (45b) showed that with very pure monomers and solvents polyamic acid with $M_n = 55,000$ and $M_w/M_n = 2.4$ resulted and that molecular weight could be varied systematically by control of stoichiometric imbalance. The temperature limitation appeared to be related to three reactions which would limit molecular weight. These were (1) partial conversion to polyimide releasing water, which could hydrolyze polyamic acid, (2) extensive conversion to polyimide above 100°C, which, in addition to hydrolysis, would result in premature precipitation of low molecular polyimide from the reaction mixture, and (3) possible transamidation with amide solvents upsetting stoichiometry.

The conversion of polyamic acid to polyimide via dehydration has been carried out both thermally (42,48) and by chemical means (43). As Sroog (45) has pointed out, the preferred methods involved the partial drying of solvated polyamic acid in fabricated form, i.e., film, fibers, etc., to high solids content (65–75%) and subsequent heating of the shaped articles at 300°C for time periods up to 1 hr. The complete conversion (44) was indicated by the disappearance of the NH band at 3.08 μ and appearance of characteristic imide bands at 5.63 and 13.8 μ (Figs. 11 and 12). The chemical conversion of polyamic acid to polyimides has been effected utilizing dehydrating agents such as acid anhydrides and catalysts such as tertiary amines. Chemical agents which have been utilized successfully for the conversion of polyamic acids to polyimides are summarized (45,53) in Table X.

Using these general procedures, the preparation of not only aromatic polyimides but also aromatic poly(amide-imides), poly(hydrazide-imides),

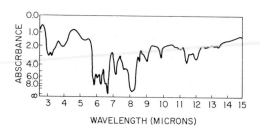

Fig. 11. Infrared spectrum of polyimide film (polyamic acid from 4,4'-diaminodiphenyl ether and pyromellitic dianhydride), 0.1 mil thick, dried 2 hr at 80°C (44).

IV. AROMATIC POLYMERS CONTAINING HETEROCYCLIC RINGS

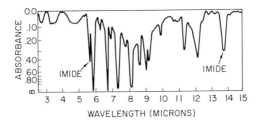

Fig. 12. Infrared spectrum of 0.1 mil polyimide film (poly[N,N'-(p,p'-oxydiphenylene)pyromellitimide]). The film was heated to 300°C over a period of 45 min, then held at 300°C for 1 hr (44).

poly(amide-ester-imides), etc., have been reported. The polymers which have been prepared are listed in Table XI.

All the polymers listed in Table XI were highly colored ranging in color from light yellow to dark brown and were characterized by excellent resistance to irradiation and to solvent attack, and by a high degree of thermal stability (44,45). They were soluble only in concentrated sulfuric and fuming nitric acids (44,45).

The aromatic polyimides have been extensively studied, with particular emphasis on thermal behavior. (41,44,45,58-62). Thermal stability of films as indicated by thermogravimetric analysis showed that polypyromellitimide films derived from aromatic diamines had very little weight loss to over 500°C *in vacuo* or inert atmospheres (41,44,45). For the aromatic polypyromellitimides in which the diamine component was p-phenylene, **11**, m-phenylene, **39**, benzidine, **41**, bis(4-aminophenyl) ether,

TABLE X
Chemical Agents for Conversion of Polyamic Acids (45,53)

Anhydrides	
Propionic anhydride	Naphthoic anhydride
Acetic anhydride	Benzoic anhydride
Butyric anhydride	Aliphatic ketenes
Valeric anhydride	

Bases	
4-Isopropylpyridine	N-Ethylmorpholine
N-Dimethylbenzylamine	N-Methylmorpholine
Isoquinoline	Diethylcyclohexylamine
4-Benzylpyridine	N,N-Dimethylcyclohexylamine
N-Dimethyldodecylamine	4-Benzoylpyridine
Trimethylamine	2,4,6-Collidine
Triethylenediamine	Pyridine

TABLE XI
Aromatic Polyimides

Polymer	Diamine	Dianhydride	PMT^a (°C)	Ref.
39	1,3-Diaminobenzene	Pyromellitic	>600	44, 45, 50, 54, 57
40	1,4-Diaminobenzene	Pyromellitic	>600	44, 45, 57
41	4,4'-Diaminodiphenyl	Pyromellitic	>600	44, 45, 50, 54
42	4,4'-Diaminodiphenyl methane	Pyromellitic	>600	44, 45, 50, 54
43	2,2-Bis(4-aminophenyl)propane	Pyromellitic	>500	44, 45, 50, 54
44	4,4'-Diaminodiphenyl sulfide	Pyromellitic	>600	44
45	4,4'-Diaminodiphenyl ether	Pyromellitic	>600	44, 45, 50, 54
46	4,4'-Diaminodiphenyl sulfone	Pyromellitic	>600	44, 45
47	3,3'-Diaminodiphenyl sulfone	Pyromellitic	>600	44, 45
48	3,4'-Diaminodiphenyl ether	Pyromellitic	>600	57
49	2,6-Diaminopyridine	Pyromellitic	>600	57
50	1,3-Diaminobenzene	Benzophenone-3,3',4,4'-tetracarboxylic acid	>600	57
51	4,4'-Diaminodiphenyl ether	Benzophenone-3,3',4,4'-tetracarboxylic acid	>600	57
52	4,4'-Diaminodiphenyl ether	Naphthalene-1,4,5,8-tetracarboxylic acid	>600	57
53	3,3'-Diaminobenzanilide	Pyromellitic	>500	50, 54

54	3,4′-Diaminobenzanilide	Pyromellitic	>500	50, 54
55	4,3′-Diaminobenzanilide	Pyromellitic	>500	50, 54
56	4,4′-Diaminobenzanilide	Pyromellitic	>500	50, 54
57	Isophthal(4-aminoanilide)	Pyromellitic	>500	50, 54
58	N,N'-1,3-Phenylene-bis(4-aminobenzamide)	Pyromellitic	>500	50, 54
59	Isophthal(3-aminoanilide)	Pyromellitic	>500	50, 54
60	N,N'-1,3-Phenylene-bis(4-aminobenzamide)	Pyromellitic	>500	50, 54
61	N,N'-Bis(3-aminobenzoyl)-2,4-diamino-diphenyl ether	Pyromellitic	>500	50, 54
62	N,O-Bis(3-aminobenzoyl)-1,4-aminophenol	Pyromellitic	>500	50, 54
63	Bis(4-aminophenyl)isophthalate	Pyromellitic	>500	50, 54
64	4,4′-Diaminophenyl benzoate	Pyromellitic	>500	50, 54
65	Resorcinol-bis(4-aminobenzoate)	Pyromellitic	>500	50, 54
66	5,5′-Bis(2-aminobenzothiazole)	Pyromellitic	>500	50, 54
67	Isophthalic dihydrazide	Pyromellitic	>500	50, 54
68	4,4′-Diamino-2,2′-dimethyldiphenyl	Pyromellitic	>500	58
69	1,2-Bis(4-aminophenyl)ethane	Pyromellitic	>500	58
70	1,3-Diaminobenzene	Trimellitic	>500	—

[a] PMT: polymer melt temperature. See R. G. Beaman and F. B. Cramer, *J. Polymer Sci.*, **21**, 223 (1956).

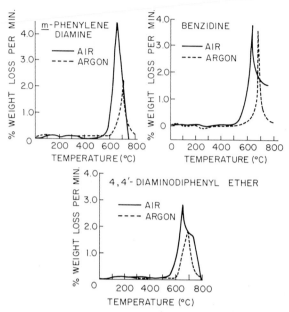

Fig. 13. TGA of polypyromellitimides (heating rate, 4°C/min) (41).

45, and 4,4'-diaminodiphenylsulfone, **47**, above this temperature there was a sharp increase in the rate of weight loss, followed by leveling out after about 35% of the starting weight had been lost. Beyond this the residues showed almost no weight loss up to 1000°C. Polyimides derived from

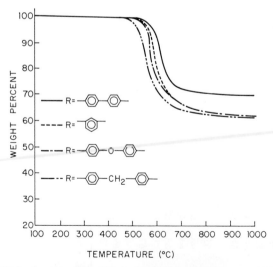

Fig. 14. TGA of pyromellitimides in dry helium (heating rate, 3°C/min) (44).

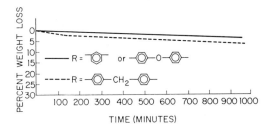

Fig. 15. Isothermal weight loss of

$$\left[-R-N \begin{array}{c} \underset{\underset{O}{\parallel}}{C} \\ \underset{\underset{O}{\parallel}}{C} \end{array} \underset{}{\bigcirc} \begin{array}{c} \underset{\underset{O}{\parallel}}{C} \\ \underset{\underset{O}{\parallel}}{C} \end{array} N- \right]_n$$

at 450°C in helium (44).

m-phenylene, **40**, benzidine, **41**, and 4,4′-diaminodiphenyl ether, **45**, exhibited under 1.5% weight loss on heating to 500°C in an inert atmosphere. By contrast upon heating in air at these temperatures, these particular aromatic polyimides slowly but completely oxidized (Figs. 13 and 14).

Thermal stability of the aromatic polyimides has also been measured by isothermal weight loss (44,45). The aromatic polypyromellitimides derived from bis(4-aminophenyl) ether showed isothermal weight loss in helium atmosphere of only 1.5% after 15 hr at 400°C, 3.0% after 15 hr at 450°C, and 7% after 15 hr at 500°C. Isothermal weight loss in helium for polypyromellitimides derived from *m*-phenylenediamine followed a similar pattern (Fig. 15). Additional weight loss measurements in inert and oxidizing atmospheres are collected in Table XII and illustrated in Figure 16.

TABLE XII
Isothermal Weight Loss of Pyromellitimides (50)

Polymer	Diamine	Wt loss at 325°C			
		100 hr	200 hr	300 hr	400 hr
34	1,3-Diaminobenzene	3.3	4.3	5.0	5.6
41	4,4′-Diaminodiphenyl	2.2	3.6	5.1	6.5
45	4,4′-Diaminodiphenyl ether	3.3	4.0	5.2	6.6
48	3,4′-Diaminodiphenyl ether	3.4	3.8	5.1	7.2
44	4,4′-Diaminodiphenyl sulfide	4.8	5.8	6.8	7.9
42	4,4′-Diaminodiphenyl methane	9.4	12.9	14.7	16.8
43	2,2-Bis(4-aminophenyl)propane	16.1	26.2	31.0	36.0

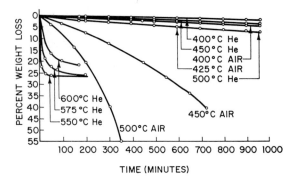

Fig. 16. Isothermal weight loss of [structure] in helium and air (44).

The thermal stability of the aromatic polyimides has also been extensively studied via differential thermal analysis, both in air and in inert atmospheres (58). Nishizaki and Fukami (58) found that in nitrogen all of the aromatic polyimides showed clearcut endothermic reactions which were believed to be decomposition. The onset temperatures of heat absorption and temperatures of maximum heat absorption are given in Table XIII. The polyimides obtained from aromatic diamines began to

TABLE XIII
Differential Thermal Analysis of the Polypyromellitimides (58)
(In N_2; Speed of temperature rise, 10°C/min)

Polymer	Diamine	Heat absorption start temp. (°C)	Heat absorption peak temp. (°C)
40	1,4-Diaminobenzene	500	610
39	1,3-Diaminobenzene	460	590
41	4,4'-Diaminodiphenyl	510	615
28	4,4'-Diamino-2,2'-dimethyldiphenyl	490	540
45	4,4'-Diaminodiphenyl ether	490	595
46	4,4'-Diaminodiphenyl sulfone	420	485
47	4,4'-Diaminodiphenyl methane	480	550
43	1,2-Bis(4-aminophenyl)ethane	470	580
49	2,2-Bis(4-aminophenyl)propane	400	430, 485

decompose at between 400 and 510°C. For those which were connected only with the benzene nucleus, *p*-substitution gave higher decomposition temperatures and better thermal stability than *m*-substitution. In those cases where the benzene rings of the diamine component were connected with functional groups, such as isopropylidine and sulfone, the decomposition temperatures were as low as 400–420°C. In those cases where the rings were connected with an ether group, a methylene group, or an ethylene group, decomposition occurred above 480°C. These workers found the following order of stability.

―⟨◯⟩―⟨◯⟩― > ―⟨◯⟩― > ―⟨◯(CH₃)⟩―⟨◯(CH₃)⟩― >

―⟨◯⟩―O―⟨◯⟩― > ―⟨◯⟩―CH₂―⟨◯⟩― >

―⟨◯⟩―CH₂―CH₂―⟨◯⟩― > ―⟨◯⟩― > ―⟨◯⟩―SO₂―⟨◯⟩― >

―⟨◯⟩―C(CH₃)(CH₃)―⟨◯⟩―

When similar experiments were carried out in air, the oxidation reaction was found to be exothermic. The order of stability in oxygen was quite different from that in an inert atmosphere. In the case of polyimides containing aliphatic chains between aromatic nuclei in the diamine portion, thermal oxidation began at temperatures as low as 230°C. For polymers containing alkyl substitution on the benzene ring, onset exothermic reaction was at temperatures of 320–330°C. Similarly, for polymers containing side chain methyl groups, such as in the case of the polyimides derived from 2,2-bis(4-aminophenyl)propane, decomposition occurred at 330°C. In those cases where the benzene rings in the diamine component were separated with functional groups, the nature of the functional group determined the stability. Whereas the polyimide containing a sulfone group as connecting group decomposed at 320°C, those containing an ether as connecting group were stable up to 400°C. By far the most stable of these polyimides in oxidizing atmospheres were those containing *p*-phenylene linkages in which the imide rings were bonded either by *p*-phenylene or 4,4′-biphenylene.

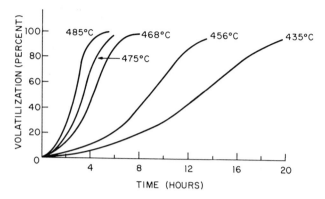

Fig. 17. Thermal degradation of polymer **45** in air (59).

The thermal degradation of polyimides has also been studied in air and vacuum in the range of 400–700°C by thermogravimetry by Bruck (59–61) and Heacock and Berr (62). They found that although the aromatic polyimides were remarkably stable in air up to approximately 420°C, at temperatures in excess of this they began to volatilize (Figs. 17 and 18).

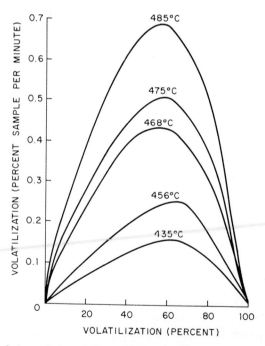

Fig. 18. Rates of thermal degradation of polymer **45** at various temperatures in air (59).

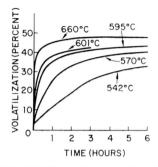
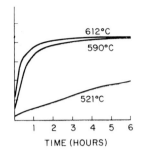

Fig. 19. Thermal degradation of polymer **45** in vacuum (59).

Fig. 20. Thermal degradation of polymer **45** in vacuum (59).

At 485°C, practically total volatilization took place within 5 hr. *In vacuo* ($\sim 10^{-3}$ mm Hg), the polymers showed even greater heat stability, with no appreciable loss even after prolonged exposure to temperatures up to approximately 500°C. Above this temperature volatilization occurred leaving brittle carbonized residues which appeared to have a limiting weight corresponding to approximately 45% of the original sample and showing no infrared absorption bands (Figs. 19–21). From thermal

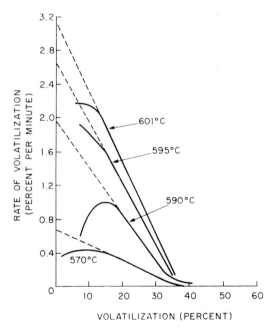

Fig. 21. Rates of thermal degradation of polymer **45** in vacuum at various temperatures (59).

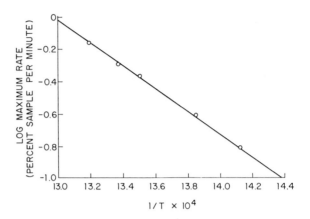

Fig. 22. Arrhenius plot for thermal degradation of polymer **45** in air (61).

degradation profiles, the rates of volatilization were calculated, and from the Arrhenius relationship (Figs. 22 and 23) an activation energy of 33 kcal/mole and 74 kcal/mole for the degradation in air and vacuum, respectively (Table XIV), were determined. The general shape of the rate

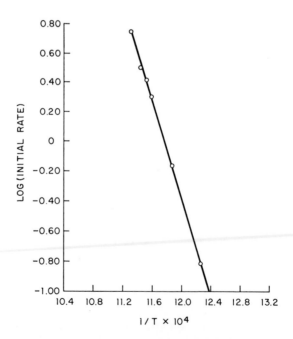

Fig. 23. Arrhenius plot for thermal degradation of polymer **45** in vacuum (61).

TABLE XIV

Rates and Activation Energies for the Thermal Degradation of Polymer **45** in Air and Vacuum (61)

Thermal degradation	Temp. (°C)	Total time (hr)	Total volatilization (%)	Rates (% sample/min)		Activation energy (kcal/mole)
				Extrapolated initial	Maximum	
Air	485	5.5	98.4	—	0.685	
	475	5.5	93.3	—	0.510	
	468	9.0	98.9	—	0.435	33
	456	13.5	94.6	—	0.252	
	435	20.0	95.6	—	0.157	
Vacuum	660	20.0	49.0	—[a]	—	
	612	20.0	45.2	5.620	—	
	601	3.0	42.3	3.180	—	
	595	10.0	43.3	2.700	—	
	590	16.0	43.4	1.990	—	74
	570	18.0	41.9	0.680	—	
	542	18.0	38.6	0.150	—	
	521	18.0	32.1	—[b]	—	

[a] Rates too fast for reliable measurement and extrapolation.
[b] Degradation complication by side reactions at this low temperature.

curves of these polymers, as the result of thermal degradation *in vacuo*, resembled those of polytrivinylbenzene. In the vacuum pyrolysis studies, based on the activation energy, bond dissociation energies, and elemental analysis of the residue remaining after pyrolysis, the primary scission process appeared to occur at the imide bonds, followed by a secondary cleavage resulting in the elimination of CO groups.

Heacock and Berr (62) have described the gaseous products of the decomposition of the polypyromellitimide derived from bis(4-aminophenyl) ether, **45**, in closed systems *in vacuo* at 540°C (Table XV). The principal decomposition products found in this work were carbon dioxide and carbon monoxide and are in harmony with the proposed mechanism of Bruck for the thermal degradation of the aromatic polyimides.

The copolyimides such as amide-imides, ester-imides, and hydrazide-imides are reported to be substantially inferior to the homopolyimides (63,64). The isothermal weight loss at 300°C for the polyamide-imide from 4,4′-diaminobenzanilide, **56**, and pyromellitic dianhydride was about twice that for the polyimide from 4,4-diaminodiphenyl) ether, **45**. Similar

TABLE XV
Mass Spectrometric Analysis of Gaseous Degradation Products of Polymer **45** (62)
(Vacuum pyrolysis 540°C, 2 hr)

Component	Mole (%)
Benzene	0.7
Carbon dioxide	35.1
Carbon monoxide	58.7
Water	1.2
Hydrogen	2.6
Ammonia	Trace
Hydrogen cyanide	1.2
Benzonitrile	0.5
Wt loss on sample	34.6%
Original sample weight	0.4382 g

behavior has also been reported for the polyhydrazide-imide and polyester-imide and polyamide-ester-imide (Table XVI). A further measure of the difference in thermal stability of the polyimides vs. the polyamide-imides is given by the thermogravimetry behavior of these polymers. Collected in Figure 24 are the TGA curves for the polyimide derived from 4,4-diaminodiphenyl ether, **45**, the polyamide-imide derived from 4,4′-diaminobenzanilide, **56**, and a polyamide-imide from trimellitic acid anhydride, **70**. It is apparent from the inspection of these data that the polyimide, in helium, showed an initial weight loss at approximately 500°C, whereas the initial break in the polyamide-imide curves occurred approximately 50–75°C lower. All structures showed the leveling out of the curves after about 35% of starting weight has been lost, and the resultant residues showed almost no weight loss up to 1000°C. In addition, the polyamide-imides which have

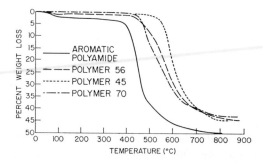

Fig. 24. TGA of copolyimides in helium (heating rate, 3°C/min) (64).

TABLE XVI
Isothermal Weight Loss of Copolypyromellitimides (63,64)

Polymer	Diamine	Wt loss at 325°C			
		100 hr	200 hr	300 hr	400 hr
56	4,4'-Diaminobenzanilide	5.7	8.4	11.9	12.1
55	4,3'-Diaminobenzanilide	4.3	7.8	10.8	11.9
54	3,4'-Diaminobenzanilide	2.0	4.2	6.9	9.8
53	3,3'-Diaminobenzanilide	3.2	6.5	9.8	11.2
57	Isophthal(4-aminoanilide)	6.9	9.4	14.4	20.4
58	N,N'-1,3-Phenylene-bis(4-aminobenzamide)	6.0	9.2	12.5	15.6
59	Isophthal(3-aminoanilide)	6.8	8.1	10.5	13.2
60	N,N'-1,3-Phenylene-bis(3-aminobenzamide)	6.2	8.3	14.0	20.3
61	N,N'-Bis(3-aminobenzoyl)2,4-diaminodiphenyl ether	24.1	31.5	38.6	44.3
62	N,O-Bis(3-aminobenzoyl)p-aminophenol	12.0	17.0	21.0	27.0
63	Bis(4-aminophenyl)isophthalate	3.6	6.7	10.9	15.0
64	4,4'-Diaminophenyl benzoate	3.3	5.5	7.6	9.7
65	Resorcinol-bis(3-aminobenzoate)	9.5	13.3	17.3	27.7

been extensively studied were reported (63,64) to be more solvent sensitive, have lower resistance to irradiation, and higher water sensitivity than the homopolyimide.

Thus, one can only conclude that aromatic polyimides, based on the foregoing data, are more thermally stable in inert atmospheres and in oxidizing atmospheres than the polybenzimidazoles and the polybenzoxazoles. Due to the conflicting data on the polybenzothiazoles, a judgment as to the relative stability of this polymer class compared with that of the polyimides is impossible.

The outstanding thermal stability and resistance to irradiation, to mechanical deformation, and to solvent attack, along with excellent balance of mechanical and electrical properties, plus their ease of processibility, have prompted considerable research in finding applications for these polymers in a number of end-uses including adhesives, laminates, sealants, coatings, foams, films, and fibers. These applications of the polyimides and their thermal stability in such applications will be discussed in Chapter VII.

E. AROMATIC POLYOXADIAZOLES

One of the more attractive heterocyclic components for the preparation of polymers with intralinear aromatic heterocyclic rings is the oxadiazole ring, both the 1,2,4-oxadiazole and the 1,3,4-oxadiazole rings. These heterocyclic compounds possess excellent thermal stability and outstanding hydrolytic stability (65,66). In addition, it has been reported that p-phenylene and the 2,5(1,3,4-oxadiazole) radicals are spectrally and electronically equivalent (67).

1. Aromatic Poly-1,2,4-Oxadiazoles

The poly-1,2,4-oxadiazoles have been prepared (68a) via the thermal cyclodehydration of precursor poly-O-acyl-amideoxime. The poly-O-acyl-amideoximes in turn have been prepared by the fusion of diamideoximes with dicarboxylic acids, esters, imidazoles, or thiol esters or by the solution polymerization of diacid chlorides with diamideoximes in suitable solvents (68a) (see reaction 17).

TABLE XVII
Poly-O-Acyl-Amideoximes (68a)

Diamidoxime	Diacid chloride	$\eta_{inh}{}^{a,b}$(dl/g)	PMT^c (°C)
Terephthalamide	Sebacyl	0.43^a	> 300
Sebacamide	Isophthaloyl	0.64^b	260
Sebacamide	Sebacyl	0.20^b	280
Oxamide	Isophthaloyl	Insol.	340
Adipamide	Isophthaloyl	0.20^b	290
Oxamide	Adipyl	0.31^b	295
Terephthalamide	Isophthaloyl	1.13^a	> 300
Terephthalamide	Terephthaloyl	Insol.	> 300

^{a} Inherent viscosities were determined at 0.2 wt % concentrations in N,N-dimethylacetamide at 30°C.
^{b} Inherent viscosities were determined at 0.2 wt % concentrations in trifluoroacetic acid at 30°C.
^{c} PMT: polymer melt temperature. See R. G. Beaman and F. B. Cramer, J. Polymer Sci., **21**, 223 (1956).

TABLE XVIII
Poly-1,2,4-Oxadiazoles (68a)

$$\left[R_1 - C \underset{N-O}{\overset{N}{\diagup\!\!\!\diagdown}} C - R_2 - C \underset{O-N}{\overset{N}{\diagup\!\!\!\diagdown}} C \right]_x$$

Structure			
R$_1$	R$_2$	$\eta_{inh}{}^a$(dl/g)	PMT^b (°C)
1,4-Tetramethylene	1,4-Tetramethylene	0.30	~350
1,3-Phenylene	1,8-Octamethylene	0.25	~350
1,3-Phenylene	1,3-Phenylene	0.38	> 400
1,3-Phenylene	1,4-Phenylene	0.45	> 400
1,4-Phenylene	1,8-Octamethylene	0.43	~350
1,4-Phenylene	1,4-Phenylene	0.42	> 400

^{a} η_{inh}: inherent viscosities at 0.5 wt % concentrations in conc. sulfuric acid at 30°C.
^{b} PMT: polymer melt temperature. See R. G. Beaman and F. B. Cramer, J. Polymer Sci., **21**, 223 (1956).

In Table XVII are listed some of the poly-*O*-acyl-amideoximes which have been prepared. These polymers have been cyclodehydrated thermally or chemically to the corresponding poly-1,2,4-oxadiazole (Table XVIII) (see reaction 18). Recently, aromatic poly-1,2,4-oxadiazoles have been

$$\left[\begin{array}{c} H_2N \\ \diagdown \\ O-N \end{array} C-R_1-C \begin{array}{c} NH_2 \\ \diagup \\ N-O-C-R_2-C \end{array} \begin{array}{c} O \\ \| \\ \end{array} \right]_x \longrightarrow$$

$$\left[R_1-C \begin{array}{c} N \\ \diagup \diagdown \\ N-O \end{array} C-R_2-C \begin{array}{c} N \\ \diagup \diagdown \\ O-N \end{array} C \right]_x \quad (18)$$

prepared by (*1*) the 1,3-dipolar cycloaddition of terephthalonitrile oxide to terephthalonitrile (68*b*), and (*2*) the solid state 1,3-dipolar cycloaddition of *p*-cyanobenzonitrile oxide (68*c*).

The all aromatic poly-1,2,4-oxadiazoles were soluble only in strong acids such as trifluoroacetic and sulfuric acid and showed extremely good hydrolytic stability in both acid and basic media and good resistance to ultraviolet-catalyzed oxidative attack (68). Although these polymers have high softening temperatures and were reported to be stable at elevated temperatures, no quantitative data as to such stability at elevated temperatures are available.

2. Aromatic Poly-1,3,4-Oxadiazoles

The first reported synthesis of aromatic poly-1,3,4-oxadiazoles was that involving the 1,3-dipolar addition of acid chlorides to tetrazoles (67). Using *p*-tetrazoloyl benzoyl chloride as an AB monomer, this preparation of poly-1,3,4-oxadiazoles yielded a 9-ring oligomer (see reaction 19).

$$Cl-\overset{O}{\underset{\|}{C}}-\underset{\bigcirc}{}-C\underset{N-NH}{\overset{N=N}{\diagup}} \longrightarrow$$

$$\underset{\bigcirc}{}-\left[C\underset{N-N}{\overset{O}{\diagup\diagdown}}C-\underset{\bigcirc}{}\right]_3 C\underset{N-N}{\overset{O}{\diagup\diagdown}}C-\underset{\bigcirc}{} \quad (19)$$

In subsequent work (69) the preparation of poly-1,3,4-oxadiazoles from *m*- and *p*-phenyleneditetrazole with terephthaloyl, isophthaloyl, oxalyl, adipyl, 4,4'-biphenyldicarbonyl, 2,2'-biphenyldicarbonyl, 3,5-pyridine-dicarbonyl, and 2,6-pyridine-dicarbonyl chlorides at elevated temperatures in pyridine has been accomplished (see reaction 20).

IV. AROMATIC POLYMERS CONTAINING HETEROCYCLIC RINGS

$$\text{(diamidrazone-phenylene-diamidrazone)} + Cl-\underset{O}{\overset{\|}{C}}-R-\underset{O}{\overset{\|}{C}}-Cl \xrightarrow{\text{Pyridine}}$$

$$\left[\text{—(oxadiazole-phenylene-oxadiazole-R)—} \right]_x \quad (20)$$

Although the poly-1,3,4-oxadiazole structure of these polymers was shown by elemental analysis and infrared spectra, the molecular weights were too low to permit fabrication of fibers or films.

The preparation of high molecular weight poly-1,3,4-oxadiazoles for use in the production of shaped articles and coatings has been reported via the reaction of bisimino ethers or their salts with dicarboxylic acid dihydrazides (70,71).

$$H_2N-NH-\underset{O}{\overset{\|}{C}}-R_1-\underset{O}{\overset{\|}{C}}-NH-NH_2 + NH=\underset{\underset{OR}{|}}{C}-R_2-\underset{\underset{OR}{|}}{C}=NH \longrightarrow$$

$$\left[-\underset{N-N}{C \diagup \overset{O}{\diagdown} C}-R_1-\underset{N-N}{C \diagup \overset{O}{\diagdown} C}-R_2- \right]_x + 2NH_3 + 2ROH \quad (21)$$

The polymerizations were carried out at elevated temperatures at 100–250°C, with a preferred range of 120–180°C, in such solvents as N,N-dimethylformamide and dimethyl sulfoxide. Using this general procedure, the poly-1,3,4-oxadiazoles from the dihydrazides of succinic, sebacic, and terephthalic acid with the methanol or ethanol iminoethers derived from oxalic, succinic, adipic, suberic, p-phenylene bisacetic, isophthalic and terephthalic acids, were prepared.

The preparation of poly-1,3,4-oxadiazoles has also been reported via the cyclodehydration of polyhydrazides. Cyclization has been effected by dehydrating agents such as acid chlorides, anhydrides, sulfuric acid, toluenesulfonic acid, and phosphorus oxychloride and by thermal treatments (72).

By far the most successful preparation of poly-1,3,4-oxadiazoles has been achieved via the thermal cyclodehydration of the corresponding polyhydrazide as outlined in equation 22 (71). As indicated in Table XIX,

$$\left[-R-\underset{O}{\overset{\|}{C}}-NH-NH-\underset{O}{\overset{\|}{C}}- \right]_x \xrightarrow{\Delta} \left[-R-\underset{N-N}{C \diagup \overset{O}{\diagdown} C}- \right]_x \quad (22)$$

TABLE XIX
Poly-1,3,4-Oxadiazoles from Thermal Cyclodehydration of Polyhydrazides (73)

$$\left(R_1-\underset{NH-NH}{\overset{O}{\underset{\|}{C}}}\quad\underset{}{\overset{O}{\underset{\|}{C}}}-R_2-\underset{NH-NH}{\overset{O}{\underset{\|}{C}}}\quad\underset{}{\overset{O}{\underset{\|}{C}}} \right)_x \longrightarrow$$

$$\left(R_1-\underset{N-N}{\overset{O}{\underset{\diagdown\ \diagup}{C\quad C}}}-R_2-\underset{N-N}{\overset{O}{\underset{\diagdown\ \diagup}{C\quad C}}} \right)_x$$

Poly-1,3,4-oxadiazoles	R_1	R_2
71	1,3-Phenylene	1,4-Phenylene
72	1,3-Phenylene	1,3-Phenylene
73	1,3-Phenylene	—
74	1,3-Phenylene	1,4-Tetramethylene
75	1,3-Phenylene	1,8-Octamethylene
76	1,4-Tetramethylene	1,4-Tetramethylene
77	1,8-Octamethylene	1,8-Octamethylene
78	1,3-Phenylene	1,4-Cyclohexylene
79	1,3-Phenylene	1,10-Bis(1,3-phenoxylene)decane
80	1,3-Phenylene	Oxydi(1,4-phenyline)
81	1,4-Tetramethylene	1,7-Heptamethylene
82	2,6-Pyrazine	—
83	1,3-Phenylene	2,6-Dichloro-1,4-phenylene

this technique has yielded a whole range of poly-1,3,4-oxadiazole structures. An examination of this table shows that along the polymer chain in addition to the 1,3,4-oxadiazole units are aromatic, heterocyclic, and aliphatic connecting units.

The thermal conversion of polyhydrazides to poly-1,3,4-oxadiazoles was usually carried out at or near the glass transition temperature of the polyhydrazide in a nitrogen atmosphere or *in vacuo*. The rate of conversion of the polyhydrazide to the poly-1,3,4-oxadiazole was dependent not only on the chemical structure but also on the physical state of the polymer (74). For example, unoriented films or bulk polymer converted from 8 to 10 times faster than oriented fibers. On first inspection, this behavior seemed somewhat unusual but when it was considered that (*1*) in the oriented state a greater percentage of the polymer chains probably were in the extended form, **84**, than in the unoriented state and, (*2*) in the folded or unextended form, **85**, the polyhydrazide structure more closely approximated the transition state, then it was not too unexpected that the rates of con-

IV. AROMATIC POLYMERS CONTAINING HETEROCYCLIC RINGS

version of unoriented structures were considerably more rapid than those of oriented structures.

$$
\begin{array}{cc}
\text{84} & \text{85} \\
\end{array}
$$

(Structure 84: cyclic with C=O, NH, NH, C=O groups; Structure 85: -C(=O)-NH-NH-C(=O)-)

The kinetics of the conversion of polyhydrazides to poly-1,3,4-oxadiazoles have been studied (74). The conversion has been found to follow first-order kinetics over the first 80–90% of reaction. These data along with thermodynamic quantities, determined in the usual manner, are summarized in Table XX.

TABLE XX
Kinetic Data and Thermodynamic Quantities for the
Conversion of Polyhydrazides to Poly-1,3,4-Oxadiazoles (74)

Polymer[a]	Temp. (°C)	k (sec^{-1})	$\Delta E\ddagger$ (kcal)	$\Delta S\ddagger$ (255°C) (eu)
71	255	6.81×10^{-6}	50.4	14.0
	265	2.03×10^{-5}		
	277	5.28×10^{-5}		
	285	8.89×10^{-5}		
72	255	3.06×10^{-6}	62.0	34.2
	265	1.36×10^{-5}		
	277	4.47×10^{-5}		
73	255	8.89×10^{-6}	53.0	16.0
	265	3.40×10^{-5}		
	277	6.81×10^{-5}		
74	222	2.01×10^{-5}	49.0	16.7
	242	7.67×10^{-5}		
	255	2.67×10^{-4}		

[a] For polymer structures, see Table XXII.

For the poly-1,3,4-oxadiazoles—**71**, **72**, and **74**—and where there was a wide variation in substituent groups on the hydrazide linkages the specific velocity constants (k) for the conversion reaction were quite different. However, there were only small differences in the $\Delta E\ddagger$ and $\Delta S\ddagger$ values, and the latter were positive. The suggested explanation of these results was related to the fact that these reactions were carried out in the solid state on highly crystalline polymers. It was argued that the energy required to go from reactants to transition state was comparable to the lattice energy of the crystallites involved. This was supported by the fact that the poly-1,3,4-oxadiazoles, although crystalline, have different crystalline patterns and were much less crystalline than the precursor polyhydrazides. Since the lattice energy of these polyhydrazides should be of the same order of magnitude, the difference in $\Delta E\ddagger$ values should be small.

Similarly, the small differences and positive values of $\Delta S\ddagger$ were explained. Even though in these reactions cyclic intermediates were being formed from noncyclic reactants, the decrease in order in the system on going from highly crystalline polyhydrazides to the transition state would be such that the $\Delta S\ddagger$ would be positive. Since this decrease in order would be comparable for all hydrazides in question, the $\Delta S\ddagger$ values would not be markedly different.

Using the same arguments the larger value of $\Delta E\ddagger$ and $\Delta S\ddagger$ for the preparation of the poly-1,3,4-oxadiazole, **72**, could be explained on the basis of increased order in the polyhydrazide due to the higher crystallinity of the homopolymer.

Recently, aromatic poly-1,3,4-oxadiazoles have been prepared from benzenedicarboxylic acids or their derivatives and hydrazine by solution polymerization in fuming sulfuric acid or polyphosphoric acid (PPA) (75) (see reaction 23).

$$H_2N-NH-\overset{O}{\underset{\|}{C}}-\!\!\left\langle\!\!\bigcirc\!\!\right\rangle\!\!-\overset{O}{\underset{\|}{C}}-NH-NH_2$$

$$X-\!\!\left\langle\!\!\bigcirc\!\!\right\rangle\!\!-X + NH_2NH_2\cdot H_2SO_4 \quad \xrightarrow[\text{Oleum}]{\text{Heat}}_{\text{or PPA}} \quad \left[-\!\!\left\langle\!\!\bigcirc\!\!\right\rangle\!\!-\overset{\overset{O}{\diagup\diagdown}}{\underset{N-N}{\overset{\|}{C}\overset{\|}{C}}}-\right] \quad (23)$$

where X represents $-CO_2H$, $-CONH_2$, or $-CN$

Polymerizations were carried out at temperatures from 85 to 185°C for time periods from 0.5 to 13 hr to yield film- and fiber-forming molecular weights. Fuming sulfuric acid was found to be superior to polyphosphoric acid. Terephthalic acid and its derivatives reacted much faster with hydrazine than the corresponding isophthalic acid derivatives.

As in the case of the polyimides, copoly-1,3,4-oxadiazoles have also been prepared. Recently, the preparation of copolyimide-1,3,4-oxadiazoles and polybenzimidazole-1,3,4-oxadiazole has been reported. In this work intermediates containing the preformed 1,3,4-oxadiazole ring were used in the preparation of these copolymers (76) (see reactions 24a and 24b).

In Table XXI are listed the properties of the poly-1,3,4-oxadiazoles and the copoly-1,3,4-oxadiazoles which have been reported. These polymers were characterized by extremely high polymer melt temperatures, medium crystallinity, outstanding resistance to hydrolytic degradation in acidic or basic media, and a high degree of thermal stability. The poly-1,3,4-oxadiazoles were soluble only in such strong acids as sulfuric, trifluoroacetic, and polyphosphoric acid and were much less colored than the polyaromatic heterocyclics previously discussed, ranging from colorless to deep yellow. The outstanding resistance of the aromatic poly-1,3,4-oxadiazoles to hydrolytic degradation in acid or basic media was demonstrated by the fact that treatment of poly-1,3,4-oxadiazole, **71**, with 20%

sulfuric acid solutions or 20% sodium hydroxide solutions for 24 hr at 100°C resulted in no change in the molecular weight of the polymer (73).

The thermal stability of the poly-1,3,4-oxadiazoles has been extensively investigated. Frazer and Sarasohn (74) have reported on the thermal behavior of the poly-1,3,4-oxadiazoles listed in Table XXII. These workers studied the thermogravimetric analysis and differential thermal analysis in air and nitrogen of these representative polymer structures (Tables XXIII and XXIV and Figs. 25 and 26). In thermogravimetric analysis, all the poly-1,3,4-oxadiazoles exhibited a small weight loss from room temperature up to 100°C which corresponded to the loss of absorbed water. The

TABLE XXI
Poly-1,3,4-Oxadiazoles and Copoly-1,3,4-Oxadiazoles

Polymers[a]	PMT[e](°C)	η_{inh}[f], dl/g	Ref.
71	>400	1.5	69, 72, 73, 75
72	>400	2.2	71, 73, 75
73	>400	0.6	73
74	>400	0.5	69, 73
75	>400	0.4	73
76	320	0.4	73
77	300	0.25	73
78	300	0.35	73
79	170	0.5	73
80	>400	0.6	73
81	305	1.45	73
82	350	0.36	73
83	350	0.25	73
86[b]	>400	3.8	73, 75
87[c]	>400	1.1	76
88[d]	>400	1.2	76

[a] For other data on polymers **71–73**, see Table XX.
[b] **86** is poly-1,4-phenylene-2,5-(1,3,4-oxadiazole).
[c] **87** is:

[d] **88** is:

[e] PMT: polymer melt temperature. See R. G. Beaman and F. B. Cramer, *J. Polymer Sci.*, **21**, 223 (1956).
[f] η_{inh}: inherent viscosities determined at 0.5 wt % concentrations in conc. sulfuric acid at 30°C.

poly-1,3,4-oxadiazoles consisting of benzene rings and 1,3,4-oxadiazole rings, **71**, **72**, and **73**, were of particular interest.

Polymer **71**, after a loss of absorbed water, showed no weight loss up to temperatures of approximately 450°C where rapid loss in weight occurred

TABLE XXII
Poly-1,3,4-Oxadiazoles (74)

$$\left(R-C\underset{N-N}{\overset{O}{\diagup\diagdown}}C-R'-C\underset{N-N}{\overset{O}{\diagup\diagdown}}C \right)_x$$

Polymers	R	R'
71	1,3-Phenylene	1,4-Phenylene
72	1,3-Phenylene	1,3-Phenylene
73	1,3-Phenylene	—
74	1,3-Phenylene	1,4-Tetramethylene
75	1,3-Phenylene	1,8-Octamethylene
76	1,4-Tetramethylene	1,4-Tetramethylene
77	1,8-Octamethylene	1,8-Octamethylene

and continued up to 560°C. The residue slowly lost weight up to 700°C. At this temperature, 51% of the original weight remained. This behavior was observed both under a nitrogen atmosphere and in air.

A similar behavior was observed for the poly-1,3,4-oxadiazole, **72**. The first significant break in the curve occurred at approximately 450°C, with rapid weight loss continuing up to 560°C. At 700°C, 54.7% of the original weight still remained under both air and nitrogen.

For the poly-1,3,4-oxadiazoles, **73**, where the polymer chain consisted of alternating benzene and bisoxadiazole rings, the first break in the curve

TABLE XXIII
A Summary of Thermogravimetric Data for Poly-1,3,4-Oxadiazoles (74)

Polymer	$(T_i/T_f)_1$ [a]	Δ_1 [b]	$(T_i/T_f)_2$ [c]	Δ_2 [d]	$(w/w_o)700$ [e]
71	447/560	37.5	—	—	51.8
72	450/560	38.8	—	—	54.7
73	275/355	26.0	355/541	36.0	26.6
74	314/532	44.8	—	—	46.8
75	372/579	64.3	—	—	34.8
76	275/343	9.4	343/508	60.2	26.2
77	290/348	2.0	348/508	86.8	11.4

[a] $(T_i/T_f)_1$ = (onset temp., °C/terminus temp., °C), Step 1.
[b] $\Delta_1 = (\Delta w/w_o)_{100}$, step 1.
[c] $(T_i/T_f)_2$ = (onset temp., °C/terminus temp., °C), Step 2.
[d] $\Delta_2 = (\Delta w/w_o)_{100}$, step 2.
[e] $(w/w_o)_{700}$ = % original weight remaining at 700°C; w = sample weight at temperature T; w_o = original sample weight (based on dry polymer).

TABLE XXIV
A Summary of Differential Thermal Data for
Poly-1,3,4-Oxadiazoles (74)[a]

Polymer	
71	−100(w), +375(m), +505(s)*
72	−112(m), +380(m), +505(s)*
73	−112(s), +288(m)*, +331(s)**
74	−123(m), +343(msh)* → +419(s)*
75	−66(s), −115(w), −258(w), +367(sh)* → +433(s)*
76	−130(s), −278(s), +333(s)*, +410(w)**
77	−82(m), +152(w), −249(s), +324(s)*, +438(s)**

[a] Endotherm peak: −; exotherm peak: +; * peak corresponding to first break in TG curve (excluding absorbed water); ** peak corresponding to second break in TG curve (excluding absorbed water); relative peak size: ww = very weak, w = weak, m = medium, s = strong, sh = shoulder; x(sh) → y or y ← z(sh) denotes a shoulder at temperature x (or z) which precedes (i.e., x) or follows (i.e., z) the main peak. The main peak itself reaches a maximum temperature at y.

occurred at approximately 275°C with a rapid loss of 26% of weight. This was followed by a second break in the curve at 355°C and rapid loss of weight up to 541°C. At 541°C the sample had lost approximately 62% of its original weight, and the weight loss continued up to 700°C where only 27% of the original weight remained.

This difference in behavior of these three poly-1,3,4-oxadiazoles suggested that the bisoxadiazole structure in **73** was much more unstable than the structure consisting of alternating intralinear benzene and 1,3,4-oxadiazole rings.

The thermogravimetry behavior of the other poly-1,3,4-oxadiazoles, **74**,

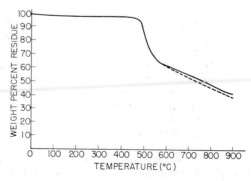

Fig. 25. TGA of poly-1,3,4-oxadiazole, **71** (74): ———, in nitrogen; - - - - -, in air.

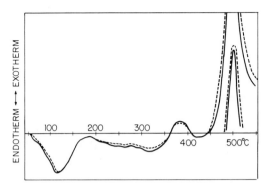

Fig. 26. DTA of poly-1,3,4-oxadiazole, 71 (74): ———, in nitrogen; -----, in air. Numbers are degrees centigrade.

75, 76, and **77,** reflected the presence of the thermally unstable aliphatic carbon–carbon linkages in the chain. In all cases, onset decomposition temperatures of below 400°C were observed.

The differential thermometry curves of the poly-1,3,4-oxadiazoles show endothermic-exothermic behavior corresponding to that found in the thermogravimetry curves. The all-aromatic poly-1,3,4-oxadiazoles, **71** and **72,** showed exothermic peaks corresponding to decomposition in the 450–540°C region where maximum weight losses were observed in the thermogravimetry curves. It was also of interest to note that the curves from both the differential thermal analysis and the thermogravimetric analysis of the poly-1,3,4-oxadiazoles were identical with those obtained in the second steps of the such curves for the corresponding polyhydrazides (see Chapter III).

Similar results were reported by Iwakura (75) and Ehlers (77) for polyphenylene-1,3,4-oxadiazoles. The poly-phenylene-1,3,4-oxadiazoles showed no loss in weight up to 450°C in air or in nitrogen with retention of over 50% of original weight up to 700°C. Iwakura (75) also reported that a film of poly-p-phenylene-1,3,4-oxadiazole, on heating at 450°C for 2 hr in air, turned black and brittle, but the infrared spectrum remained unchanged.

The practically identical thermal behavior of the poly-1,3,4-oxadiazoles in nitrogen and air suggested that the decomposition process was a thermolysis reaction. Recent mass spectral data (78) on the thermal decomposition of 3,5-diphenyl-1,2-oxadiazole and 2,5-diphenyl-1,3,4-oxadiazole showed different decomposition pathways, and, in the case of the 1,3,4-oxadiazole derivative, the main driving force for decomposition was the formation of molecular nitrogen. If the thermal decomposition of the polymer followed a similar pathway, the intermediate product would be a polyether. This argument was supported by the work of Ehlers (77) who found that in the thermogravimetric analysis of aromatic polyethers and aromatic poly-1,3,4-oxadiazoles similar responses to the atmosphere of measurement were observed, and similar curves were obtained.

Based on the foregoing data the aromatic poly-1,3,4-oxadiazoles exhibited approximately the same level of thermal stability in inert atmospheres as the previously discussed polyaromatic heterocyclics. However, in air or oxidizing atmospheres, the poly-1,3,4-oxadiazoles were far superior to those aforementioned polymer systems. Similarly, the copolyaromatic heterocyclics containing the 1,3,4-oxadiazole ring were superior to other copolyaromatic heterocyclics in air or oxidizing atmospheres. This outstanding thermal stability in air and superior resistance to hydrolytic attack in acid and basic media has prompted an extensive study of the aromatic poly-1,3,4-oxadiazoles in fiber and adhesive applications. These will be discussed in detail in Chapter VII.

F. AROMATIC POLYPYRAZOLES

Pyrazoles, which are synthesized from hydrazine derivatives and 1,3-dicarbonyl compounds (see reaction 25), are isomeric with imidazoles (79).

$$R_1-NH-NH_2 + R_2-\overset{O}{\overset{\|}{C}}-CR_3R_4\overset{O}{\overset{\|}{C}}-R_5 \longrightarrow \underset{R_2-C\diagdown_N\diagup N-R_1}{R_3-C=C-R_5} \qquad (25)$$

Although not as high melting or as resistant to attack by acidic, basic, and oxidizing agents as the imidazole ring (80,81), the pyrazole ring has sufficient chemical and thermal stability to be of interest as a component in polymer structures consisting of intralinear aromatic and heterocyclic rings.

Korshak and co-workers (82,83) first reported the preparation of

$$H_2NNH-\overset{O}{\overset{\|}{C}}-R_1-\overset{O}{\overset{\|}{C}}-NH-NH_2 + \left(R_2-\overset{O}{\overset{\|}{C}}-CH_2-\overset{O}{\overset{\|}{C}}-R_3 \right)_{/2} \longrightarrow$$

$$\left[\underset{R_3-C\diagdown_N\diagup N-C_1-R_1-C-N\diagdown_N\diagup C}{\overset{R_1}{\underset{HC=C}{}}\overset{O}{\underset{\|}{}}\quad \overset{O}{\underset{\|}{}}\overset{R_2}{\underset{C=CH}{}}} \right]_x \qquad (26)$$

IV. AROMATIC POLYMERS CONTAINING HETEROCYCLIC RINGS

aromatic polypyrazoles by the reaction of dihydrazides with bis-β-diketones (1) by melt polymerization or (2) by a two-stage polycondensation–cyclization reaction.

In the melt polymerization, the polypyrazoles were prepared directly as outlined in reaction 26.

In the two-stage polycondensation–cyclization, the initial reaction was a solution polycondensation to yield a polyhydrazone which was subsequently thermally cyclized in the melt (see reaction 27).

$$H_2NHN-\overset{O}{\underset{\|}{C}}-R_1-\overset{O}{\underset{\|}{C}}-NH-NH_2 + (R_2-\overset{O}{\underset{\|}{C}}-CH_2-\overset{O}{\underset{\|}{C}}-)_2R_3$$

↓ Solution

1st stage:

$$\left[\begin{array}{c} -R_3-C=N-NH-\overset{O}{\underset{\|}{C}}-R_1-\overset{O}{\underset{\|}{C}}-NH-N=C- \\ H_2C CH_2 \\ C=O O=C \\ R_2 R_2 \end{array} \right]_x \quad (27)$$

↓ Δ

2nd stage:

$$\left[\begin{array}{c} R_2 R_2 \\ HC=C O O C=CH \\ -R_3-C \diagdown N-\overset{\|}{C}-R_1-\overset{\|}{C}-N \diagup C- \\ N N \end{array} \right]_x$$

The polypyrazoles initially prepared were of little interest as high temperature polymers due to the presence of the carbonyl group in the polymer chain arising from the use of dihydrazides as one of the reactants.

In subsequent work, Korshak (84) has reported the preparation of polypyrazoles from the reaction of bisdiazo compounds with bisacetylenes (see reaction 28). The polymers derived from bis(diazo)xylene and *p*-diethynylbenzene were high melting, colored solids in the 10,000 molecular weight range.

Recently the preparation of polyphenylpyrazoles has been reported (85)

$$N_2CH-R-CHN_2 + CH\equiv C-R_1-C\equiv CH \longrightarrow$$

$$\left[\begin{array}{c} -R-C-\!\!\!-\!\!\!-C-R_1-C-\!\!\!-\!\!\!-C- \\ \|\|\|\| \\ N CH HC N \\ \diagdown_{NH}\diagup \diagdown_{NH}\diagup \end{array} \right]_x \quad (28)$$

$$\text{(reaction scheme)} \quad (29)$$

via the reaction of 1,3-dihydrazinobenzene with 1,1,2,2-tetraacylethanes (see reaction 29).

Polymers were prepared by heating equivalents of monomers under a nitrogen atmosphere up to temperatures of 200°C to yield prepolymer which was subsequently polymerized to high molecular weight by further heating up to temperatures of 350°C *in vacuo* (ca. 0.1 mm). Based on spectrophotometric and mass spectral data the resulting polymers were shown to have the bispyrazole structure.

The polypyrazoles which have been prepared are listed in Table XXV. The aromatic polypyrazoles had polymer melt temperatures in excess of 400°C, were highly colored, and were soluble only in concentrated sulfuric acid. Their thermal stability has been the subject of only limited investigation.

The polypyrazole derived from *p*-bis(diazo)benzene and *p*-diethylnylbenzene was reported not to melt up to temperatures of 500°C and to be

TABLE XXV
Polypyrazoles

Polymer	Reactants		PMT[a](°C)	Ref.
	Diazo or dihydrazino	Diethynyl or dicarbonyl		
89	1,6-Bis(diazo)hexane	1,4-Diethynylbenzene	440	84
90	1,6-Bis(diazo)hexane	Diacetylene	~360	84
91	1,4-Bis(diazo)xylene	1,4-Diethynylbenzene	>500	84
92	1,3-Dihydrazinobenzene	1,1,2,2-Tetraacetylethane	>500	85

[a] PMT: polymer melt temperature. See R. G. Beaman and F. B. Cramer, *J. Polymer Sci.*, **21**, 223 (1956).

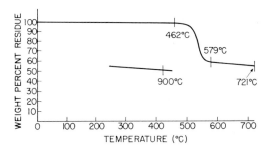

Fig. 27. TGA plot for polybispyrazole, **92**, in nitrogen (heating rate, 3°C/min).

stable at this temperature (84). No differential thermal or thermogravimetric analysis of this polymer was reported.

The thermogravimetric analysis of the polybispyrazole derived from the 1,3-dihydrazinobenzene and 1,1,2,2-tetraacetylethane has been reported (85) (Figs. 27 and 28). Under a nitrogen atmosphere this polymer showed no weight loss up to 462°C, where a rapid loss in weight occurred and continued up to 579°C. The residue slowly lost weight up to 900°C. At this temperature, approximately 52% of the original weight remained. In air, no weight was lost up to 410°C. At this temperature, rapid loss of weight was initiated and continued up to 541°C, with total volatilization of the sample at 690°C.

The differential thermal analysis of this polymer (Figs. 29 and 30) in air and nitrogen showed exothermic behavior, probably indicating decomposition, corresponding to the major weight losses observed in thermogravimetry.

Based on this rather scanty thermal data, it would appear that the aromatic polybispyrazoles are of the same order of stability as the previously discussed polyaromatic heterocyclics, showing the typical susceptibility to oxidative attack observed in the case of the polyimides, polybenzimidazoles, and polybenzoxazoles.

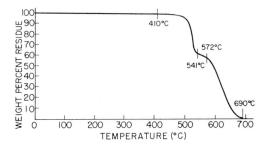

Fig. 28. TGA plot for polybispyrazole, **92**, in air (heating rate, 3°C/min).

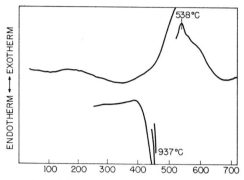

Fig. 29. DTA plot of polybispyrazole, **92**, in nitrogen (heating rate, 10°C/min).

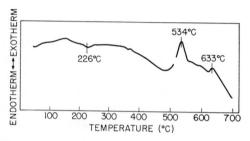

Fig. 30. DTA plot of polybispyrazole, **92**, in air (heating rate, 10°C/min).

G. POLYQUINOXALINES

Quinoxalines or benzopyrazines, which are synthesized from o-phenylenediamines and 1,2-dicarbonyl compounds (see reaction 30), are generally high melting, stable, crystalline solids which exhibit unusual chemical resistance to acid, basic, and oxidizing agents (86). Like the benzimidazoles, these have been of interest as structural units in high temperature resistant polymers.

$$\tag{30}$$

The preparation of polyquinoxalines in high molecular weight has recently been reported (87–90) via the reaction of tetracarbonyl compounds such as 1,4-diglyoxalbenzene with tetramines such as 3,3′-diaminobenzidine by melt polymerization or by solution polymerization at elevated temperatures (see reaction 31).

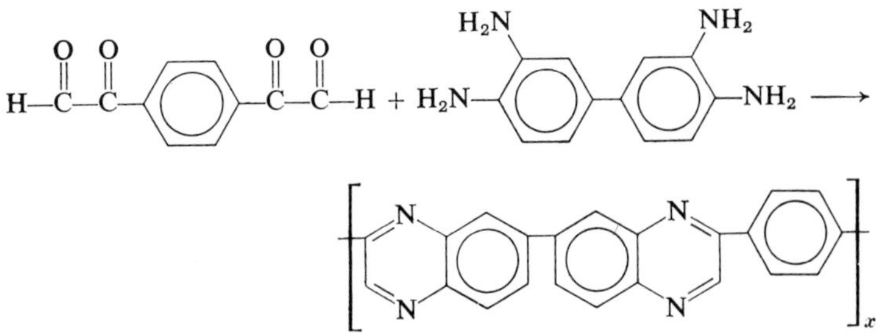

(31)

In the melt polymerization procedure, the two monomers were heated to temperatures as high as 375°C in a two-stage polymerization cycle quite similar to that employed in the preparation of polybenzimidazoles. In the solution polymerization procedure, the monomers were allowed to react in hexamethylphosphoramide at temperatures of 100–200°C for periods up to 3 hr, after which the isolated polymer or prepolymer was heated at temperatures between 350 and 400°C under reduced pressure.

The polymers prepared by these methods are listed in Table XXVI. All of the polyquinoxalines were highly colored, showed only trace crystallinity, and were soluble only in concentrated sulfuric acid. The polyquinoxalines containing flexible groups such as the ether group in the polymer chain were also soluble in hexamethylphosphoramide.

The thermogravimetric analysis (87–90) in air and nitrogen of polyquinoxalines showed behavior typical of most of the polyaromatic heterocyclics (Figs. 31–36). In nitrogen, there was essentially no weight loss up to 400–500°C, where a slow loss in weight occurred and continued up to 900°C. At this temperature, 60–80% of original weight still remained. In air, similar behavior was observed up to 400–500°C, where a precipitous loss in weight occurred and continued up to the 550–700°C region, with complete volatilization of the polymer.

TABLE XXVI
Polyquinoxalines

Polymer	PMT^a(°C)	Ref.
Poly-2,2'-(1,4-phenylene)-6,6'-diquinoxaline	>500	87, 88, 90
Poly-2,2'-(1,4-phenylene)-6,6'-oxydiquinoxaline	>500	89, 90
Poly-2,2'-(1,3-phenylene)-6,6'-diquinoxaline	>500	89, 90
Poly-2,2'-(1,3-phenylene)-6,6'-oxydiquinoxaline	>400	89, 90
Poly-2,2'-(4,4'-oxydiphenylene)-6,6'-diquinoxaline	>400	89, 90
Poly-2,2'-(4,4'-oxydiphenylene)-6,6'-oxydiquinoxaline	>400	89, 90

^{a} PMT: polymer melt temperature. See R. G. Beaman and F. B. Cramer, J. Polymer Sci., **21**, 223 (1956).

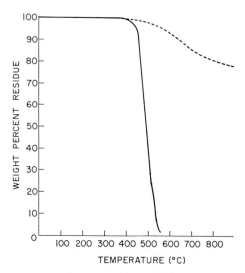

Fig. 31. TGA plot of poly-2,2'-(1,4-phenylene)-6,6'-diquinoxaline in air (——), and nitrogen (-----). Heating rate, 1.5°C/min (88).

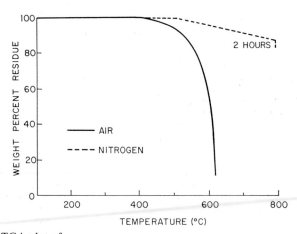

Fig. 32. TGA plot of

(heating rate, 2°C/min) (89).

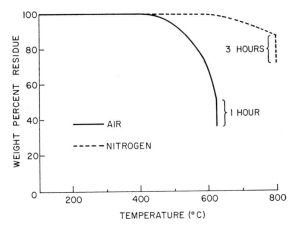

Fig. 33. TGA plot of (heating rate, 2°C/min) (89).

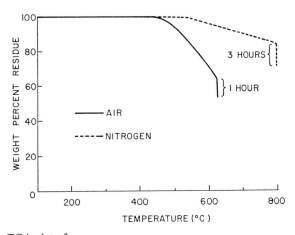

Fig. 34. TGA plot of heating rate, 2°C/min) (89).

Fig. 35. TGA plot of

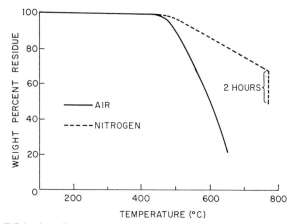

(heating rate, 2°C/min) (89).

Fig. 36. TGA plot of

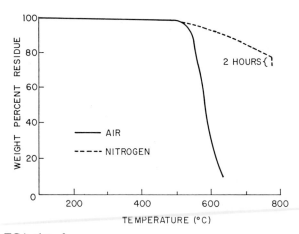

(heating rate, 2°C/min) (89).

IV. AROMATIC POLYMERS CONTAINING HETEROCYCLIC RINGS 197

Thus, the stability of the polyquinoxalines is of the same order as that observed in the previously discussed polyaromatic heterocyclics with the exception of the poly-1,3,4-oxadiazoles.

H. AROMATIC POLY-1,3,4-THIADIAZOLES

Like the 1,3,4-oxadiazole ring, the 1,3,4-thiadiazole ring has been of interest as a component in polymers containing intralinear aromatic and heterocyclic rings. Like its oxygen analog, this heterocyclic ring is reported to possess excellent thermal stability and outstanding hydrolytic stability (91,92).

The poly-1,3,4-thiadiazoles have been prepared (93) via the cyclodehydration of polyoxathiahydrazides

$$\left[\begin{array}{c} O \\ \| \\ R-C-NH-NH-C \\ \| \\ S \end{array}\right] \xrightarrow{-H_2O} \left[\begin{array}{c} S \\ R-C \diagdown C \\ \| \quad \| \\ N-N \end{array}\right] \quad (32)$$

or the cyclodehydrosulfurization of polydithiahydrazides.

$$\left[\begin{array}{c} S \\ \| \\ R-C-NH-NH-C \\ \| \\ S \end{array}\right] \xrightarrow{-H_2S} \left[\begin{array}{c} S \\ R-C \diagdown C \\ \| \quad \| \\ N-N \end{array}\right] \quad (33)$$

The cyclodehydration or cyclodehydrosulfurization reaction was carried out at elevated temperatures in a nitrogen atmosphere or *in vacuo*. The rate of conversion of the polyoxathiahydrazide or polydithiahydrazide to the poly-1,3,4-thiadiazole structure was considerably more rapid than the polyhydrazide → poly-1,3,4-oxadiazole conversion. For example, whereas at temperatures of 250°C the conversion of the polyhydrazide to poly-1,3,4-oxadiazole required hours, the conversion of the aforementioned precursor to the 1,3,4-thiadiazole structure occurred in a matter of minutes (93).

In Table XXVII are listed the poly-1,3,4-thiadiazoles which have been prepared. These polymers were characterized by extremely high polymer melt temperatures, medium crystallinity, outstanding resistance to hydrolytic degradation in acidic or basic media, and a high degree of thermal stability. The poly-1,3,4-thiadiazoles were soluble in such strong acids as sulfuric and methanesulfonic acid and, like the poly-1,3,4-oxadiazoles, were much less colored than the polyaromatic heterocyclics previously discussed, ranging in color from colorless to deep yellow (93).

TABLE XXVII
Poly-1,3,4-Thiadiazoles (93)

$$-R-C\underset{N-N}{\overset{S}{\underset{\|}{C}}}C-R-C\underset{N-N}{\overset{S}{\underset{\|}{C}}}C-$$

Code	Structure R	Structure R'	PMT (°C)[a]	η_{inh}[b] (dl/g)
93	1,3-Phenylene	1,5'-Biphenylene	300	0.26
94	1,3-Phenylene	1,4-Cyclohexylene	>375	0.70
95	1,3-Phenylene	2,4-Pyridylene	>375	0.24
96	1,3-Phenylene	3,5-Pyridylene	>375	0.56
97	1,3-Phenylene	2,5-Dichloro-1,4-phenylene	>375	0.28
98	1,3-Phenylene	—	~300	0.14
99	1,3-Phenylene	1,3-Phenylene	>375	0.60
100	1,3-Phenylene	1,4-Phenylene	>375	0.50
101	1,3-Phenylene	1,4-Tetramethylene	>375	—
102	1,3-Phenylene	1,18-Octadecylene	>375	—
103	1,4-Phenylene	1,4-Phenylene	>375	0.14
104	2,5-Dichloro-1,4-phenylene	2,5-Dichloro-1,4-phenylene	>375	0.28
105	1,4-Cyclohexylene	1,4-Cyclohexylene	>375	0.35
106	1,4-Tetramethylene	1,4-Tetramethylene	>375	0.93
107	1,4-Phenylene	1,4-Phenylene	>375	0.20

[a] PMT: polymer melt temperature. See R. G. Beaman and F. B. Cramer, *J. Polymer Sci.*, **21**, 223 (1956).

[b] η_{inh}: inherent viscosity determined at 0.5 wt % concentration in methanesulfonic acid at 30°C.

The outstanding resistance of the aromatic poly-1,3,4-thiadiazoles to hydrolytic degradation in acid or basic media was demonstrated by the fact that the aromatic poly-1,3,4-thiadiazole, **99**, showed no change in molecular weight after treatment with 20% sulfuric acid solution or 20% sodium hydroxide solution for 24 hr at 100°C (93).

The thermal stability of the poly-1,3,4-thiadiazoles was quite similar to that of the poly-1,3,4-oxadiazoles. Listed in Table XXVIII are the results of the thermogravimetric analysis of these polymers in air (93). It is quite apparent from inspection of these data and a comparison with similar data of the poly-1,3,4-oxadiazoles (Tables XXIII and XXIV) that these polymers exhibited almost identical behavior. The aromatic poly-1,3,4-thiadiazoles showed no weight loss up to temperatures of approximately 450°C where rapid loss in weight occurred and continued up to the 600°C region. The residue slowly lost weight up to 700°C. At this temperature between 50 and 56% of the original weight remained. Like the poly-

TABLE XXVIII
Thermogravimetric Data for Poly-1,3,4-Thiadiazoles in Air (93,94)

Polymer[a]	(T_i/T_f)[b]	Δ[c]	(w/w_o) 700°C[d]
93	450/610	36	52
94	380/500	37	36
95	450/600	38	53
96	456/585	38	52
97	440/595	39	50
98	360/520	47	25
99	460/580	37	53
100	450/590	35	53
101	380/480	50	26
102	360/490	43	30
103	360/520	45	30
104	450/600	36	51
105	381/499	48	32
106	360/490	60	10
107	460/610	35	56

[a] See Table XXVII for structures.
[b] (T_i/T_f) = (onset temp., °C/terminus temp., °C).
[c] $\Delta = (\Delta w/w_o)_{100}$.
[d] $(w/w_o)_{700°C}$ = % original weight remaining at 700°C; w = sample weight at 700°C, w_o = original sample weight.

1,3,4-oxadiazoles, this behavior was observed both under a nitrogen atmosphere and in the air.

Thus, the poly-1,3,4-thiadiazoles, like the poly-1,3,4-oxadiazoles, appear to be far superior to the previously discussed polyaromatic heterocyclic systems in air or oxidizing atmospheres. This outstanding thermal stability in air and superior resistance to hydrolytic attack in acid and basic media has prompted an extensive study of the aromatic poly-1,3,4-thiadiazoles in fiber applications. These will be discussed in detail in Chapter VII.

I. AROMATIC POLYTHIAZOLES

The thiazole ring, like the benzothiazole ring, has been of interest as a heterocyclic component in aromatic heterocyclic polymers. Like the benzothiazoles, the thiazoles are high melting, possess both acid and basic characteristics, and exhibit chemical resistance to acid, basic, and oxidizing reagents (95,96). In addition, the thiazole nucleus is more resistant to electrophilic substitution, such as sulfonation, than benzene itself (97).

The preparation of polythiazoles was first reported (98) by Mulvaney and Marvel from the reaction of bis-α-haloketones with dithioamides in acetic acid or N,N-dimethylformamide (see reaction 34). Although this

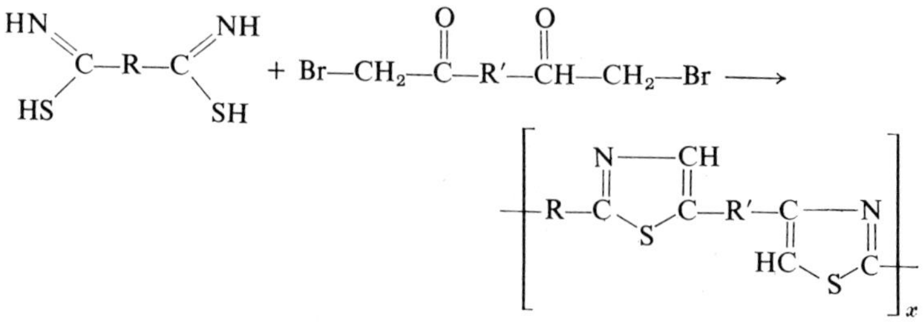

$$\underset{HS}{\overset{HN}{\diagdown}}C-R-C\underset{SH}{\overset{NH}{\diagdown}} + Br-CH_2-\overset{O}{\underset{\|}{C}}-R'-\overset{O}{\underset{\|}{C}}H-CH_2-Br \longrightarrow$$

$$\left[-R-\underset{S}{\overset{N-CH}{\underset{\|}{C}}}\overset{}{\underset{}{C}}-R'-\underset{HC}{\overset{C}{\underset{\|}{\diagdown}}}\underset{S}{\overset{N}{\diagdown}}C- \right]_x \quad (34)$$

procedure yielded aliphatic–aromatic polythiazoles in film-forming molecular weights, all attempts to prepare all-aromatic polythiazoles of high molecular weight were unsuccessful.

Using formic acid and o-dichlorobenzene as reaction media in this reaction, Craven (99) has prepared all-aromatic polythiazoles, **108**, and aromatic–cycloaliphatic polythiazoles, **109**, in film-forming molecular weights.

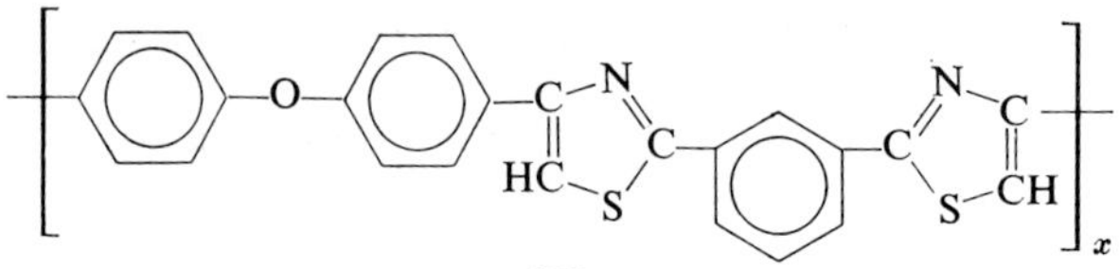

108

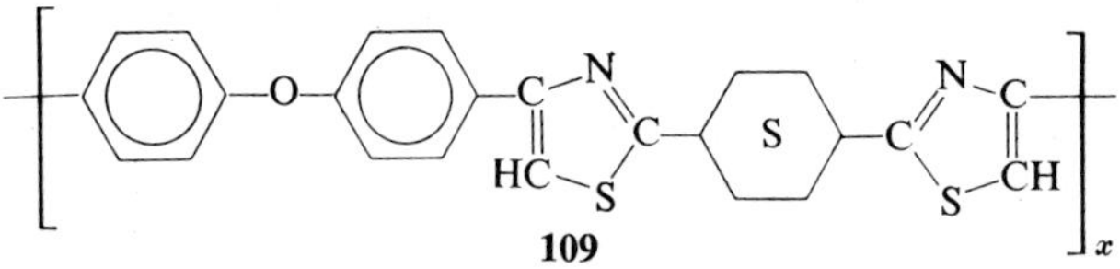

109

The preparation of aromatic polydithiazoles has been reported (100) from the reaction of aryl bis(bromomethylketones) with dithiooxamide in N,N-dimethylformamide (see reaction 35). The all-aromatic polydithia-

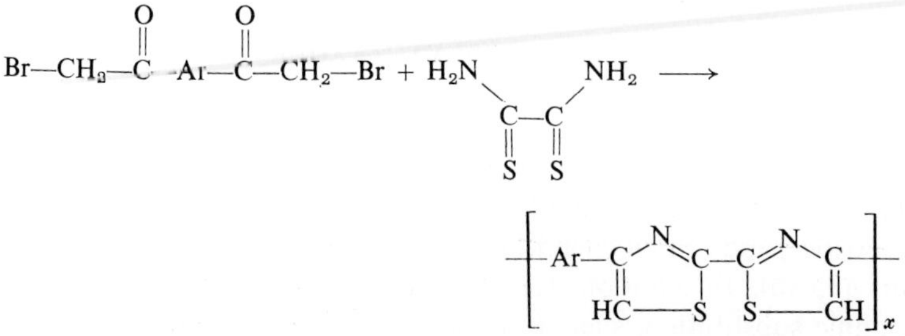

$$Br-CH_2-\overset{O}{\underset{\|}{C}}-Ar-\overset{O}{\underset{\|}{C}}-CH_2-Br + H_2N\underset{\overset{\|}{S}}{\overset{}{\diagdown}}\underset{\overset{\|}{S}}{\overset{C-C}{\diagup}}NH_2 \longrightarrow$$

$$\left[-Ar-\underset{HC-S}{\overset{N}{\underset{\|}{C}}}\overset{}{\underset{}{C}}-\underset{S-CH}{\overset{N}{\underset{\|}{C}}}\overset{}{\underset{}{C}}- \right]_x \quad (35)$$

TABLE XXIX
Aromatic Polythiazoles and Polydithiazoles

$$-R_1-C\underset{HC-S}{\overset{N}{\diagdown}}C-R_2-C\underset{S-CH}{\overset{N}{\diagdown}}C-$$

Polymer	Structure R₁	R₂	PMT (°C)[a]	Ref.
110	1,4-Tetramethylene	1,4-Phenylene	~300	98, 101
111	1,4-Phenylene	1,4-Phenylene	>400	97, 101
108	1,3-Phenylene	1,4-Phenylene	>400	97, 101
109	1,4-Cyclohexylene	Bis-(4-phenylene)ether	~230	98, 101
112	1,3-Phenylene	Bis-(4-phenylene)ether	>400	98, 101
113	1,4-Phenylene	—	>400	100, 101
114	4,4′-Biphenylene	—	>400	100
115	Bis(4-phenylene)methane	—	>400	100
116	Bis(4-phenylene)ether	—	>400	100
117	Bis(4-phenylene)methane	1,4-Phenylene	>400	101
118	2,6-Pyridylene	1,4-Phenylene	>400	101
119	2,6-Pyridylene	Bis(4-phenylene)ether	>400	101

[a] PMT: polymer melt temperature. See R. G. Beaman and F. B. Cramer, *J. Polymer Sci.*, **21**, 223 (1956).

zoles were so intractable that complete characterization and fabrication of these polymers into useful articles was not achieved.

The polythiazoles and polydithiazoles which had been prepared are listed in Table XXIX. All of the aromatic polythiazoles and polydithiazoles

TABLE XXX
Polythiazoles and Polydithiazoles
Percent Weight Loss at Various Temperatures by Thermogravimetry[a,b]

Polymer	400°C Air	400°C Inert atm.	500°C Air	500°C Inert atm.	600°C Air	600°C Inert atm.	700°C Air	700°C Inert atm.	Ref.
112	2	—	50	2	—	20	—	40	99
113	—	—	—	10	—	23	—	38	100
111	—	33	—	90	—	—	—	—	101
110	20	—	50	—	—	—	—	—	99

[a] See Table XXIX for polymer structure.
[b] Heating rate, 6°C/min.

TABLE XXXI
Polythiazoles and Polydithiazoles (100)

Polymer[a]	% wt loss at various temps. by static measurements[b]				
	300°C	400°C	500°C	600°C	Total
113	—	—	1	18	19
114	7	13	—	—	20
115	2	21	9	—	32
116	1	2	14	28	42

[a] See Table XXIX for polymer structure.
[b] Samples under a nitrogen atmosphere held at 300, 400, 500, and 600°C converting for 1 hr each.

were highly colored, exhibited high crystallinity, and were soluble only in such strong acids as concentrated sulfuric acid (98–101).

The thermal stability of these polymers has been studied by thermogravimetric analysis and static weight loss measurements (99–101). These data are collected in Tables XXX and XXXI. Under an inert atmosphere, the polythiazoles and polydithiazoles exhibited no loss in weight up to approximately 500°C, where rapid loss in weight occurred up to approximately 600°C. In air, there was loss of 50% of original weight at temperatures as low as 500°C.

Based on these limited data, it would appear that the aromatic polythiazoles and polydithiazoles have thermal stability comparable to that of most of the other polyaromatic heterocyclics.

J. AROMATIC POLYTETRAAZOPYRENES

The model compound for aromatic polytetraazopyrene, 2,7-diphenyl-1,3,6,8-tetraazopyrene, is a high melting aromatic structure which was resistant to acid, base, and oxidizing agents (103). This suggested that polymers of this structure would be of interest for high temperature resistant polymers.

Polytetraazopyrenes have been prepared via the polycondensation of 1,4,5,8-tetraaminonaphthalene with diphenyl esters (103) as shown in reaction 36.

These polymers were brown or black powders not melting below 400°C, with inherent viscosities in concentrated sulfuric acid ranging from 0.15 to 0.70; their infrared and ultraviolet spectra showed absorptions which were characteristic of the 2,7-diphenyl-1,3,6,8-tetraazopyrene ring system (103).

The thermal stability of these polymers in air and nitrogen has been

IV. AROMATIC POLYMERS CONTAINING HETEROCYCLIC RINGS

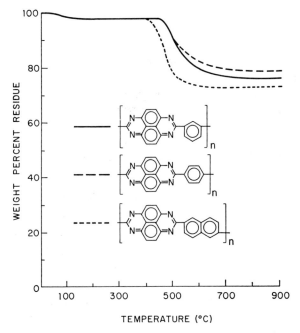

where R represents isophthalic, terephthalic, 2,6-naphthalenic, or oxybisbenzoic nucleus

studied by thermogravimetric analysis (103). As is apparent from Figure 37, in nitrogen, the polymers synthesized from diphenyl isophthalate or terephthalate have almost identical behavior showing a similar initial loss of about 20% in weight up to 410°C, followed by 22% loss up to 840°C. The polymer synthesized from diphenyl-2,6-naphthalenedicarboxylate showed less than 1% weight loss up to 170°C, followed by 26% weight loss

Fig. 37. TGA plot in nitrogen (heating rate, 6°C/min) (103).

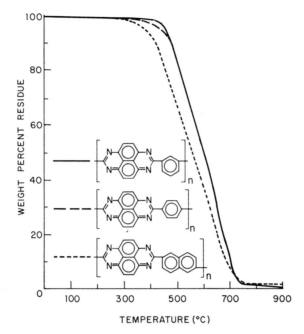

Fig. 38. TGA plot in air (heating rate, 6°C/min) (103).

between 390 and 850°C. In air (Fig. 38) they showed quite different behavior. The isophthalate and terephthalate derived polymers had initial loss of 1% from room temperature up to 340°C and a rapid loss of 98% of weight between 340 and 710°C. Similarly, the polymer derived from diphenyl-2,6-naphthalenedicarboxylate had less than 1% weight loss up to 270°C, with 99% weight loss between 360 and 780°C.

This markedly different behavior in air and nitrogen is similar to that observed with other polyaromatic heterocyclics and indicates that these polymers are of the same order of thermal stability.

K. AROMATIC POLY-4-PHENYL-1,2,4-TRIAZOLES

The 4-phenyl-1,2,4-triazole ring, like the 1,3,4-oxadiazole and the 1,3,4-thiadiazole rings, has been of interest as a component in polymers containing intralinear aromatic and heterocyclic rings. Like these other heterocyclic rings, this heterocyclic ring is reported to possess excellent thermal stability and outstanding hydrolytic stability (102).

The first reported synthesis of aromatic poly-4-phenyl-1,2,4-triazoles was by a modification of the reaction of tetrazoles with imide chlorides (69). In this procedure, *m*-phenylene bistetrazole was made to react

$$\underset{\text{HN}-\text{N}}{\overset{\text{N}=\text{N}}{\underset{|}{\text{C}}}}\!\!-\!\!\bigcirc\!\!-\!\!\underset{\text{N}-\text{NH}}{\overset{\text{N}=\text{N}}{\underset{|}{\text{C}}}} + \text{Cl}-\overset{\phi-\text{N}}{\underset{\|}{\text{C}}}\!\!-\!\!\bigcirc\!\!-\!\!\overset{\text{N}-\phi}{\underset{\|}{\text{C}}}\!\!-\!\!\text{Cl} \longrightarrow$$

$$\left[\bigcirc\!\!-\!\!\underset{\text{N}-\text{N}}{\overset{\overset{\phi}{|}}{\underset{}{\text{C}}}}\!\!\!\!\!\!\!\!\!\!\!\!\right]_x \quad (37)$$

at elevated temperatures with N,N'-diphenylisophthalimide chloride in pyridine solution (see reaction 37).

The polymers which were obtained were soluble only in concentrated sulfuric acid and were in the molecular weight range of from 5000 to 6000. Although in this molecular weight range fibers and films could not be

$$\left(\text{R}-\overset{\text{O}}{\underset{\|}{\text{C}}}-\text{NH}-\text{NH}-\overset{\text{O}}{\underset{\|}{\text{C}}}\right)_x + \phi-\text{NH}_2 \xrightarrow{\text{PPA}} \left[-\text{R}-\underset{\text{N}-\text{N}}{\overset{\overset{\phi}{|}}{\underset{}{\text{C}}}}\!\!\!\!\!\!\!\!\!\!\!\!\right]_x \quad (38)$$

fabricated, the polymers did exhibit excellent thermal stability as evidenced by thermogravimetric analysis in nitrogen (69).

High molecular aromatic poly-4-phenyl-1,2,4-triazoles have been reported (104,105) via the reaction of high molecular weight aromatic polyhydrazides with aniline at elevated temperatures in polyphosphoric acid

TABLE XXXII
Properties of Phenylenetriazole Polymers (104,105)

Structure	$\eta_{\text{inh}}{}^a$(dl/g)	Film
Poly(m-phenylene)4-phenyltriazole	0.280	Poor Brittle Clear
Poly(m-, p-phenylene)4-phenyltriazole	1.004	Good Clear Flexible
Poly[(1,3-phenylene), (2,6-naphthylene)]-4-phenyltriazole	0.704	Fair Opaque Flexible
Poly[(1,3-phenylene), (4,4'-biphenyl)]-4-phenyltriazole	Insoluble	Poor Opaque Brittle

^{a} η_{inh}: inherent viscosity determined at 0.5 wt % concentration in formic acid at 30°C.

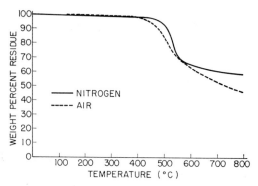

Fig. 39. TGA plot of poly(*m*-, *p*-phenylene)-4-phenyl-1,2,4-triazole (heating rate, 10°C/min) (104).

(PPA). Using this procedure, a whole series of high molecular weight aromatic poly-4-phenyl-1,2,4-triazoles have been prepared as shown in Table XXXII. These polymers were characterized by high polymer melt temperatures and solubility only in strong acids such as formic and sulfuric acid (104,105).

The thermal stability of the poly-4-phenyl-1,2,4-triazoles has been demonstrated (104) by the behavior of poly(*m*, *p*-phenylene)-4-phenyl-1,2,4-triazole in thermogravimetric analysis and differential thermal analysis (Figs. 39 and 40). Under a nitrogen atmosphere, this polymer was stable up to 450°C where rapid weight loss occurred and continued up to 550°C. The residue slowly lost weight up to 800°C with 60% original weight remaining at this temperature. In air, weight loss began at a lower temperature, slightly above 400°C, and was typical of oxidative decomposition, with continuing weight loss up to 800°C. The differential thermal analysis (Fig. 40) curve in nitrogen was in agreement, with the maximum of

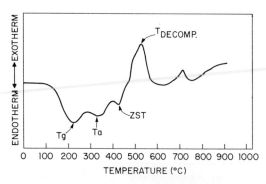

Fig. 40. DTA plot of poly(*m*-, *p*-phenylene)-4-phenyl-1,2,4-triazole in nitrogen (heating rate, 10°C/min) (104).

the large exotherm at 525°C comparing well with the temperature of maximum weight loss.

Thus the aromatic poly-4-phenyl-1,2,4-triazoles exhibited thermogravimetry behavior similar to that of the other polyaromatic heterocyclics which have been discussed previously, showing outstanding resistance to thermal attack in nonoxidizing atmospheres and indicating a susceptibility to air oxidation, at temperatures in excess of 400°C.

With high molecular weight polyhydrazides, it has been possible to prepare fibers and films of the aromatic poly-1-phenyl-1,3,4-triazoles via the reaction with aniline. The properties of the fibers and films will be discussed in detail in Chapter VII.

L. POLY(QUINAZOLINEDIONES) AND POLY(BENZOXAZINONES)

The 2-aryl quinazolinediones and the 2-aryl benzoxazinone, like the previously discussed aromatic heterocyclic ring systems, have been of

(39)

interest as component in polymers for high temperature resistance uses. These aromatic heterocyclic ring systems are reported to possess excellent thermal stability and outstanding hydrolytic stability (106).

The fully aromatic poly(quinazolinediones) of high molecular weight were prepared in three successive steps as shown in reaction 39 (107). A polycondensation of this type is especially noteworthy since it involved the formation of high molecular weight tractable polyurea acid, **117**, followed by cyclodehydration to obtain poly(2-imino-4H-3,1-benzoxazin-4-one), **118**, and then underwent intramolecular rearrangement along the polymer chain upon heating to form the thermodynamically stable poly(quinazolinedione), **119**.

The poly(benzoxazinone) was prepared by the cyclopolycondensation of 4,4-diaminobiphenyl-3,3-dicarboxylic acid with an aromatic dicarboxylic acid halide in two successive steps as shown in reaction 40 (108).

The cyclopolycondensation reaction proceeded through the formation of a high molecular weight polyamide-acid, **120**, in the first step, an open chain, tractable precursor, which subsequently underwent thermal cyclodehydration along the polymer chain, in the second step, to yield a fully aromatic poly(benzoxazinone), **121**.

The thermal stability of the poly(quinazolinediones) and the poly(benzoxazinones) has been demonstrated by their behavior in thermogravimetric analysis (107,108). Inspection of Figures 41 and 42 clearly

IV. AROMATIC POLYMERS CONTAINING HETEROCYCLIC RINGS

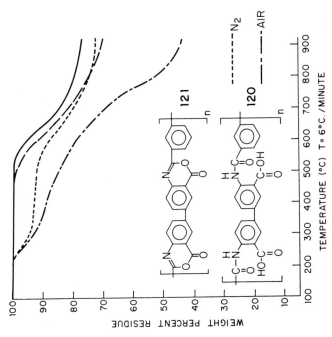

Fig. 42. TGA curves for polyamide-acid, **120**, and polybenzoxazinone, **121**, in nitrogen and air.

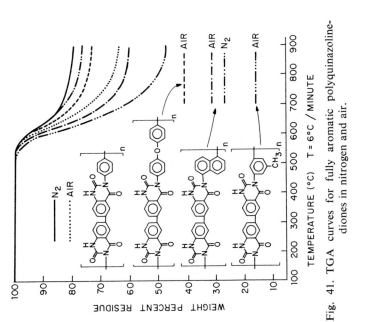

Fig. 41. TGA curves for fully aromatic polyquinazolinediones in nitrogen and air.

show these fully aromatic heterocyclic polymers have thermal stabilities comparable to that of the aromatic polyimides, polybenzimidazoles, and polybenzoxazoles.

References

1. K. Hofmann, *Imidazole and Its Derivatives*, Interscience, New York, 1953.
2. E. A. Steck, F. C. Nichold, G. W. Ewing, and N. H. Gorman, *J. Am. Chem. Soc.*, **70**, 3406 (1948).
3. J. W. Dale, I. B. Johns, E. A. McElhill, and J. O. Smith, WADS Tech. Report 59–95, Jan. 1959.
4. K. C. Brinker and I. M. Robinson, U.S. Pat. 2,895,948 (June 1959).
5. H. A. Vogel and C. S. Marvel, *J. Polymer Sci.*, **50**, 511 (1961).
6. (*a*) D. N. Gray and G. P. Shulman, *Am. Chem. Soc., Div. Polymer Chem., Preprints*, **6**, No. 2, 778 (1965), *J. Macromol. Sci.* (*Chem.*), **A-1**, No. 3, 395 (1967). (*b*) D. N. Gray, L. L. Rouch, and E. L. Strauss, *Am. Chem. Soc., Div. Polymer Chem., Preprints*, **8**, No. 2, 1138 (1967).
7. H. Vogel and C. S. Marvel, *J. Polymer Sci. A*, **1**, 1531 (1963).
8. L. Plummer and C. S. Marvel, *J. Polymer Sci. A*, **2**, 2559 (1964).
9. R. T. Foster and C. S. Marvel, *J. Polymer Sci. A*, **3**, 417 (1965).
10. C. S. Marvel and H. A. Vogel, U.S. Pat. 3,174,947 (March 1965).
11. Y. Iwakura, K. Uno, and Y. Imai, *J. Polymer Sci. A*, **2**, 2605 (1964).
12. Y. Iwakura, K. Uno, and Y. Imai, *Makromol. Chem.*, **77**, 33 (1964).
13. (*a*) Y. Iwakura, K. Uno, and Y. Imai, *Makromol. Chem.*, **77**, 41 (1964). (*b*) Y. Iwakura, Y. Imai, S. Inone, and K. Uno, *Makromol. Chem.*, **95**, 236 (1966).
14. French Pat. 1,363,757 (May 1964).
15. K. Mitsuhashi and C. S. Marvel, *J. Polymer Sci. A*, **3**, 1661 (1965).
16. H. H. Levine, C. B. Delano, and K. J. Kjoller, *Am. Chem. Soc., Div. Polymer Chem., Preprints*, **5**, No. 1, 160 (1964).
17. R. Phillips and W. W. Wright, *J. Polymer Sci. B*, **2**, 47 (1964).
18. W. W. Wright, *Soc. Chem. Ind.* (*London*) *Monograph*, **13**, 248 (1961).
19. A. H. Frazer, J. J. Kane, and F. T. Wallenberger, ASD, TDR-62-679, July 1962.
20. A. Ladenburg, *Ber.*, **9**, 1524 (1876).
21. W. Dech and D. Dains, *J. Am. Chem. Soc.*, **55**, 4986 (1933).
22. H. Bogert and J. Maimans, *J. Am. Chem. Soc.*, **57**, 1529 (1935).
23. R. Stickings, *J. Chem. Soc.*, **1928**, 3131.
24. K. C. Brinker, D. D. Cameron, and I. M. Robinson, U.S. Pat. 2,904,537 (Sep. 1959).
25. W. W. Moyer, C. Cole, and T. Anyos, *J. Polymer Sci. A*, **3**, 2107 (1965); French Pat. 1,365,114 (1964); Brit. Pat. 1,026,633 (1966); W. W. Moyer, U.S. Pat. 3,230,196 (Jan. 1966).
26. T. Kubota and R. Nakanishi, *J. Polymer Sci. B*, **2**, 655 (1964).
27. Y. Imai, I. Taoka, K. Uno, and Y. Iwakura, *Makromol. Chem.*, **83**, 167 (1965).
28. (*a*) Y. Imai, I. Taoka, K. Uno, and Y. Iwakura, *Makromol. Chem.*, **83**, 179 (1965). (*b*) French Pat. 1,413,116 (1965). (*c*) V. V. Korshak, G. M. Tseitlin and A. L. Rusenov, *Vysokomolekul. Soedin.*, **8**, 1599 (1966). (*d*) R. M. Gitina, G. I. Braz, and E. S. Kronganz, *Vysokomolekul. Soedin.*, **8**, 1535 (1966).
29. H. Bogert and R. Stull, *J. Am. Chem. Soc.*, **47**, 3078 (1925).
30. H. Bogert and W. Husted, *J. Am. Chem. Soc.*, **54**, 3395 (1932).
31. I. Hunter and R. Jones, *J. Chem. Soc.*, **1930**, 941.

32. J. Rudner, R. Brumfeld, and J. O. Smith "Synthesis of Poly-Hetero-Cyclics," ASD TDR-62-249, March 1963.
33. A. Morton, J. B. Littlefield, and W. D. Mecum, U.S. Pat. 3,047,543 (June 1963).
34. V. Kiprianov and K. Mushkalo, *J. Gen. Chem. (USSR)*, **32**, 4040 (1962).
35. H. H. Levine, P. M. Hergenrother, and W. Wrasidlo, *Am. Chem. Soc., Div. Polymer Chem., Preprints*, **5**, No. 1, 153 (1964).
36. (*a*) H. H. Levine, P. M. Hergenrother, and W. Wrasidlo, *J. Polymer Sci. A*, **3**, 1665 (1965). (*b*) P. M. Hergenrother and H. H. Levine, *J. Polymer Sci. A-1*, **4**, 2341 (1966).
37. T. M. Bogert and R. R. Renshaw, *J. Am. Chem. Soc.*, **30**, 1140 (1908).
38. W. M. Edwards and I. M. Robinson, U.S. Pat. 2,710,853 (June 1955).
39. W. M. Edwards and I. M. Robinson, U.S. Pat. 2,867,609 (Jan. 1959).
40. W. M. Edwards and I. M. Robinson, U.S. Pat. 2,880,230 (March 1959).
41. J. I. Jones, F. W. Ochynski and F. A. Rackley, *Chem. Ind., London*, **1962**, 1686.
42. W. M. Edwards, U.S. Pat. 3,179,614 (April 1965).
43. W. M. Edwards, U.S. Pat. 3,179,634 (April 1965).
44. (*a*) C. E. Sroog, A. L. Endrey, S. V. Abramo, C. E. Berr, W. M. Edwards, and K. L. Oliver, *Am. Chem. Soc., Div. Polymer Chem., Preprints*, **5**, No. 1, 132 (1964). (*b*) C. E. Sroog, A. L. Endrey, S. V. Abramo, C. E. Berr, W. M. Edwards, and K. L. Olivier, *J. Polymer Sci. A*, **3**, 1373 (1965).
45. (*a*) C. E. Sroog, *J. Polymer Sci. C*, **16**, 1191 (1967). (*b*) M. L. Wallach, *Am. Chem. Soc., Div. Polymer Chem., Preprints*, **6**, No. 1, 53 (1965); *ibid.*, **8**, No. 1, 656 (1967); *ibid.*, **8**, No. 2, 1170 (1967); *J. Polymer Sci. A-2*, **5**, 653 (1967). (*c*) R. Ikeda, *J. Polymer Sci. B*, **4**, 353 (1966).
46. L. E. Amborski, *Am. Chem. Soc., Div. Polymer Chem., Preprints*, **4**, No. 1, 175 (1963).
47. Brit. Pat. 903,271 (1963).
48. W. M. Edwards and A. L. Endrey, U.S. Pat. 3,179,631 (April 1965).
49. French Pat. 1,365,545 (May 1964).
50. G. M. Bower and L. W. Frost, *J. Polymer Sci. A*, **1**, 3135 (1963).
51. A. L. Endrey, U.S. Pat. 3,179,633 (April 1965).
52. Brit. Pat. 942,025 (1963).
53. A. L. Endrey, U.S. Pat. 3,179,630 (April 1965).
54. L. W. Frost and I. Kesse, *J. Appl. Polymer Sci.*, **8**, 1039 (1964).
55. R. J. Angelo, U.S. Pat. 3,073,785 (June 1963).
56. Belg. Pat. 589,179 (1960).
57. H. H. Levine, P. M. Hergenrother, and W. J. Wrasidlo, "High Temperature Adhesives," Contract No. w-63-0420-C, April 1964.
58. S. Nishizaki and A. Fukami, *Chem. Soc. Japan*, **66**, 382 (1963).
59. S. D. Bruck, *Am. Chem. Soc., Div. Polymer Chem., Preprints*, **5**, No. 1, 148 (1964).
60. S. D. Bruck, *Polymer*, **5**, 435 (1964).
61. S. D. Bruck, *Polymer*, **6**, 49 (1965).
62. J. F. Heacock and C. E. Berr, *SPE Trans.*, **5**, No. 2, 1 (1965).
63. (*a*) J. H. Freeman, E. J. Traynor, J. Miglarese, and R. H. Lunn, *SPE Trans.*, **2**, No. 3, 216 (1962). (*b*) J. H. Freeman, L. W. Frost, G. M. Bower, and E. J. Taylor, *ibid.*, **5**, No. 4, 75 (1965). (*c*) T. Unishi, *J. Polymer Sci. B*, **3**, 679 (1965). (*d*) D. F. Loncrini, W. B. Walton, and R. B. Hughes, *J. Polymer Sci., A-1*, **4**, 440 (1966). (*e*) D. F. Loncrini, *ibid.*, **4**, 1531 (1966). (*f*) Y. Iwakura, Y. Imai, and K. Uno, *Makromol. Chem.* **94**, 114 (1966).

64. R. J. Angelo, private communication.
65. F. Tiemann and P. Kruger, *Ber.*, **17**, 1693 (1884).
66. R. Stolle, *J. Prakt. Chem.*, Ser. 2, **68**, 30 (1903).
67. J. Sauer, R. Huisgen, and H. J. Strum, *Tetrahedron*, **11**, 214 (1960).
68. (*a*) D. C. Bloomstrom, U.S. Pat. 3,044,994 (1962). (*b*) C. G. Overberger and F. Fujimoto, *J. Polymer Sci. B*, **3**, 735 (1965). (*c*) M. Akiyama, Y. Iwakura, S. Shirashi, and Y. Imai, *J. Polymer Sci. B*, **4**, 305 (1966). (*d*) V. V. Korshak, E. S. Kronganz, and A. L. Rusenov, *Dokl. Nauk SSSR*, **166**, 356 (1966). (*e*) Y. Iwakura, K. Uno, Y. Imai, and M. Akiyama, *Makromol. Chem.*, **95**, 275 (1966).
69. E. J. Abshire and C. S. Marvel, *Makromol. Chem.*, **44–46**, 388 (1961).
70. Belg. Pat. 628,775 (1963).
71. Can. Pat. 673,091 (1963).
72. E. J. Vandenberg and C. G. Overberger, *Science*, **141**, 176 (1963).
73. A. H. Frazer, W. Sweeny, and F. T. Wallenberger, *J. Polymer Sci. A*, **2**, 1157 (1953).
74. A. H. Frazer and I. M. Sarasohn, *J. Polymer Sci. A-1*, **4**, 1649 (1966).
75. Y. Iwakura, K. Uno, and S. Hara, *J. Polymer Sci. A*, **3**, 45 (1965); *Makromol. Chem.*, **101**, 2500 (1967).
76. (*a*) J. Preston and W. B. Black, *Am. Chem. Soc., Div. Polymer Chem., Preprints*, **6**, No. 2, 757 (1965); *J. Polymer Sci. A-1*, **4**, 5410 (1967). (*b*) K. Sato and K. Santome, *Makromol. Chem.*, **100**, 91 (1966). (*c*) Y. Iwakura, Y. Imai, K. Uno, and Y. Takse, *Makromol. Chem.*, **94**, 114 (1966).
77. G. F. L. Ehlers, ASD-TR-61-622, Feb. 1962.
78. (*a*) J. L. Cotter, *J. Chem. Soc.*, **1964**, 5491. (*b*) J. L. Cotter and G. J. Knight, *Chem. Commun.*, **1966**, 336.
79. R. Knoor, *Ber.*, **16**, 2597 (1883).
80. W. Buckner, *Ann.*, **4191** (1919).
81. R. Curtus and W. Wirsing, *J. Prakt. Chem.*, Ser. 2, **50**, 531 (1894).
82. V. V. Korshak, Y. S. Kronganz, and A. M. Berlin, *Vysokomolekul. Soedin.*, **6**, 1078 (1964).
83. V. V. Korshak, Y. S. Kronganz, A. M. Berlin, and A. P. Travnikova, *Vysokomolekul. Soedin.*, **6**, 1087 (1964).
84. V. V. Korshak, E. S. Kronganz, and A. M. Berlin, *Dokl. Akad. Nauk SSSR*, **152**, 1108 (1963); *J. Polymer Sci. A*, **3**, 2425 (1965).
85. J. P. Schaefer and J. L. Bertram, *J. Polymer Sci. B*, **3**, 95 (1965).
86. J. C. E. Simpson, *Condensed Pyradazine and Pyrazine Rings*, Interscience, New York, 1953.
87. J. K. Stille and J. R. Williamson, *J. Polymer Sci. B*, **2**, 209 (1964).
88. J. K. Stille and J. R. Williamson, *J. Polymer Sci. A*, **2**, 3867 (1964).
89. (*a*) J. K. Stille, J. R. Williamson, and F. E. Arnold, *J. Polymer Sci. A*, **3**, 1013 (1965). (*b*) J. K. Stille and F. E. Arnold, *J. Polymer Sci. A-1*, **4**, 551 (1966).
90. G. De Gandemaris, B. Sillion, and J. Preve, *Bull. Soc. Chem.*, **1964**, 1763.
91. R. Stolle, *Ber.*, **32**, 797 (1899).
92. O. Simroth and R. De Montmollin, *Ber.*, **43**, 2904 (1910).
93. A. H. Frazer, W. P. Fitzgerald, and T. A. Reed, ML-TDR-285, Aug. 1964.
94. A. H. Frazer, W. P. Fitzgerald, T. A. Reed, and L. W. Wilson, AFML-TR-65-221, Pt. 1, Aug. 1965.
95. E. Gabriel, *Ber.*, **43**, 134 (1910).
96. R. Robinson, *J. Chem. Soc.*, **95**, 2167 (1907).

IV. AROMATIC POLYMERS CONTAINING HETEROCYCLIC RINGS

97. R. H. Wiley, *Organic Chemistry*, Vol. IV, H. Gilman, Ed., Wiley, New York, 1953.
98. J. E. Mulvaney and C. S. Marvel, *J. Org. Chem.*, **26**, 95 (1961).
99. J. M. Craven and T. M. Fischer, *J. Polymer Sci. B*, **3**, 35 (1965).
100. D. T. Longone and H. H. Un, *Am. Chem. Soc., Div. Polymer Chem., Preprints*, **4**, No. 2, 49 (1963); *J. Polymer Sci. A*, **3**, 3117 (1965).
101. W. C. Sheehan, T. B. Cole, and L. G. Picklesimer, *J. Polymer Sci. A*, **3**, 1443 (1965).
102. K. T. Potts, *Chem. Rev.*, **61**, 87 (1961).
103. F. Dawans, B. Reichel, and C. S. Marvel, *J. Polymer Sci. A*, **2**, 5005 (1964).
104. M. R. Lilyquist and J. R. Holsten, *Am. Chem. Soc., Div. Polymer Chem., Preprints*, **6**, No. 2 (1963); *J. Polymer Sci. A*, **3**, 3905 (1965).
105. Belg. Pat. 645,926 (1964); Ger. Pat. 1,224,491 (1966).
106. P. Friedlander and S. Wleugel, *Ber.*, **16**, 2227 (1883).
107. N. Yoda, R. Nakanishi, M. Kurihara, Y. Bamba, S. Tohyama, and K. Ikeda, *J. Polymer Sci. B*, **4**, 11 (1966); *J. Polymer Sci. A-1*, **5**, 1780, 5405 (1967).
108. N. Yoda, M. Kurihara, K. Ikeda, S. Tohyama, and R. Nakanishi, *J. Polymer Sci. B*, **4**, 551 (1966); *J. Polymer Sci. A-1*, **5**, 5406 (1967).

V. INORGANIC POLYMERS

The present interest in inorganic polymers as high temperature resistant polymers is based on the prospect of new materials superior in some ways to any of the numerous and widely different organic plastics and the ever-present interest in the relationship of chemical and physical properties to chemical bonding. Research on inorganic polymers has been carried out on three major classes:

1. Stable inorganic polymeric materials having linear chains consisting of such typical repeating units as silicon–nitrogen, boron–nitrogen, phosphorus–nitrogen, etc.

2. Inorganic-organic (semiorganic) polymers having inorganic chains framed by organic substituents such as in silicones, or organic units as the members of the backbone chain. The major emphasis in this work has been aimed at the replacement of the silicon in silicone-like structures by elements such as aluminum, titanium, tin, and boron, in order to increase the already high thermal stability of the silicone structure.

3. Metal chelate or coordination polymers. Such polymers have been prepared by (*a*) linking polydentate ligands by metal ions, (*b*) forming of the polymer with simultaneous incorporation of the metal ions, (*c*) incorporating metal ions into existing polymers, (*d*) forming the polymer by reaction with chelate containing functional groups, and (*e*) a special case, ferrocene-containing polymers.

Within the last few years, several excellent books dealing with inorganic polymers have appeared (1–4). It is not the intent of this chapter to exhaustively or comprehensively discuss the preparation of all such inorganic polymers. The emphasis in this discussion will be on the properties, specifically the thermal and hydrolytic stability, of the inorganic polymers which have been prepared.

A. BORON-CONTAINING POLYMERS

Elementary boron has yielded three dimensional polymers, one form of which has a melting point of 2300°C and is nearly as hard as diamond. Boron formed a number of polymeric materials with many other elements, involving particularly bonds of boron with hydrogen, carbon, oxygen, nitrogen, and phosphorus. The high bond energies of some of these bonds, e.g., boron–oxygen, 119.3; boron–nitrogen, 104.3; boron–carbon, 89; and boron–boron, 80 kcal/mole, indicate the strength of the bonds formed. Unfortunately, the bonds which boron formed with other elements were usually fairly susceptible to attack by oxygen and water. In addition, many syntheses of boron polymers yielded thermodynamically favored small ring products rather than linear macromolecules.

1. Addition Polymers of Borozens

Borozen polymers have been considered as derived from reactions of the type

$$X \diagdown B-\ddot{N} \diagup \longrightarrow \left[\bar{B}-\overset{+}{N} \right]_x \quad (1)$$

in which each structural unit was joined to the next by a coordinate link. This type of system was analogous to the addition polymerization of olefins.

Although considerable effort has been expended in this area of research, only one example of a polymeric borozen has been reported. Dimeric dimethylaminoborane on heating at 150°C under 3000 atm pressure yielded amorphous, insoluble, infusible polymer which appeared to be stable to hydrolysis, but on heating or standing for a few months at room temperature, reverted back to the dimers.

2. Addition Polymers of Borozins (Condensed Systems)

The condensed borozins have been considered to be derived from the hypothetical reaction 2, in which each structural unit was joined to the

$$X-(-B=\ddot{N}-) \longrightarrow \left[-B-\ddot{N}- \right]_x \quad (2)$$

next by a covalent link. Trimers of this type, i.e., borazoles, have been studied extensively. Although cyclic tetramers (6,7), the borazocines, **1**,

$$\begin{array}{c} \text{Y} \quad \text{X} \\ \text{N}-\text{B} \\ \text{XB} \qquad \ddot{:}\text{NY} \\ | \qquad | \\ \text{YN}\ddot{:} \qquad \text{B}-\text{X} \\ \text{B}-\ddot{\text{N}} \\ \text{X} \quad \text{Y} \end{array}$$

1

have been prepared, no well-defined high molecular weight products have been obtained. Even the products described as polymeric materials, on further heating, reverted back to the borazole derivative.

3. Addition Polymers of Borozins (Fused Systems)

Included in this classification of boron polymers were the types of structures shown in structure **2**, in which each structural unit was joined to the next by two covalent links to make a fused system. The simplest compound of this type was the naphthalene analog shown in **2**. Included

$$\begin{array}{c} \text{H} \quad\quad \text{H} \\ \text{HN:}\overset{\underset{|}{B}}{\diagdown}\text{N}\overset{\underset{|}{B}}{\diagdown}\text{:NH} \\ | \quad\quad | \quad\quad | \\ \text{HB}\overset{}{\diagdown}\underset{|}{N}\overset{}{\diagdown}\underset{|}{B}\overset{}{\diagdown}\underset{|}{N}\overset{}{\diagdown}\text{BH} \\ \text{H} \quad\quad \text{H} \end{array}$$

2

in the classification, was boron nitride in the cubic or "diamond" form which has been shown to have a zinc blende structure (8) and to be unaffected by heating *in vacuo* up to 2000°C. It was not attacked by common mineral acids and was only slowly oxidized in the atmosphere at 2000°C (8). It was a good electrical insulator and hard enough to scratch diamonds. All of these polymers in this thermally stable form were highly crosslinked.

4. Polymeric Boron Complexes

The polymeric materials included in this classification were those which may be regarded as derived from monomeric units by intermolecular coordination and have other atoms as well as boron and nitrogen in the main polymer chain (see reaction 3). The monomeric species were substances

$$\text{>B}\!\sim\!\ddot{Z}\!\sim\!\longrightarrow\;\longrightarrow\;{}^{-}\text{B}\!\!\left[\overset{|}{-}Z^{+}\!\!-\!\text{B}\!\diagup\right]_{x}\!\overset{+}{\underset{\diagdown}{Z}} \quad\quad (3)$$

which possessed both acceptor (three-coordinate boron atoms) and donor (indicated as Z) in the various sites.

Examples of these types of polymers were polymeric cyanoborane and polymeric boron isocyanates. The polymeric cyanoborane was hydrolyzed slowly with 10% sodium hydroxide but pyrolyzed rapidly at 300–330°C to yield butane (9). Although the isocyanates showed less tendency to hydrolyze than cyanides, they, too, underwent depolymerization on heating at temperatures in excess of 300°C (10,11).

5. Polymers Having Only Boron and Nitrogen in the Main Chain

These polymers were derived from the aminoborane–amine elimination reaction as shown in reaction 4 (12). This general reaction provided a

$$2\;\text{>B·NH—R}\;\xrightarrow{\text{heat}}\;\text{>B—N(R)—B<}\;+\;\text{RNH}_2 \quad\quad (4)$$

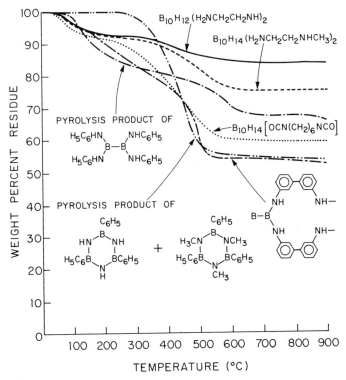

Fig. 1. TGA plot of boron–nitrogen polymers in nitrogen (heating rate, 3°C/min) (16).

means of joining two boron atoms through a nitrogen bridge and has been utilized to prepare a number of boron–nitrogen polymers (13–15). Although the structure and the molecular weight of the polymers have not been clearly defined, thermogravimetric analysis of some of these polymers has been carried out by Ehlers (16). These boron–nitrogen polymers (Fig. 1) all showed marked weight loss and decomposition at temperatures well below 300°C. It is apparent that polyaromatic heterocyclics discussed in Chapter IV are far more thermally stable.

6. Polymers Having Only Boron, Nitrogen, and Elements Other than Carbon in the Main Chain

These polymers were those in which successive boron–nitrogen units were linked together through oxygen, sulfur, or phosphorus.

Polymers linking boron–nitrogen units through oxygen have been derived from the reaction of hydroxylamine, N-methyl- and N,N-dimethylhydroxylamine with diborane. Such polymers were white solids, insoluble in inert organic solvents, and probably had the structure shown in structure 3 (17).

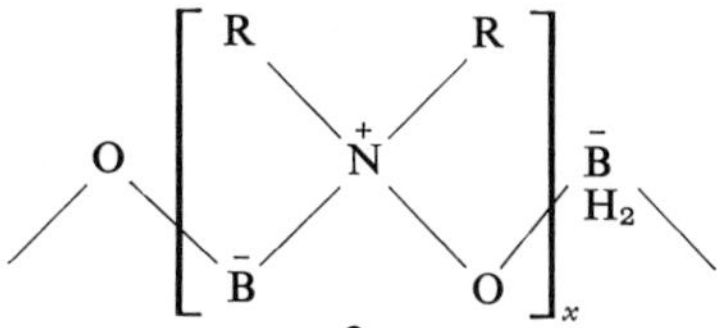

$$\left[\begin{array}{c} R \quad\quad R \\ \diagdown \;\;\;\diagup \\ O \quad N^+ \quad \bar{B} \\ \diagup \;\;\;\diagdown \quad H_2 \\ \bar{B} \quad\quad O \end{array} \right]_x$$

3

They have been obtained from the molecular addition compounds (RR'NOH, BH$_3$; R=H=R'; R=Me=R'; or R=H; R'=Me) via dehydrogenation at 25°C. All such polymers, on heating, decomposed to monomer.

Although the polymer prepared by linking two boron–nitrogen units through sulfur has not been reported, the model reaction has been carried out with sulfamide and o-phenylene chloroboronate (18).

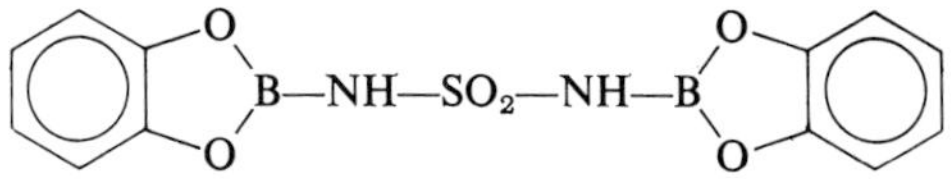

$$\text{(o-C}_6\text{H}_4\text{O}_2\text{)B—NH—SO}_2\text{—NH—B(O}_2\text{C}_6\text{H}_4\text{-}o\text{)}$$

4

The polymer linking two boron–nitrogen units through phosphorus has probably been formed from the reaction of β-trichloroborazole with trialkyl phosphates (19). In this reaction sequence, the intermediate,

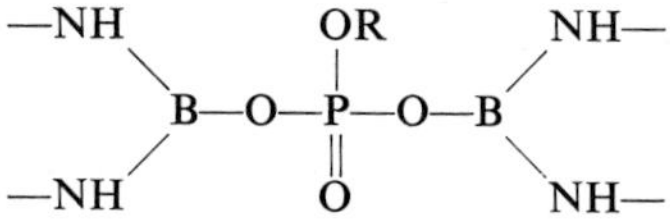

$$\begin{array}{ccc} \text{—NH} & \text{OR} & \text{NH—} \\ \diagdown & | & \diagup \\ \text{B—O—P—O—B} & & \\ \diagup & \| & \diagdown \\ \text{—NH} & \text{O} & \text{NH—} \end{array}$$

5

B-phosphatoborazole, 6 (15,19), on heating under reduced pressure at 200°C yielded trialkyl phosphate and an insoluble resin. This polymeric

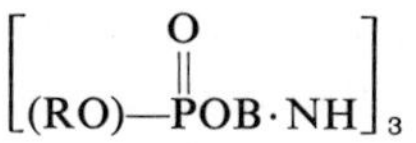

$$\left[(RO)\text{—}\overset{\overset{\displaystyle O}{\|}}{P}OB\cdot NH \right]_3$$

6

material was slowly hydrolyzed with water and, on heating, decomposed at temperatures in excess of 400°C to a hard, brittle, substantially inorganic product (15).

7. Polymers having Boron, Nitrogen, and Carbon in the Main Chain

In this class of compounds were the polymers derived from diamines and boron trichloride or diboronic acids. Of particular interest were those derived from (*1*) the reaction of phenylene diamines with boron trichloride (20-23), **7**, and (*2*) bis(*o*-phenylenediamines) with diboronic acid, (HO$_2$)B—X—B(OH)$_2$ (21,24-26) **8**. Although a number of the polyboro-

$$\begin{bmatrix} & C_6H_4 & \\ & | & \\ & N & \\ Cl-B & & B-Cl \\ & & \\ -N & & N-C_6H_4- \\ & B & \\ & | & \\ & Cl & \end{bmatrix}_x$$

7

$$\begin{bmatrix} H & & & H \\ | & & & | \\ N & & & N \\ -B & \langle\text{benzo}\rangle-\langle\text{benzo}\rangle & B-R- \\ N & & & N \\ | & & & | \\ H & & & H \end{bmatrix}_x$$

8

amides of the type **8** have been prepared and some have been reported to be stable at 300°C or above, all were readily hydrolyzed by room temperature water. However, the boron analog of the aromatic polybenzimidazoles, the polybenzoborimidazoline, has been prepared and the thermal behavior of this polymer studied (16) by thermogravimetry (Fig. 2). A comparison of this curve with that of the corresponding aromatic polybenzimidazole clearly showed that significant weight loss for the boron containing polymer occurred at a temperature at least 100°C lower.

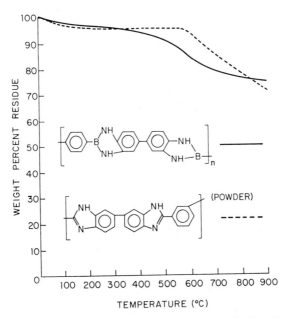

Fig. 2. TGA plot of polybenzimidazole and polybenzoborimidazoline in nitrogen (heating rate, 3°C/min) (16).

8. Polymers Derived from Isocyanates

The reaction of boronic acids with organic diiso- and diisothiocyanates at temperatures between 35 and 200°C has been reported to yield poly(boroureas) with loss of carbon dioxide (27). Subsequent work has shown that the product was not a poly(borourea) but a boron-free polymer (28).

9. Polymers containing Boron, Oxygen, and Carbon in the Main Polymer Chain

In this section, the polymers which will be considered are: polyesters from boric acid, polyesters from boronic acid, and polymers derived from hexaalkylborozole with bisphenols.

a. Polyesters Derived from Boric Acid

Although there were numerous references (18,29–36) in the literature to polymers derived from boric acid and its anhydride with glycols to yield products ranging from viscous liquids to vitreous solids, all of these products were easily hydrolyzed and decomposed at elevated temperatures to crosslinked structures.

b. Polymers from Boronic Acid and Dihydroxy Compounds

As in the case of boric acid, polymers derived from boronic acid have been the subject of detailed studies (37–39). In general, the polyesters from short-chain aliphatic boronic acids were usually readily oxidized in air whereas those from the aromatic acids were more stable. Most of the polyesters prepared from boronic acids with aliphatic hydroxy compounds were readily hydrolyzed on exposure to air.

The oxygen analogs of polybenzoborimidazoline have been prepared from p-phenylene diboronic acid and bis[o-diphenols]. Polymers, **9**,

9

showed a behavior in thermogravimetric analysis similar to that of the polybenzoborimidazoline, being less thermally stable than the corresponding polybenzimidazole (40).

c. Polymers Derived from Hexaalkylborazoles and Bisphenols

The polymers in this classification which have been most extensively studied were those derived from hexamethylborazole and resorcinol, **10**, hydroquinone, **11**, and 2,2-bis-(4-hydroxyphenyl)propane, **12**.

$$(\text{Me}-\text{B}-\text{Me})_3 + \text{HO}-\text{Ar}-\text{OH} \longrightarrow \left[\begin{array}{c} \text{Me} \\ | \\ -\text{B}-\text{O}-\text{Ar}-\text{O}- \end{array} \right]_x \quad (5)$$

where

Ar = (*m*-phenylene) **10**

= (*p*-phenylene) **11**

= -C₆H₄-C(CH₃)₂-C₆H₄- **12**

The polymer, **10**, which was an excellent adhesive to glass and metals, softened at 200°C and showed only a 7% weight loss in thermogravimetric analysis on heating to 500°C. The related hydroquinone derivative (**2**) was somewhat higher melting, melting between 350 and 400°C, and showed similar thermal stability by thermogravimetry. The related compounds derived from the bisphenol, **3**, melted at 250°C and exhibited 22% weight loss when heated to 500°C (40).

Similar polymers have been prepared from bisphenols and alkyl or aryl di-*n*-alkoxyboranes. Surprisingly, the phenyl-substituted polymers were reported to be less stable than the alkyl-substituted polymers (40).

10. *p*-Vinylphenylboronic Acid and Related Polymers

These polymers which contain boron in the polymer side chain were derived from *p*-vinylphenylboronic acid and its derivatives. They did not appear to be significantly different from other styrene derivatives (28,42–50) in thermal stability. Although they did not melt below 300°C, in air, such polymers show obvious darkening and decomposition well below this temperature.

11. Polymers Containing Boron–Phosphorus Bonds

Although polymers of this type have been the subject of extensive investigation, in only a limited number of cases have well-characterized high molecular weight polymers been obtained.

Linear borophanes, phosphinoborane polymers, have been prepared in high molecular weight via the pyrolysis of methyl-*n*-alkylphosphine boranes with 20 mole % of triethylamine (51). Similarly, the polymeric phosphinoboranes have been reported from the reaction of pentaborane

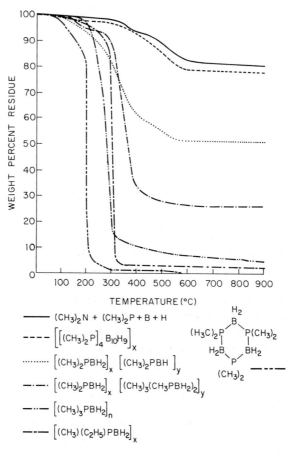

Fig. 3. TGA plot of phosphorus–boron polymers in nitrogen (heating rate, 3°C/min) (16).

with (1) dimethylaminodimethylphosphine (52), (2) tetramethylphosphine (53), and (3) aminodiphosphine (54). All of these phosphinoborane polymers, especially the poly p-dialkylborophanes, decomposed at temperatures in excess of 200°C to the p-hexaalkyltriborophane (51). All evidence suggested that these polymers depolymerized by an unzipping of monomer units which recombined to form the thermodynamically preferred cyclic products.

Similar thermal stability has been reported based on the behavior of these boron–phosphorus polymers in thermogravimetric analysis (16). As shown in Fig. 3, it is apparent that these phosphorus–boron polymers were of the same order of stability as the boron–nitrogen polymers and were considerably less thermally stable than the polyaromatic heterocyclics discussed in Chapter IV.

B. PHOSPHORUS-CONTAINING POLYMERS

1. Phosphorus–Nitrogen Polymers

a. *Phosphonitriles*

The current interest in phosphorus–nitrogen polymers arose from the early observations that trimeric phosphonitrilic chloride could be polymerized at temperatures above 250°C to a rubberlike solid which was stable to 350°C. Unfortunately, these polymeric phosphonitrilic compounds were hydrolytically unstable. Major efforts, therefore, have been exerted to prepare phosphorus–nitrogen polymers in which the bonding system could be characterized by the repeating unit, $-PR_2\!\!=\!\!N-$ (55).

Polyphosphonitriles have been prepared by the reaction of PCl_5 or substituted phosphorus chloride with ammonia or amines and ammonium chloride (56,57).

$$PCl_5 + NH_4Cl + NH_3 \longrightarrow (-NPCl_2-)_x \tag{6}$$

From this synthesis, depending on temperature, ratio of reactants, reaction times, etc., a wide range of products has been obtained. The proposed pathway for this reaction (58) is outlined in reaction 7.

(7)

The rubberlike solids which presumably were high molecular weight polymers were swollen by solvents and underwent reactions such as alcohololysis and hydrolysis. Although little is known of the nature of these polymers, they were presumed to be crosslinked, and at temperatures over 350°C, cyclic monomers were regenerated from the polymers (59). When amino or anilido derivatives were used, modified polyphosphonitriles were obtained with some decrease in thermal stability (61).

Recently, the preparation of a large variety of symmetrically substituted polyphosphonitriles (60–62) has been achieved by an extension of the classic synthesis in the following manner:

$$R_2PCl_3 + NH_3 \text{ and/or } NH_4Cl \longrightarrow -PR_2=N- \qquad (8)$$

Similarly, mixed symmetrically substituted organochlorophosphonitriles have been prepared by this reaction (63–65):

$$RPCl_4 + NH_4Cl \longrightarrow -PR(Cl)=N- \qquad (9)$$

A different approach to the synthesis of phosphonitriles polymers involved the reaction of an organodiphosphine with an organic bisazide (66) (see reaction 10). All of these modified phosphonitriles showed considerably improved hydrolytic stability (60–66).

$$\begin{array}{c} N_3-R_1-N_3 + R_2-P-R_3-P-R_2 \\ \downarrow \\ [-R_1-N=PR_2-R_3-R_2P=N-]_x \end{array} \qquad (10)$$

The only detailed study of the thermal stability of these polymers was that carried out by Ehlers (16) who studied the behavior of polyphosphonitriles derived from diamines by thermogravimetric analysis (Fig. 4). Inspection of these data clearly showed that these particular polymers had rapid and continuing weight loss at onset temperatures of less than 200°C.

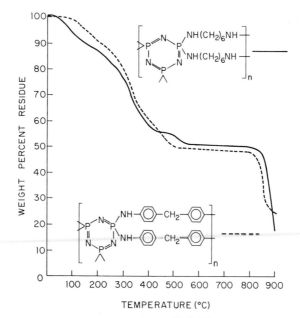

Fig. 4. TGA plot of phosphorus–nitrogen polymers in nitrogen (heating rate, 3°C/min) (16).

Thus it would appear that the upper thermal stability limit of 350°C was achieved only with the phosphonitrilic chloride. Any modifications of this basic polymer structure improved hydrolytic stability with a concomitant loss in thermal stability.

b. Polymers with Phosphorus in a Lower Oxidation State

Related to the phosphonitrile polymers are these polymers in which phosphorus is in a lower oxidation state. When dialkylammonium phosphorus dichlorides were reacted with ammonia, polymeric substances such as those shown in structures **13** and **14** were produced (67). All

$$\left[\begin{array}{c} H \\ | \\ -N-P- \\ | \\ N-R \\ | \\ R \end{array} \right]_x \qquad \left[\begin{array}{c} H \\ | \\ -N-P^- \\ | \\ N^+-R \\ | \\ R \end{array} \right]_x$$

$$\quad\quad\quad 13 \quad\quad\quad\quad\quad\quad\quad\quad 14$$

substances containing phosphorus in the oxidation state of three were unstable, decomposing in moist air, releasing phosphine (67).

2. Phosphoryl Amides and Derivatives

The main structural unit, **15**, in polyphosphate which will not be discussed here is shown below. In addition, in the polyphosphimates there is an additional unit, **16**, also shown below. Polymers containing phos-

$$\left[\begin{array}{c} O \\ \| \\ -P-O- \\ | \\ R \end{array} \right] \qquad \text{and} \qquad \left[\begin{array}{c} O \\ \| \\ -P-NH- \\ | \\ R \end{array} \right]$$

$$\quad\quad\quad 15 \quad\quad\quad\quad\quad\quad\quad\quad 16$$

phimate units have been prepared by heating the triamide of phosphoric acid (see reaction 11).

$$H_2N-\underset{\underset{NH_2}{|}}{\overset{\overset{O}{\|}}{P}}-NH_2 \longrightarrow H_2N-\underset{\underset{NH_2}{|}}{\overset{\overset{O}{\|}}{P}}-NH-\underset{\underset{NH_2}{|}}{\overset{\overset{O}{\|}}{P}}-NH_2 \longrightarrow H_2N-\left[\underset{\underset{NH_2}{|}}{\overset{\overset{O}{\|}}{P}}-NH\right]_x-H \quad (11)$$

These products on heating at temperatures up to 600°C became insoluble, and the resulting polymer which was believed to have the PON structure was obtained (68,69).

$$H_2N-\left[\underset{\underset{NH_2}{|}}{\overset{\overset{O}{\|}}{P}}-NH\right]_x \longrightarrow \left[\underset{\underset{O}{|}}{\overset{|}{P}}-N\right]_x \quad \text{and} \quad \left[\underset{\underset{O}{|}}{\overset{|}{P}}=N\right]_x \quad (12)$$

3. Phosphonic Acids, Phosphonic Amides, and Derivatives

Polymer structures having the repeating units shown below have been prepared via the self-condensation of phenylphosphonic diamides at temperatures of 200–300°C (70) (see reaction 13). Similar polymer

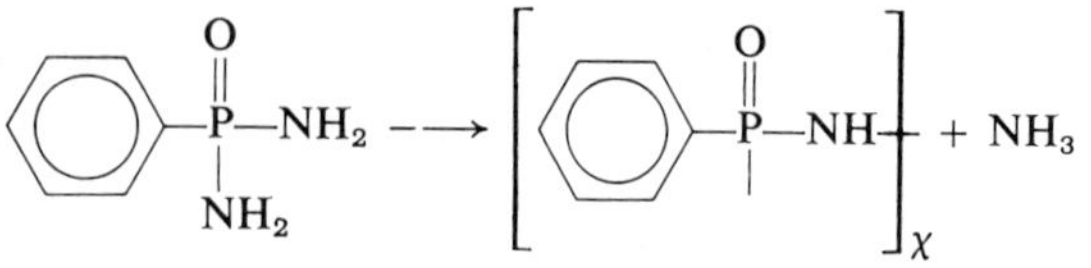

$$\text{Ph-P(=O)(NH}_2\text{)NH}_2 \longrightarrow \left[\text{Ph-P(=O)(NH}_2\text{)-NH} \right]_x + NH_3 \quad (13)$$

structures have been obtained via the reaction of phenylphosphinic dichloride with diamines (72) and urea (71) (see reaction 14). Similarly, the

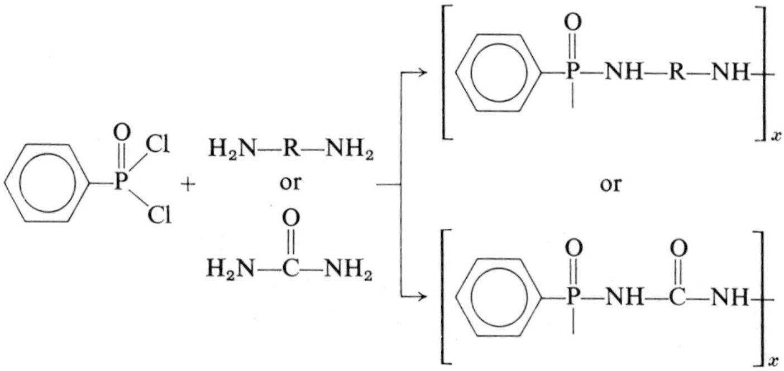

$$\text{Ph-P(=O)Cl}_2 + \begin{array}{c} H_2N-R-NH_2 \\ \text{or} \\ H_2N-C(=O)-NH_2 \end{array} \longrightarrow \begin{array}{c} \left[\text{Ph-P(=O)(NH}_2\text{)-NH-R-NH} \right]_x \\ \text{or} \\ \left[\text{Ph-P(=O)(NH}_2\text{)-NH-C(=O)-NH} \right]_x \end{array} \quad (14)$$

replacement of the imido group by oxygen led to structures such as **17** (74).

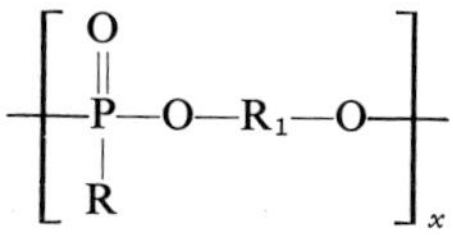

$$\left[-P(=O)(R)-O-R_1-O- \right]_x$$
17

All of the above structures had extremely low melting points and were susceptible to hydrolysis (72).

Dichlorophosphorus compounds and diimidazolyl phosphorus compounds have been reacted with suitable diamines, in some cases aromatic diamines, to yield high molecular weight polymers (73) (see Table I and reaction 15). In all cases the resultant phosphorus-containing polymer melted at least 100°C lower than the corresponding carboxylic polyamide (73).

The thermal stability of this type of polymer as determined by thermogravimetric analysis (16) (Fig. 5) was considerably poorer than that of the carbon analog showing significant weight loss at temperatures of 400°C or less.

TABLE I
Polyamides Containing Phosphorus (73)

Phosphorus compound	Diamine	η_{inh} (dl/g)[a]	PMT (°C)[b]	Decomp. pt. (°C)[c]
$C_6H_5P(O)Cl_2$	$NH_2(CH_2)_6NH_2$	0.14	90–100	300
$C_6H_5P(O)Cl_2$	$NH_2(CH_2)_6NH_2$	0.05		
	$C_4H_9NH(CH_2)_6NH_2$	0.04	75	320
	$NH_2(CH_2)_3NH_2$	0.03	60	260
	$NH_2(CH_2)_3O(CH_2)_3NH_2$	0.10	60	260
	$NH_2(CH_2)_3NH(CH_2)_3NH_2$	0.03	70	295
	$NH_2(CH_2)_2NH(CH_2)_2NH_2$		100	285
	$NH_2CH_2C_6H_4CH_2NH_2$	0.06	120	270
	$NH_2C_2H_4$–(tetramethylphenyl)–$C_2H_4NH_2$	0.15	210	320
	bis(o-aminoanil) of biphenyl-based dialdehyde (HC=N/HN– biphenyl –NH/N=CH)		280	275
$C_6H_5P(S)Cl_2$	$NH_2(CH_2)_6NH_2$	0.05	80	225
$C_6H_5OP(O)Cl_2$	$NH_2(CH_2)_6NH_2$	0.17	70	325
	$NH_2C_2H_4$–(tetramethylphenyl)–$C_2H_4NH_2$	0.21	170	325
$C_6H_5N(CH_3)P(O)Cl_2$	bis(o-aminoanil) of biphenyl-based dialdehyde	0.02	>440	430–450
$(C_6H_5)_2NP(O)Cl_2$	bis(o-aminoanil) of biphenyl-based dialdehyde		>430	430

(continued)

TABLE I (continued)
Polyamides Containing Phosphorus (73)

Phosphorus compound	Diamine	η_{inh} (dl/g)^a	PMT (°C)^b	Decomp. pt. (°C)^c
$(C_6H_5)_2NPCl_2$	HC=N–C$_6$H$_4$–C$_6$H$_4$–N=CH, HN–...–NH		225	320
$C_6H_5P(O)Im_2$	$NH_2(CH_2)_6NH_2$	0.08	80–85	250
$C_6H_5P(O)Im_2$	$NH_2C_6H_4NH_2$	0.05	245–255	350
	$NH_2C_6H_4CH_2C_6H_4NH_2$	0.07	185–190	370–390
	$NH_2C_6H_4C_6H_4NH_2$		210–215	330
	piperazine (HN(CH$_2$CH$_2$)$_2$NH)	0.08	185–190	210

^{a} η_{inh}: inherent viscosity determined at 0.5 wt % concentrations in conc. sulfuric acid at 30°C.

^{b} PMT: polymer melt temperature. See R. G. Beaman and F. B. Cramer, *J. Polymer Sci.*, **21**, 223 (1956).

^{c} Decomp. pt.: decomposition temperature determined by differential thermal analysis.

$$H_2N-R-NH_2 + \begin{array}{c} R_1-\overset{O}{\underset{\|}{P}}-Cl_2 \\ \text{or} \\ R_1-\overset{O}{\underset{\|}{P}}\left[N\begin{array}{c}HC=N\\ \| \\ HC=CH\end{array}\right]_2 \end{array} \longrightarrow \left[\begin{array}{c}O\\ \|\\ P-NH-R-NH\\ |\\ R_1\end{array}\right]_x \quad (15)$$

C. SILICON-CONTAINING POLYMERS

1. Silicones or Polysiloxanes

Since silicone chemistry and technology have been described in numerous books, in this discussion only some of the more recent advances in this field related to thermal stability will be considered. The commercially

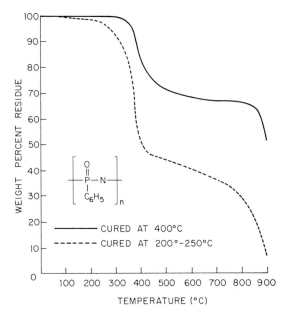

Fig. 5. TGA plot of phosphorus–nitrogen polymer in nitrogen (heating rate, 3°C/ min) (16).

available silicones, usually alkyl- or phenyl-substituted polysiloxanes, have five properties which have made them applicable in a large number of uses. These are:

1. They possess heat stability and maintained their physical properties over a wide range of temperatures (-70 to 250 or even 300°C).

2. They are resistant to oxidation and are inert to chemical attack.

3. They exhibit water repellency. This is due in part to their ability to bond themselves to a wide range of materials so that the water shedding organic groups are exposed.

4. They have excellent dielectric properties.

5. They exhibit certain specific surface-active effects that give rise to applications such as antifoaming.

Only those properties pertaining to heat stability and resistance to oxidation which have been extensively studied (74–78) will be discussed. In the absence of oxygen and at temperatures greater than 350°C, linear polydimethylsiloxanes readily yielded volatile cyclic products. The analysis of both the starting materials and the by-products corresponded to $[(CH_3)_2SiO]$. In addition, the absence of charred coloration from the products (3–9) suggested that the Si—C and C—H linkages were stable at this temperature. Branched-chain polysiloxanes indicated no Si—C cleavage at temperatures in excess of 400°C, while tetramethylsilane was stable in the vapor phase above 600°C. Thus, in the absence of oxygen, the degradation occurred by cleavage of the Si—O bond.

The thermal stability of tetraphenylsilane in the presence of oxygen has been known for some time. A comparison of the stability of polydimethylsiloxane and polyphenylmethylsiloxane under oxidation conditions confirmed the greater stability of the silicon–phenyl bond relative to that of the silicon–methyl bond (3-9). Silicon–methyl cleavage occurred with subsequent crosslinking of the polymer, under conditions in which no phenyl cleavage occurred. Thus, it would appear that:

1. The oxidative degradation reaction occurred at a much lower temperature than the polymer rearrangement reaction.

2. The polyphenylsiloxanes were more stable to oxidative degradation than the polymethylsiloxanes.

Various heat stability additives have been suggested for extending the life of the silicones at 250°C under oxidative conditions. Compounds of iron (79), phosphorus (80), tin (79), and cerium (81) have all proved to be effective when used with phenylmethylsiloxane, the effectiveness decreasing with increasing phenyl content.

Carbon-functional silicones have been utilized in the preparation of polymer and copolymers. Such polymeric compositions were prepared in order to obtain products with improved thermal stability. The ≡Si—H group has been useful in the preparation of ≡Si—Cl, ≡Si—OH, ≡Si—R, ≡Si—OR, etc. which have been used in the preparation of "carbon-functional" silicones (82). In every case, no improvement in the thermal stability of the silicone was achieved, and if anything, these polymers and copolymers were less stable than the unmodified silicones (82).

Of the many attempts to improve the thermal durability of the silicone polymers recent attempts have involved the incorporation of a highly stable aromatic ring in the backbone of the chain. Such aromatic polysiloxanes have been prepared from the reaction of bisphenols with suitable silicon derivatives (83) (see reaction 16).

$$\begin{matrix} R_2Si(OR)_2 \\ \text{or} \\ R_2SiCl_2 \end{matrix} + HO-Ar-OH \longrightarrow \left[\begin{matrix} R \\ | \\ Si-O-Ar-O \\ | \\ R \end{matrix} \right] \quad (16)$$

From the reaction of bisphenols with bis(anilino) diphenylsilane (84) (reaction 17) the polymers listed in Table II, some of which have extremely high polymer melt temperatures, were prepared.

$$\phi\text{—NH—}\underset{\underset{\phi}{|}}{\overset{\overset{\phi}{|}}{\text{Si}}}\text{—NH—}\phi + \text{HO—Ar—OH} \longrightarrow \left[\underset{\underset{\phi}{|}}{\overset{\overset{\phi}{|}}{\text{Si}}}\text{—O—Ar—O}\right] + \phi\text{—NH}_2 \quad (17)$$

The thermal stability of these polymers, as shown by thermogravimetric analysis under nitrogen (Figs. 6 and 7), was superior to that of the usual

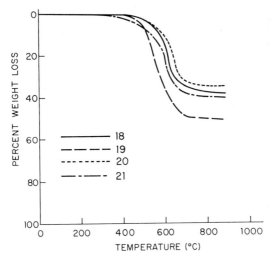

Fig. 6. TGA plot of polymers **18, 19, 20,** and **21** from Table II (heating rate, 6°C/ min) (84).

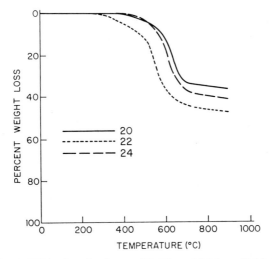

Fig. 7. TGA plot of polymers **20, 22,** and **24** from Table II (heating rate, 6°C/min) (84).

TABLE II
Silicon Polymers of Aryl Diols (84)

Polymer	Polymer repeating unit	Description of polymer[a]	PMT (°C)[b]
18	[−Si(C₆H₅)(C₆H₅)−O−C₆H₄−O−]ₙ	Hard, slightly brittle, amber solid	>300
19	[−Si(C₆H₅)(C₆H₅)−O−C₆H₄−O−]ₙ (ortho)	Hard, slightly brittle, amber solid	240–245
20	[−Si(C₆H₅)(C₆H₅)−O−C₆H₄−C₆H₄−O−]ₙ	Hard, tough, semiflexible light amber solid	>300
21	[−Si(C₆H₅)(C₆H₅)−O−naphthyl−O−]ₙ	Hard, brittle dark brown solid	123–125

V. INORGANIC POLYMERS

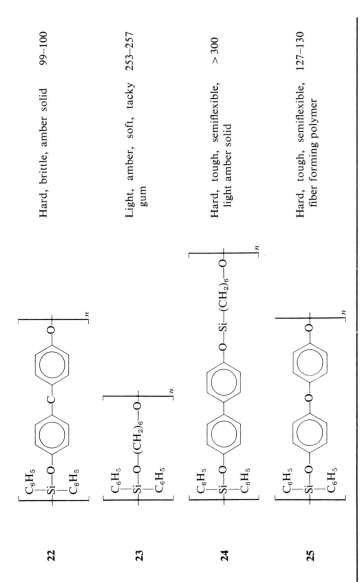

22 Hard, brittle, amber solid 99–100

23 Light, amber, soft, tacky gum 253–257

24 Hard, tough, semiflexible, light amber solid >300

25 Hard, tough, semiflexible, fiber forming polymer 127–130

[a] All polymers with the exception of 24 were soluble in N,N-dimethylformamide, tetrahydrofuran, and dimethyl sulfoxide.
[b] PMT: polymer melt temperature. See R. G. Beaman and F. B. Cramer, J. Polymer Sc., 21, 223 (1956).

siloxane polymer (84). In fact, polymer **20** showed only a 10% weight loss even at temperatures as high as 600°C. Interestingly enough these polymers were soluble in tetrahydrofuran, N,N-dimethylformamide, and dimethyl sulfoxide.

2. Modified Silicones or Polymetallosiloxanes

Although the silicones have satisfactory thermal life at 250°C, at very high temperatures (350°C) these polymers undergo extensive rearrangement to form low molecular weight cyclic products. Since the stability of the Si—O linkage largely determines the resistance of silicone polymers to thermal degradation, the thermal stability of these polymers should be improved by modifying the electronic character of this bond. It is generally recognized that as bonds become more ionic or as covalent bonds become more polar, greater thermal stability ensues. Replacement of some (or all) of the Si—O linkages in the polymer chain with M—O, where M represents a metal, would, in those cases where M is more electropositive than silicon, provide a more polar bond, and consequently, a more ionic polymer, which would be expected to have greater thermal stability. Again, it is well known that the intermolecular forces are particularly weak in the organopolysiloxanes. A higher coordination tendency produced by the introduction of a suitable metal atom into the chain would result in greater intermolecular interaction, and this might also reduce the propensity to degrade by cyclic-compound formation. These modifications are likely to lead to higher second-order transition temperatures in the polymer, but at the same time, some loss of mechanical properties is inevitable owing to the increased stiffness of the chains. Current research interest in the polymetallosiloxanes as thermally stable materials stems largely from these considerations.

a. Silicon–Oxygen–Aluminum Polymers

These polymers have been prepared via the controlled hydrolysis of tris(trialkylsiloxy)aluminums (85–89) as outlined in reaction 18. Similarly,

$$[(R)_3\text{—SiO}]_3\text{Al} + H_2O \longrightarrow (R_3)\text{SiO—Al—O—Al—O—Al—OSi—}(R)_3 \quad (18)$$
$$\underset{\text{OSi}(R)_3}{|}$$

the synthesis of polyorganometallosiloxanes has been effected via (*1*) the cohydrolysis in alkaline medium of salts or alkoxy derivatives of aluminum and the alkyl- or arylhalosilanes (88–91), and (*2*) the double decomposition of monosodium derivatives of alkyl- or arylsilanetriols with aluminum sulfate (88–91).

(*1*) $\quad RSiCl_3 + NaOH + Al_2SO_4 \longrightarrow [RSi(OH)_2O]_3Al + Na_2SO_4 + NaCl$

(*2*) $\quad RSi(OH)_2ONa + Al_2SO_4 \longrightarrow [RSi(OH)_2O]_3Al + NaSO_4$

(19)

Since water was present in both systems, further reactions between the aluminosiloxane and silanetriol yielded products which simultaneously condensed through the hydroxyl groups to give polymers (see reaction 20).

$$R\text{—}Si(OH)_2ONa + H_2O \rightleftarrows RSi(OH)_3 + NaOH$$

$$RSi(OH)_2ONa + RSi(OH)_3 \longrightarrow \quad (20)$$

$$[R\text{—}Si(OH_2O)_2Al\text{—}O\text{—}\underset{OH}{SiR}\text{—}O\text{—}\underset{OH}{SiR}\text{—}OH \xrightarrow{etc.}$$

The polyphenylaluminosiloxane containing four silicon atoms to one aluminum atom has been found to be a colorless, transparent, brittle resin, readily soluble in benzene and other similar organic solvents, but insoluble in petroleum ether. It formed clear, brittle films which retained solubility in toluene even after 100 hr heating at 150°C. Some loss in weight

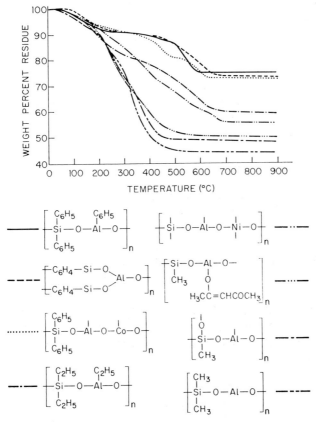

Fig. 8. TGA plot of —Si—O—Al—O— polymers in nitrogen (heating rate, 3°C/ min) (16).

occurred due to condensation involving free hydroxyl groups, and at higher temperatures, 200–500°C, the polymer gradually lost its solubility in organic solvents. Short heat treatments at 400°C rendered the materials completely insoluble. This phenomenon was undoubtedly associated with crosslinking rearrangements occurring in the chains with cleavage of siloxane and aluminosiloxane linkages and their conversion into three-dimensional structures. Based on chemical analysis, the repeating unit was that shown in structure **25**. The outstanding characteristics of the poly-

$$\left[\begin{array}{ccccc} & C_6H_5 & C_6H_5 & C_6H_5 & C_6H_5 \\ -O- & Si-O- & Si-O- & Si-O- & Si-O-Al- \\ & | & | & | & | & | \\ & OH & OH & O_{0.5} & O_{0.5} & O_{0.5} \end{array} \right]$$

25

phenylaluminosiloxanes and related polyalkylaluminosiloxanes were their complete infusibility; neither showing any tendency to melt or sinter at temperatures up to 500°C. These polymers were readily soluble in organic solvents, but their melting points were above their decomposition temperatures (88–92).

In spite of the lack of fusibility of this class of polymers, a comparison of the thermal stability of these modified silicones with the unmodified silicones by thermogravimetric analysis (16) (Fig. 8) clearly showed that the unmodified silicones were much more thermally stable.

b. Silicon–Oxygen–Titanium Polymers

Polyorganotitanosiloxanes have been prepared by the cohydrolysis of alkyl- or arylchlorosilanes with butyl titanate (90,93) (see reaction 21).

$$R_2SiCl_2 + Ti(OC_4H_9)_4 \xrightarrow[OH^-]{H_2O} \left[\begin{array}{c} R \\ | \\ -O-Si- \\ | \\ R \end{array} \right]_n \left[\begin{array}{c} OC_4H_9 \\ | \\ -O-Ti-O- \\ | \\ OC_4H_9 \end{array} \right]_m \quad (21)$$

Similar experimental procedures have yielded a whole variety of polymers containing silicon, oxygen, and titanium (93–99). Some of the polyorganotitanosiloxanes which have been prepared are as shown in structures **25**, **26**, and **27**.

$$\left[\begin{array}{ccccc} C_2H_5 & C_2H_5 & C_2H_5 & C_2H_5 & C_2H_5 \\ | & | & | & | & | \\ -O-Si-O-Si-O-Si-O-Si-O-Ti- \\ | & | & | & | & | \\ OH & OH & O_{0.5} & O_{0.5} & O_{0.5} \end{array} \right]_x$$

25

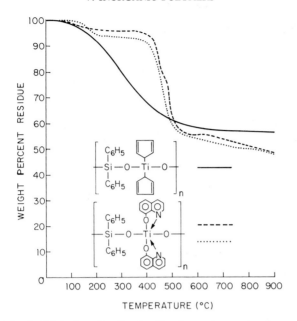

Fig. 9. TGA plot of —Si—O—Ti—O— polymers in nitrogen (heating rate, 3°C/min) (16).

$$\left[\begin{array}{cc} R & OSi(R)_3 \\ | & | \\ -Si-O-Ti-O- \\ | & | \\ R & OSi(R)_3 \end{array} \right]_x$$
26

$$\left[\begin{array}{c} OSi(R)_3 \\ | \\ -Ti-O- \\ | \\ OSi(R)_3 \end{array} \right]_x$$
27

Thermogravimetric analysis of the silicon–oxygen–titanium polymers showed that the unmodified silicon polymers were much more thermally stable (16) (Fig. 9). Moreover, on heating tetrakis(triphenylsiloxy) derivatives of titanium and silicon, it was observed that the former discolored and showed signs of decomposition at 460–470°C, whereas, the latter showed little change even at 605°C (100).

c Silicon–Oxygen–Tin Polymers

The synthesis of polyorganostannosiloxanes has been based on the cohydrolysis of dialkyldichlorostannane with dialkyldichlorosilane (92) (see reaction 22). Modifications of the above procedure were the condensation of (1) organotin oxides with silanols (93,101–103), (2) organotin

$$(R_1)_2SiCl_2 + (R_2)_2\text{—}SnCl_2 \xrightarrow[NH_2]{H_2O} \left[\begin{array}{c} R_1 \\ | \\ -Si\text{—}O- \\ | \\ R_1 \end{array}\right] \left[\begin{array}{c} R_2 \\ | \\ -Sn\text{—}O- \\ | \\ R_2 \end{array}\right]_y \quad (22)$$

chlorides with alkali metal silanolates (101,104), and (3) acyloxysilanes with alkoxytin compounds (105) (see reactions 23–25).

$$R_2SnO + R'_2Si(OH)_2 \longrightarrow \left[\begin{array}{c} R \\ | \\ -Sn\text{—}O- \\ | \\ R \end{array}\right]_x \left[\begin{array}{c} R' \\ | \\ -Si\text{—}O- \\ | \\ R' \end{array}\right]_y \quad (23)$$

$$R_2SnCl_2 + R'_2Si(ONa)_2 \longrightarrow \left[\begin{array}{cc} R & R' \\ | & | \\ -Sn\text{—}O\text{—}Si\text{—}O- \\ | & | \\ R & R' \end{array}\right]_x \quad (24)$$

$$R_2Sn(OCOR)_2 + R'_2\text{—}Si(OR'')_2 \longrightarrow \left[\begin{array}{cc} R & R' \\ | & | \\ -Sn\text{—}O\text{—}Si\text{—}O- \\ | & | \\ R & R' \end{array}\right]_x \quad (25)$$

For the most part, these polymers were of relatively low molecular weight and showed thermal stabilities based on thermogravimetric analysis comparable to or slightly less than that of the corresponding polyorganosiloxane (16) (Fig. 10).

d. Silicon–Oxygen–Arsenic Polymers

Cohydrolysis of arsenic chloride with phenyltrichlorosilane has yielded polymeric materials with the approximate composition (91) (see structure **28**) which were soluble in organic solvents, but neither softened or melted

$$\left[\begin{array}{ccccc} \phi & \phi & \phi & \phi & \phi \\ | & | & | & | & | \\ -O\text{—}Si\text{—}O\text{—}Si\text{—}O\text{—}Si\text{—}O\text{—}Si\text{—}O\text{—}Si\text{—} \\ | & | & | & | & | \\ O & O & O & O & OH \\ | & | & | & | & \\ -O\text{—}Si\text{—}O\text{—}Si\text{—}O\text{—}As\text{—}O\text{—}Si\text{—} \\ | & | & | & & \\ \phi & \phi & \phi & & \end{array}\right]$$

28

on heating. Modifications of this general procedure have yielded a whole variety of silicon–oxygen–arsenic polymers. The materials were analogous to the polyphenylaluminosiloxanes, but have been obtained in only low

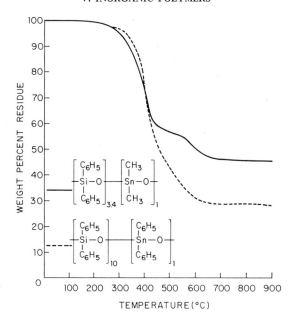

Fig. 10. TGA plot of —Si—O—Sn—O— polymers in nitrogen (heating rate, 3°C/min) (16).

molecular weights. These low molecular weight products by thermogravimetry appeared to be considerably less stable than the unmodified siloxanes (16).

e. Silicon–Oxygen–Boron Polymers

This particular class of polymers has been prepared in numerous ways. The reactions which have been utilized were:

1. organosilicon acetates and alkoxyboron compounds (106)
2. organosilicon alkoxides and acetoxyboron compounds (107)
3. organosilicon alkoxides and boric acid (108)
4. organosilicon halides and boric acid (108)
5. organosilicon hydroxides and boric acid (109)

For the most part, however, the polymers were low melting and, based on thermogravimetry, were less stable than corresponding unmodified polysiloxanes (16).

The polyorganoborosiloxanes have acquired some prominence by virtue of the extraordinary physical properties of "bouncing putty." Based on the patent literature (109–112), bouncing putty appears to possess what seem to be mutually contradictory properties, flow and bounce.

f. Silicon–Oxygen–Phosphorus Polymers

These polymers have been prepared via the reaction of polyfunctional alkoxysilanes with phosphorus pentoxide, phosphoric acid, or phosphoryl chloride (113–119) (see reaction 26).

$$\begin{array}{c}
\mathrm{-O-\underset{R}{\overset{R}{Si}}-OH + HO-\underset{R}{\overset{R}{Si}}-O-} \\
+ \mathrm{P_2O_5} \longrightarrow \\
\mathrm{-O-\underset{R}{\overset{R}{Si}}-OH + HO-\underset{R}{\overset{R}{Si}}-O-}
\end{array}
\quad
\begin{array}{c}
\mathrm{-O-\underset{R}{\overset{R}{Si}}-O-\underset{\underset{O}{|}}{\overset{O}{\overset{\|}{P}}}-O-\underset{R}{\overset{R}{Si}}-O-} \\
\\
\mathrm{-O-\underset{R}{\overset{R}{Si}}-O-\underset{\underset{O}{\|}}{\overset{|}{P}}-O-\underset{R}{\overset{R}{Si}}-O-}
\end{array}
\qquad (26)$$

In many cases, polymers prepared by this and modifications of this procedure were complex three-dimensional structures which were difficult to characterize. In addition, the susceptibility of the silicon–oxygen–phosphorus linkage to hydrolysis has limited interest in this area of research.

g. Silicon–Oxygen–Sulfur Polymers

Polymers of this type have been reported from the reaction of methyltrichlorosilane with sulfuric acid (118) $[(CH_3Si)_2(SO_4)_3]_n$. Little work has been done on the characterization of such polymers since they were so readily hydrolyzed by water (119).

h. Silicon–Oxygen–Antimony Polymers

Polyorganoantimonysiloxanes have been prepared by the alkoxide-acyloxide reaction (120),

$$\mathrm{R-O-\underset{OR}{\overset{OR}{Sb}}-OR + R'CO_2-\underset{R'}{\overset{R'}{Si}}-O_2CR} \longrightarrow \text{Polymer} \qquad (27)$$

The supposed polymers have not been fully characterized, again due to their ease of hydrolysis.

i. Silicon–Oxygen–Germanium Polymers

These polymers were prepared by the cohydrolysis of dimethyldichlorosiloxane with dimethyldichloro- or dimethyldibromogermanium (121–123). On further polymerization in sulfuric acid, a rubberlike, high molecular weight compound similar to polydimethylsiloxane was obtained. The thermal stability of this germanosiloxane at 250–300°C was comparable to that of a standard silicone rubber, but somewhat inferior at higher temperatures (16).

j. Silicon–Oxygen–Lead Polymers

Polymers of this structure were reported to be obtained from the reaction of a methylalkoxysilane with an aqueous solution of sodium plumbite. These polyplumboxanoorganosiloxanes, which were presumed to have the following structure, **29**, have been little characterized, so no indication of

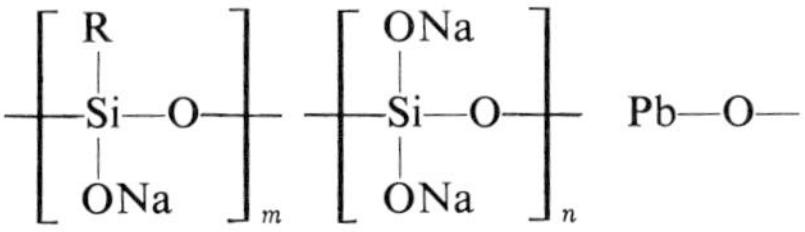

$$\left[\begin{array}{c} R \\ | \\ -Si-O- \\ | \\ ONa \end{array}\right]_m \left[\begin{array}{c} ONa \\ | \\ -Si-O- \\ | \\ ONa \end{array}\right]_n Pb-O-$$
29

their thermal stability can be deduced (123).

D. OTHER ORGANOMETALLIC POLYMERS

1. Aluminum–Oxygen Polymers

A number of aluminum–oxygen polymers have been prepared by various condensation polymerization methods. The simplest aluminum–oxygen polymers were prepared by the partial hydrolysis-condensation reaction as shown in reaction 28, but all such polymers showed poor thermal stability and disproprotionated on heating (40). Related aluminum–oxygen polymers with mixed substituents have been prepared by reorganization and

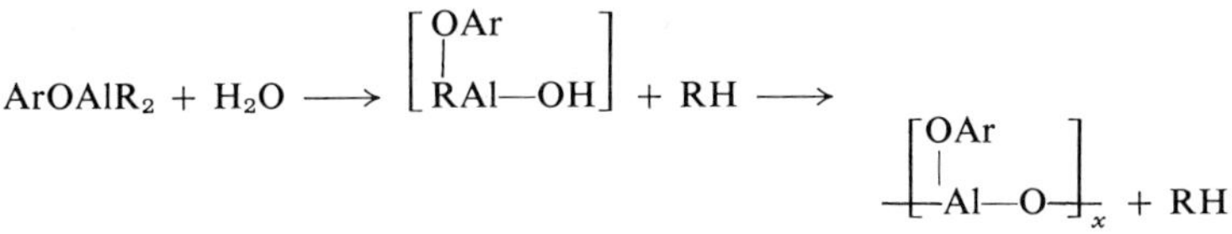

$$ArOAlR_2 + H_2O \longrightarrow \left[\begin{array}{c} OAr \\ | \\ RAl-OH \end{array}\right] + RH \longrightarrow \left[\begin{array}{c} OAr \\ | \\ -Al-O- \end{array}\right]_x + RH \quad (28)$$

$$Al(OR)_3 + Al(OX)_3 + Al(OAr)_3 \longrightarrow \left(\begin{array}{cc} OX & OAr \\ | & | \\ -Al-O-Al-O- \end{array}\right)_x + ROH + \text{olefin} \quad (29)$$

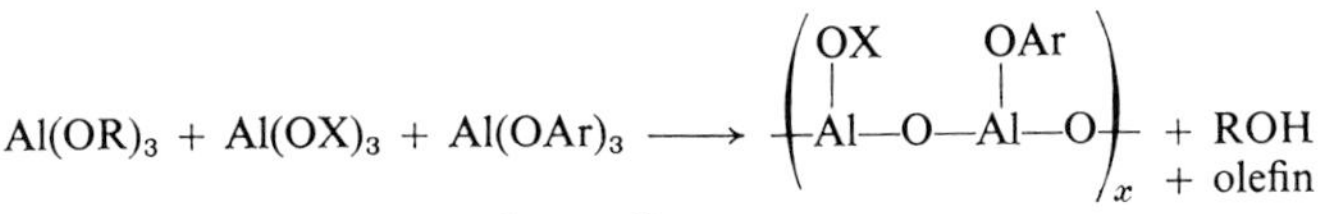

where OX = oxinate,

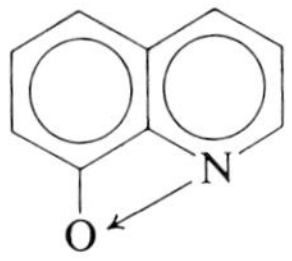

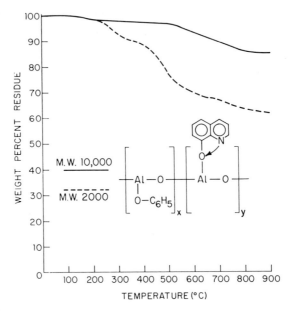

Fig. 11. TGA plot of —Al—O— polymers in nitrogen (heating rate, 3°C/min) (16).

pyrolysis reaction (40) (see reaction 29). These products were more soluble than simple aryloxyaluminoxanes and were prepared in molecular weights of 2000–2500. Thermogravimetric analysis of these polymers showed 29% weight loss at 500°C (40) (Fig. 11).

2. Aluminum–Nitrogen Polymers

Aluminum–nitrogen polymers have been prepared either by (1) the reaction of alkyl aryloxyaluminums with amines (41), or (2) the reaction of aluminum alkoxides with suitable diamines (124) (see reactions 30 and 31).

$$\text{ArOAlR}_2 + \text{ArNH}_2 \longrightarrow \left[\begin{array}{cc} \text{OAr} & \text{Ar} \\ | & | \\ \text{Al} & \text{N} \end{array} \right]_{\bar{x}} \quad (30)$$

$$(\text{RO})_3\text{Al} + \text{RNH}_2 \longrightarrow \left[\begin{array}{cc} \text{OR} & \text{R} \\ | & | \\ \text{Al} & \text{N} \end{array} \right]_{\bar{x}} \quad (31)$$

All of these aluminum–nitrogen polymers decomposed more readily than the corresponding aluminum–oxygen polymers, decomposition becoming apparent at temperatures below 200°C, with major decomposition occurring at 300°C with the formation of modified alumina (124).

3. Germanium–Oxygen Polymers

Polymers of this type have been prepared via the dehydration of organodihydroxygermaniums at elevated temperatures (125) (see reaction 32).

$$\text{HO–Ge(R)(R)–OH} \longrightarrow \left[\text{–Ge(R)(R)–O–} \right]_x + H_2O \qquad (32)$$

Although these polymers did not melt below 300°C, they underwent slow decomposition at that temperature yielding crosslinked materials (125).

4. Tin–Oxygen Polymers

These tin analogs of the silicones, which were formed by the dehydration of organotrihydroxy derivatives of tin, were high melting, crystalline, brittle solids which had the structure of **30** (126,127). The thermal stability

$$HO{\left(\begin{array}{c} R \\ | \\ Sn-O \\ | \\ OH \end{array}\right)}_x$$

30

of the organotin polymers was reported to resemble that of the silicones (126,127).

5. Polymers Containing Group IV Atoms and Carbon Chains

In this broad class of polymers were those derived from the polyaddition of organometallohydrides to carbon–carbon unsaturated compounds (128), or the Wurtz-type condensation involving organometallohalides (128) (see reactions 33 and 34).

$$R_2MH_2 + CH_2=CH-R-CH=CH_2 \longrightarrow \left[\text{–M(R)(R)–CH}_2\text{–CH}_2\text{–R–CH}_2\text{–CH}_2\text{–} \right]_x \qquad (33)$$

$$\text{–M–R–Cl} + \text{Cl–M–} \longrightarrow \left[\text{–M–R–M–} \right]_x \qquad (34)$$

Using this general technique, a large number of high molecular weight polymers containing Group IV atoms have been prepared (128) (Table III).

TABLE III-A
Polymers Containing Group IV Atoms and Carbon Chains (128)

Polymer	R	R′	$\overline{M}_\omega{}^b$
31	Phenyl	—⟨C6H4⟩—	33,000
32	Phenyl	spiro bis-dioxolane phenyl	19,000
33	Phenyl	—⟨C6H4⟩—Ge(φ)(φ)—⟨C6H4⟩—	7,000
34	Propyl	—⟨C6H4⟩—Ge(φ)(φ)—⟨C6H4⟩—	—
35	Phenyl	—⟨C6H4⟩—Sn(φ)(φ)—⟨C6H4⟩—	48,000
36	Propyl	—⟨C6H4⟩—Sn(φ)(φ)—⟨C6H4⟩—	—
37	Butyl	—⟨C6H4⟩—Sn(φ)(φ)—⟨C6H4⟩—	—
38	Phenyl	—⟨C6H4⟩—Pb(φ)(φ)—⟨C6H4⟩—	14,000

(continued)

TABLE III-A (continued)
Polymers Containing Group IV Atoms and Carbon Chains (128)

		Structure[a]		
Polymer	R		R′	$\overline{M}_\omega{}^b$
39	Propyl		$-\bigcirc\!-\!\underset{\underset{\phi}{\mid}}{\overset{\overset{\phi}{\mid}}{Pb}}\!-\!\bigcirc\!-$	—

[a] Structure, general formula:

$$\left[-\underset{\underset{R}{\mid}}{\overset{\overset{R}{\mid}}{Sn}}\!=\!CH_2\!-\!CH_2\!-\!R'\!-\!CH_2\!-\!CH_2\!- \right]$$

[b] $\overline{M}_\omega$: weight average molecular weight by light scattering measurements in chloroform solutions at 30°C.

TABLE III-B
Polymers Containing Group IV Atoms and Carbon Chains (128)

		Structure[a]	
Polymer	R	R′	$\overline{M}_\omega{}^b$
40	Phenyl	—(CH$_2$)$_2$—	75,000
41	Methyl	—(CH$_2$)$_2$—	—
42	Ethyl	—(CH$_2$)$_2$—	—
43	Propyl	—(CH$_2$)$_2$—	50,000
44	Butyl	—(CH$_2$)$_2$—	50,000
45	Phenyl	—(CH$_2$)$_5$—	100,000
46	Butyl	—(CH$_2$)$_5$—	45,000
47	Phenyl	—⟨◯⟩—	65,000

[a] Structure, general formula:

$$\left[-\underset{\underset{R}{\mid}}{\overset{\overset{R}{\mid}}{Sn}}\!-\!CH\!=\!CH\!-\!R'\!-\!CH\!=\!CH\!- \right]$$

[b] $\overline{M}_\omega$: weight average molecular weight by light scattering measurements in chlorofom solutions at 30°C.

TABLE III-C
Poly-*p*-Phenylene Silanes (128)

Polymer	Structure[a]			$\overline{M}_\omega$[b]
	R_1	R_2	R_3	
48	Methyl	Methyl	Methyl	3800
49	Methyl	Phenyl	—	2400
50	Methyl	Methyl	Phenyl	2900
51	Methyl	Phenyl	Phenyl	2600
52	Phenyl	Phenyl	Phenyl	2100

[a] Structure, general formulas:

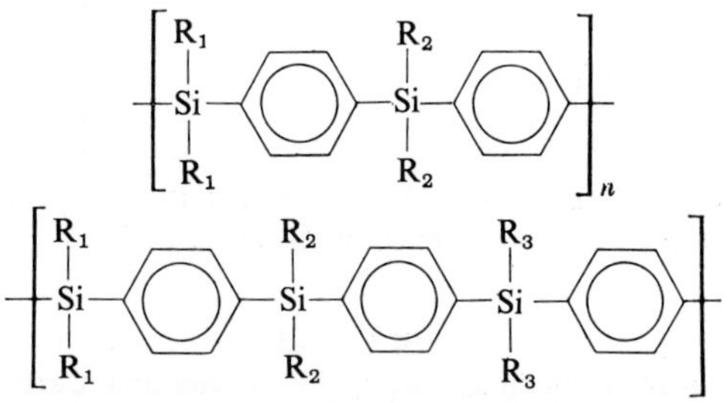

[b] $\overline{M}_\omega$: weight average molecular weight by light scattering measurements on benzene solutions at 30°C.

TABLE IV
Thermogravimetric Analysis of Group IV Metal-Containing Polymers (128)

Polymer[a]	Wt % loss at temperature (°C)[b]					
	300	350	400	450	500	600
31	5	22	50	60	65	72
33	2	7	20	50	60	70
40	10	50	58	68	70	72
48	—	1				
49	2	9	16	23	25	53
50	1	7	15	21	30	47
51	1	5	10	22	44	46
52	—	2	5	7	30	59

[a] See Table II for polymer structures.
[b] Thermogravimetric analysis under nitrogen at heating rate of 3°C/min.

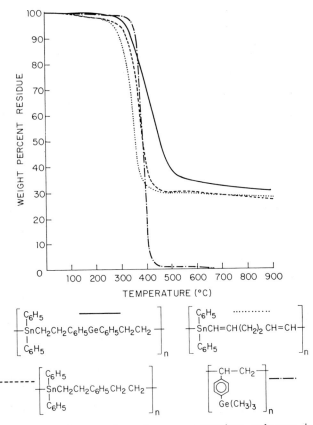

Fig. 12. TGA plot of organotin and organogermanium polymers in nitrogen (heating rate, 3°C/min) (16).

These organometallic polymers, as illustrated in Table IV and Fig. 12, in thermogravimetric analysis showed 50% weight loss at temperatures ranging from 355 to 460°C with 5% weight loss at temperatures as low as 300–350°C (128). Thus, it would appear that the carbon–metal bond was so weak as to preclude extended use of such polymers at temperatures as high as 400°C.

6. Polymers Containing Arsenic and Carbon Chains

This class of polymers has been prepared by the general reaction involving the reaction of a dilithio compound with the corresponding organodihaloarsine as outlined in reaction 35 (129). Using this general procedure, a large number of this type of polymer has been prepared as shown in Table V. Thermogravimetric analysis of these polymers as also

TABLE V
Arsenic-Containing Polymers (129)

Polymer	PMR[a] Temp. of initial (°C)	wt loss (°C)	Wt % loss at temperature[b] (°C)							Solvent
			200	300	400	500	600	700	800	
Poly-p-phenylene arsine	400	280	—	1	9	34	39	64	67	Conc. H_2SO_4
Poly-4,4-diphenyloxydiphenylarsine	180	270	—	10	25	61	65	70	74	Chloroform
Poly-m-2-methoxy-5-bromophenylenephenylarsine	300	310	—	—	23	40	49	55	58	
Poly-m-4,6-dimethoxyphenylenephenylarsine	220	250	—	12	28	40	49	53	56	
Poly-9,10-anthracenephenylarsine	270	100	3	14	56	73	75	77	77	
Poly-p-phenylene-9,10-arsanthrene	280	80	6	17	71	80	83	85	87	
Poly-p-phenylene-o-phenylenediarsenous acid anhydride	280	80	2	14	46	66	70	76		Benzene or chloroform
Poly-2,5-thiophenphenylarsine	90	230	—	13	65	83	85	86		
Poly-p-phenylenearsenic acid	400	Visual evidence of decomposition above 280°C								
Poly-4,4-diphenylphenylarsine	400	60	5	8	12	17	33	43	47	Aqueous alkali

[a] PMT: polymer melt temperature. See R. G. Beaman and F. B. Cramer, *J. Polymer Sci.*, **21**, 223 (1956).
[b] Thermogravimetric analysis under nitrogen at heating rate of 3°C/min.

$$\text{Li-R-Li} + \text{X-}\underset{\underset{\text{R}'}{|}}{\text{As}}\text{-X} \longrightarrow -\left[-\text{R-}\underset{\underset{\text{R}'}{|}}{\text{As}}-\right]_x- \quad (35)$$

shown in this table clearly demonstrated that the upper temperature limit was in the 300–400°C range.

E. METAL CHELATE POLYMERS

As stated earlier the preparative methods for such polymers may be classified in the following way:

1. The linking of polydentate ligands by metal ions.
2. The formation of polymer with simultaneous incorporation of metal ions.
3. The incorporation of metal ions into existing polymers.
4. The formation of polymer by reactions with chelates containing functional groups.
5. The preparation of polyferrocenes.

1. Linking of Ligands with Ions

The thermal stability of metal acetylacetonates has made them the obvious subject of investigation, and various workers (130,131) have studied the possibility of making polymers such as those illustrated in structures 53 and 54. The preparation of such polymers has been attempted

by (*1*) treatment of an alkaline solution of bis-β-ketone with metal ions, (*2*) the reaction of the ketone, metal salt, and excess urea, and (*3*) melt polymerization. Using these techniques, polymers with molecular weights of 4000 which yielded brittle fibers and films have been prepared (130,131). Most of these appeared to be less thermally stable than some of the previously discussed organic polymers. This is amply demonstrated by the data listed in Table VI for the sebacoyldiacetophenone metal chelate polymers (132).

Recent work has shown that beryllium chelates were thermally degraded

TABLE VI
Sebacoyl Diacetophenone–Metal Polymers (132)

Metal	Color	PMT[a] (°C)	Approximate[b] wt loss % in 3 hr at 300°C in air
Beryllium	Brown	~70	20
Aluminum	Yellow Orange	230–250	26
Copper	Brown	~100	16
Nickel	Green	230–250	43
Cobalt	Black	~230	20
Zinc	Brown	150	16
Iron	Black	300	30

[a] PMT: polymer melt temperature. See R. G. Beaman and F. B. Cramer, *J. Polymer Sci.*, **21**, 223 (1956).
[b] Static weight loss measurement.

at 150–200°C *in vacuo* to give cyclic monomers or alternatively dimers (133). Although many different chelating agents have been utilized (134) (see Table VII), all chelate polymers of this type in thermogravimetric analysis showed marked decomposition at temperatures of 300–350°C (16) (Fig. 13).

Metal-containing polymers have been obtained by heating *in vacuo* equimolar quantities of 4,4-bis(α-thiopicolinamido)biphenyl and metal acetylacetonate (135). These were colored, insoluble polymers containing

55

copper, zinc, or nickel. The most thermally stable withstood heating in air up to 380–390°C and decomposed over 400°C.

The reaction of metal acetylacetonates with bis(8-hydroxyquinolines) in *N,N*-dimethylformamide has yielded low molecular weight products with the structure **56** (136).

56

TABLE VII
Bis-Chelating Groups Used in Polymer Formation (134)

Bis(β-diketones)	$RCOCH_2CO-X-COCH_2COR$
Bis(α-amino acids)	$(RCO)_2CH-X-CH(COR)_2$
Bis(salicylaldehyde-diimines)	$HOOC-CH(NH_2)-X-CH(NH_2)-COOH$
Bis(8-hydroxyquinolines)	(2-HO-C$_6$H$_4$)-CH=N-X-N=CH-(C$_6$H$_4$-OH-2)
Bis(dithiocarbamates)	8-hydroxyquinoline—X—8-hydroxyquinoline (HO and OH on rings)
Bis(xanthates)	$(HS)(S=)C-N(R)-X-N(R)-C(=S)(SH)$
Bis(o-hydroxyaldehydes)	$(HS)(S=)C-O-X-O-C(=S)(SH)$
Bis(nitrosophenols)	4,6-diformylresorcinol-type and bis(hydroxy-formyl-phenyl)-X; also 1,4-dihydroxy-2,5-dinitrosobenzene
Bis(glyoximes)	$RS(:NOH)C(:NOH)-X-C(:NOH)C(:NOH)R$

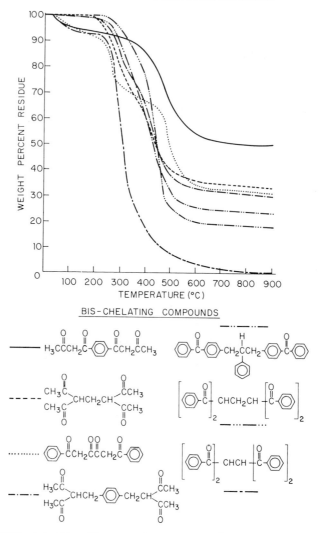

Fig. 13. TGA plot of beryllium chelates in nitrogen (heating rate, 3°C/min) (16).

The Zn- and Cu-chelate polymers showed marked decomposition at temperatures of 240–300°C (136).

Similar metal chelate polymers have been prepared from metal acetylacetonates with the bis-bifunctional Schiff base ligand, 5,5′[p-phenylenebis(methylidynenitrilo)]di-8-quinolinol (137).

The thermal stability of these polymers in which M is manganese, cobalt, nickel, copper, or zinc, has been studied by thermogravimetric

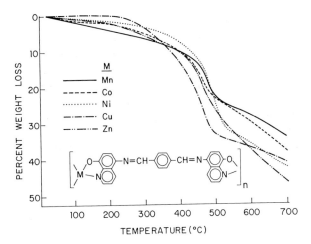

Fig. 14. TGA plot of Schiff base coordination polymers of 5,5'-[p-phenylenebis-(methylidynenitrilo)]di-8-quinolinol (heating rate, 6°C/min) (137).

analysis. An inspection of the data compiled in Fig. 14 clearly show that onset decomposition temperatures occurred in the 350°C range with well over 20% of weight loss by the time 500°C was reached. Considering the percentage of metal ion in these compositions, this represents well over 50% weight loss of the organic portion of the polymer at 500°C and demonstrated that these chelate polymers were considerably less stable than the wholly aromatic heterocyclic polymers discussed in the previous chapter.

57

Attempts to prepare polymers with bisaminophenols (136) and bissalicyclic acids (136) have yielded only low molecular weight products.

Polymers have also been prepared from tetracyanoethylene and copper acetylacetonate (138). These black, insoluble, infusible products have not been characterized.

2. Polymer Formation in the Presence of Metals

The high thermal stability of metal phthalocyanines has been recognized for many years, and attempts have been made to prepare poymers containing such groups.

Polymeric phthalocyanines have been obtained by (*1*) reacting 3,3',4,4'-tetracyanobiphenyl or 1,2,4,5-tetracyanobenzene with salts of copper, iron, and other metals (139), (*2*) reacting pyromellitic acid and phthalic acid with urea and copper salts at 170–200°C (139), and (*3*) fusing 3,3',4,4'-tetracyanodiphenyl ether, phthalonitrile, and copper dichloride at 275°C (140).

The polymeric phthalocyanines prepared by the above methods were only partially soluble in *N,N*-dimethylformamide and gave nonviscous solutions in this solvent. The molecular weight of the product varied,

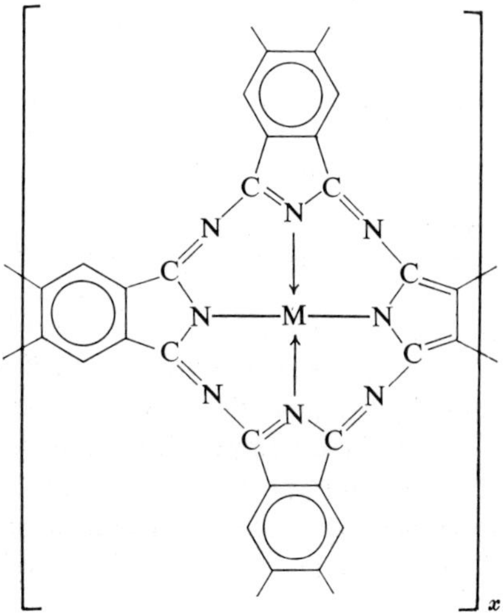

58

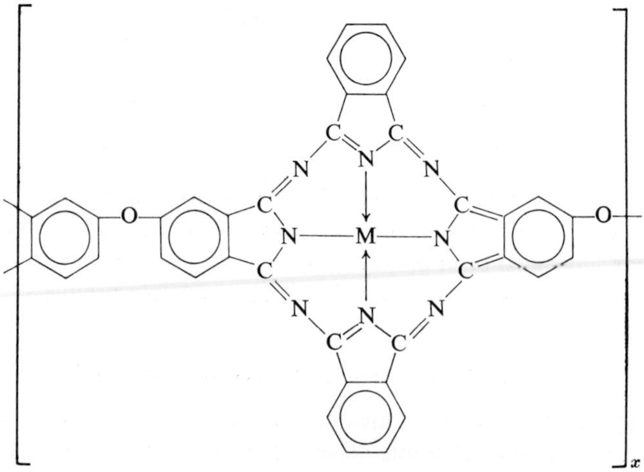

59

depending on reaction conditions, with the intrinsic viscosity ranging from 0.19 to 1.8. These highly colored polymeric phthalocyanines withstood heating in air up to 350°C but decomposed slowly at temperatures in excess of 400°C.

Similarly, polymeric chelate compounds have been prepared by the reaction of tetracyanoethylene or mixtures of tetracyanoethylene and phthalonitrile with metals or metal-containing compounds at 160–300°C (141,142) (see reaction 36). These polymeric chelate complexes of tetracyanoethylene were crystalline, chemically stable, noncombustible materials which withstood prolonged heating at 500°C.

(36)

3. Incorporation of Metal Ions in Preformed Polymers

One of the most successful routes to metal chelate polymers involved the synthesis of an organic polymer by a standard method and then the introduction of the metal ion.

Linear polymers have been prepared from Schiff bases by mixing a solution of the polymer with a soluble metal salt, the insoluble metal-containing polymer being precipitated from solution.

Polymeric Schiff bases prepared from 5,5'-methylenebis-(salicylaldehyde) and o-phenylenediamine have formed innercomplex compounds with divalent metal ions such as zinc, cadium, cobalt, nickel, iron, and copper. These polymers which were insoluble in organic solvents and were stable to 250–300°C may be represented as shown in structure **60** (143). Unfor-

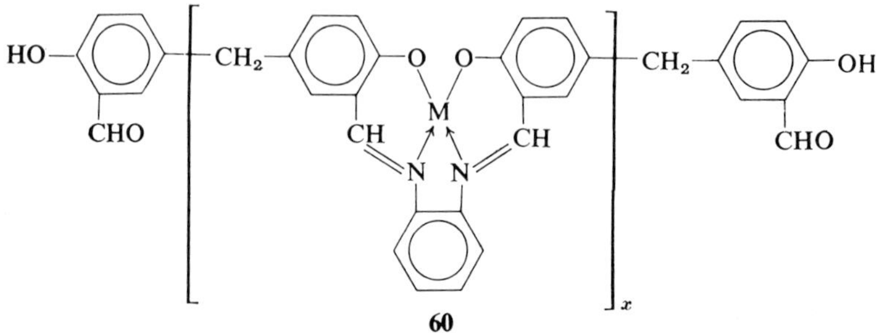

60

tunately, these polymers have been prepared only in very low molecular weight.

Metal chelate polymers have been prepared from the reaction of 8-hydroxyquinoline–formaldehyde polymers and metal ions such as zinc, nickel, aluminum, and iron (144).

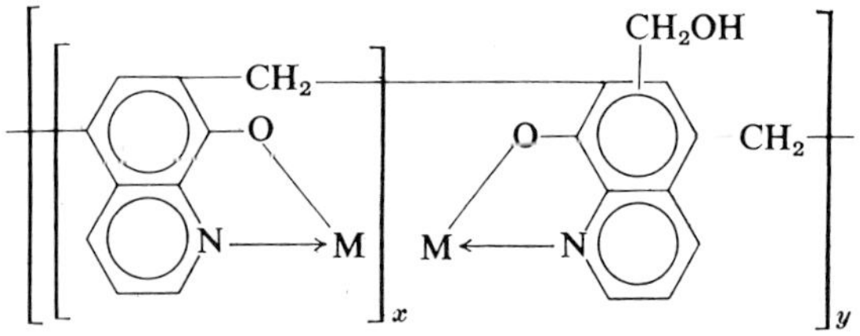

61

The thermal stability of these polymers has been studied by thermogravimetric analysis. These data are collected in Figs. 15 and 16 and represent the thermograms for the chelates of 8-hydroxyquinoline–formaldehyde polymer with aluminum, iron, nickel, and zinc. In every case the unchelated polymer was more thermally stable than the chelated polymers. The unchelated polymers showed an onset decomposition temperature between 350 and 400°C, whereas the chelated polymers decomposed approximately 50°C lower.

Metal chelates of polythiosemicarbazides have been reported (145,146). The polythiosemicarbazides derived from N,N'-diaminopiperazine with

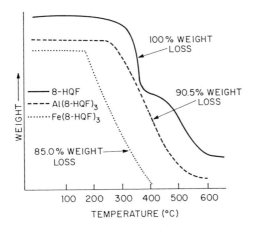

Fig. 15. Thermal decomposition of 8-hydroxyquinoline–formaldehyde polymer and its chelates with aluminum and iron (III) (144).

methylene-bis(4-phenylisothiocyanate) have been reacted with metal ions such as nickel, cobalt, iron, copper, silver, and mercury to yield the polymeric metal chelates, **62**.

Although no thermogravimetric analysis of these polymers has been carried out, based on melting phenomena, it was quite apparent that they offered no improvement in thermal stability over the unchelated polythiosemicarbazides.

The metal chelates of aromatic polyhydrazides have also been prepared

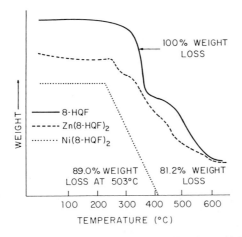

Fig. 16. Thermal decomposition of 8-hydroxyquinoline–formaldehyde polymer and its chelates with zinc and nickel (144).

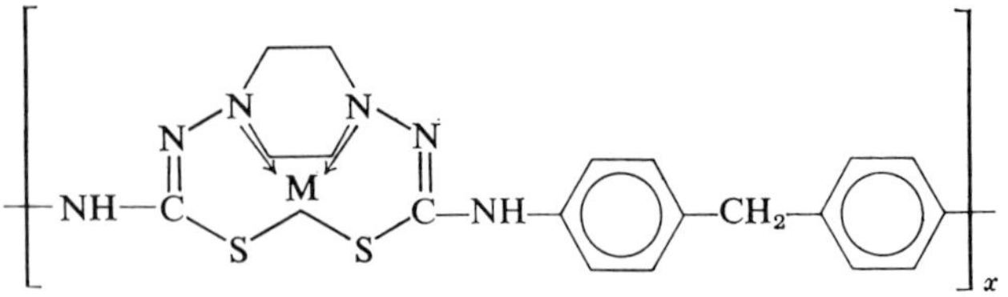

[Structure 62]

(147). Aromatic polyhydrazides derived from isophthalic and terephthalic acid have been prepared with nickel, calcium, silver, lead, mercury, zinc, cadmium, and copper.

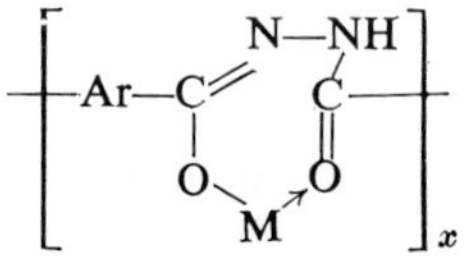

[Structures 63 and 64]

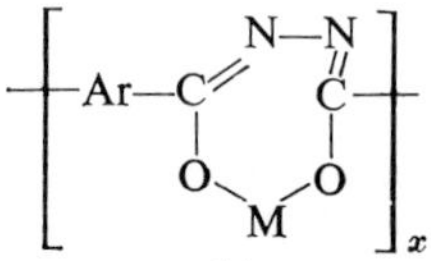

In all cases these metal chelates analyzed for a 1:1 ratio of hydrazide link per metal ion. Even though films and fibers have been prepared from such metal chelates, the thermal stability of the metal chelate of the polyhydrazide was inferior to that of the unchelated polymer.

Metal-containing polymers have been prepared from poly(aminoquinones) of the type shown in structure **65** (148). From metal ions such

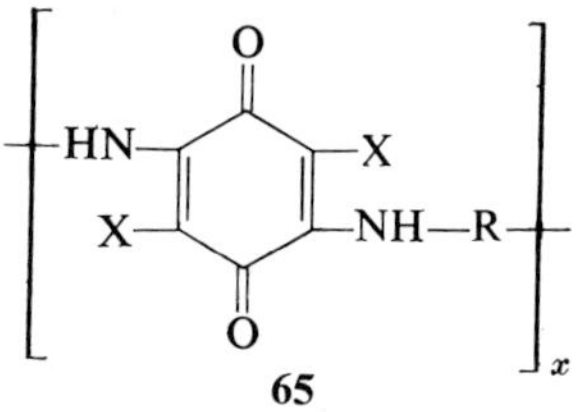

[Structure 65]

as copper, cadmium, and magnesium, insoluble complexes containing as much as 12% metal have been prepared. Unfortunately, these polymers have not been characterized, and no assessment of thermal stability could be made.

4. Reaction with Chelates Containing Functional Groups

This method was probably the most versatile and provided the best way of preparing high molecular weight material. Both radical and condensation polymerization have been employed, and both methods have been used to polymerize basic beryllium carboxylates (149). In these unusual compounds, beryllium atoms occupied the corners of the tetrahedron, at the center of which was an oxygen atom. The carboxylate groups lay along the tetrahedron edges as shown in **66**.

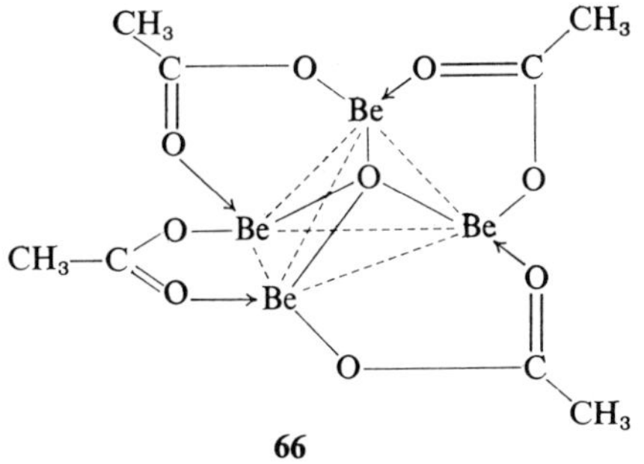

66

Polymers, obtained from compounds containing difunctional unsaturated acids, did not lose volatile materials at temperatures as high as 400°C, but disproportionated even at room temperature (see reaction 37).

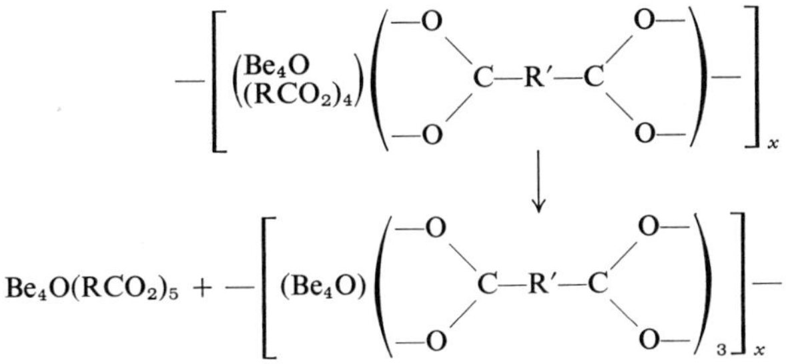

$$-\left[\begin{pmatrix}Be_4O\\(RCO_2)_4\end{pmatrix}\begin{pmatrix}-O\\-O\end{pmatrix}C-R'-C\begin{pmatrix}O-\\O-\end{pmatrix}\right]_x \quad (37)$$

$$\downarrow$$

$$Be_4O(RCO_2)_5 + -\left[(Be_4O)\begin{pmatrix}-O\\-O\end{pmatrix}C-R'-C\begin{pmatrix}O-\\O-\end{pmatrix}_3\right]_x-$$

Aluminum alkoxide derivatives of acetylacetone and ethyl acetoacetate were particularly suitable for interchange reactions.

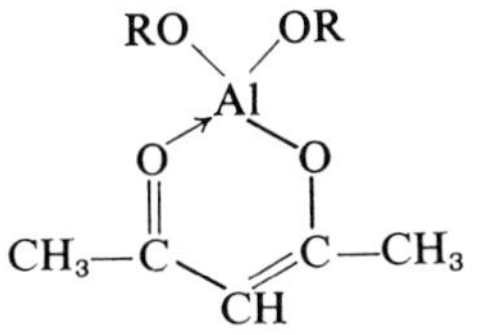

67 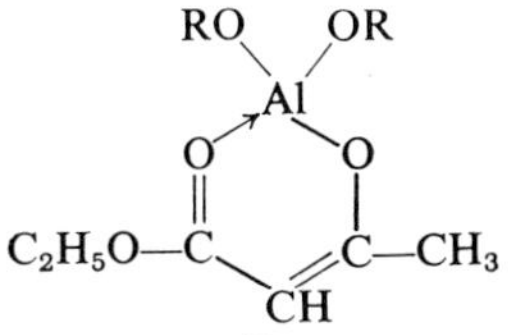   **68**

A number of polymeric substances so obtained have been described (150,151). Attempts have been made to prepare the polyterephthalate and polyurethanes from the type of complex shown below, but the products were thermally unstable (152).

The nickel derivative of β-hydroxyethylglycine was reported to be stable at 320°C, but although condensation occurred with terephthalic acids, only trimers have been obtained. β-Hydroxyethylethylenediamine gave

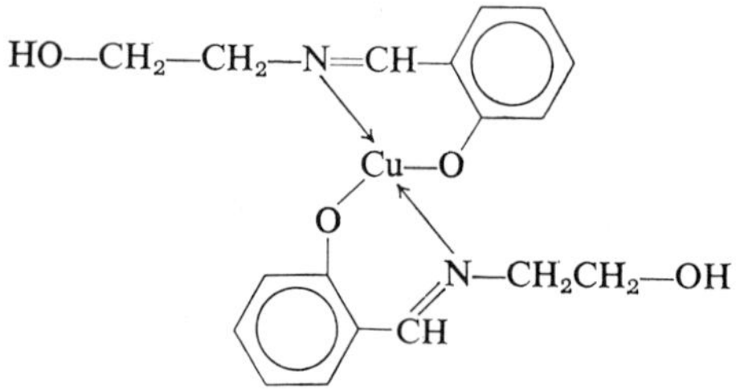

69

stable cobalt complexes, but attempts to carry out condensation reaction have all proved unsuccessful. A similar approach has been used (134), involving hydroxy derivatives of the Schiff base complex shown in structure **70**. Such a complex on heating with diphenyl carbonate or treatment

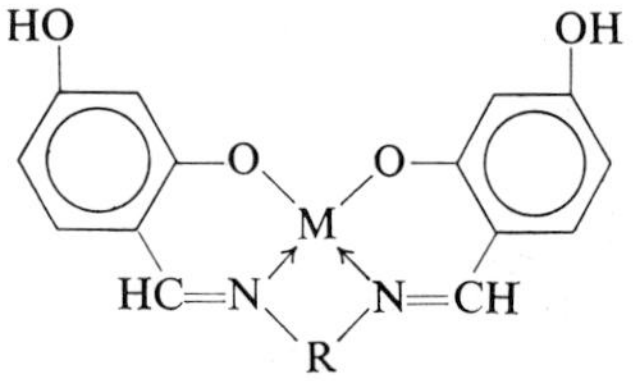

70

with phosgene in alkaline solution led only to low molecular weight polymer. Reaction with toluene diisocyanate resulted only in the formation of cyanurates.

One advantage of this method was that it could be extended to melt polymerization, provided the complexes themselves were stable at their

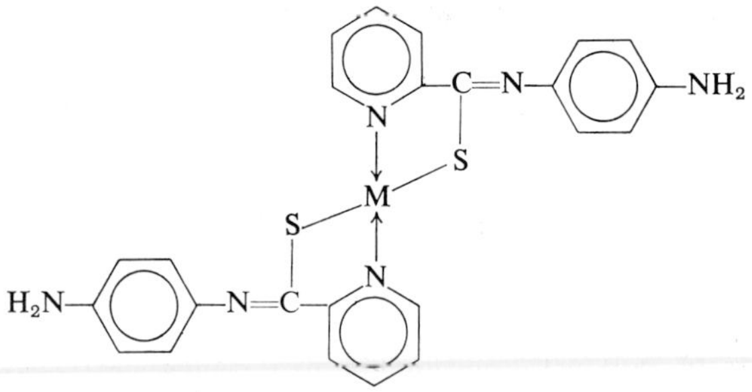

71

melting points. Using this approach (134), metal thiopicolinamides (see structure **71**) containing free amino groups were reacted with bis-acid chloride at 200°C. Some of the metal thiopicolinamides were stable at 400°C, and the polymers from bisthiopicolinamide, some of which were stable at 350°C, have been prepared (153). In both cases, only infusible, insoluble powders were obtained.

In all these examples, the ligand provided the functional group for carrying out further reactions. There is no reason why the metal ion should not also do so. It has been shown (156) that manganese phthalocyanine groups may be linked by oxygen atoms bonded to tetravalent manganese atoms. The recently reported chloro-, phenoxy-, and siloxygermanium phthalocyanines (154) may lead to phthalocyanine polymers in which the metal atom is hexaco-ordinate, instead of tetraco-ordinate and planar, and may lead to more thermally stable structures.

5. Ferrocenes

The ferrocenes are sufficiently different from compounds just described to justify separate discussion. However, as their chemistry is highly specialized and some excellent reviews (155,156) have been published, only a few examples will be mentioned here. The interest in such compounds stems from the fact that biscyclopentadienyliron is stable up to 470°C. There have been a few reports of the preparation of stable polymers containing ferrocene groups.

Vinylferrocene formed homo- and copolymers. *trans*-Cinnamoylferrocene has been shown to copolymerize readily with a number of unsaturated compounds (157). The treatment of lithioferrocene and lithiodimethylferrocene with carbon dioxide gave polymers with molecular weights of 15,000 and 6,000, respectively (158,159). Thermal stabilities were not reported, but these compounds were susceptible to oxidative attack. Siloxanylferrocenes have been prepared, but these contain only two ferrocene groups and attempts to make higher molecular weight compounds were unsuccessful (160). 1,1'-Bis(chloroformyl)ferrocene condensed with a number of polymethylenediamines and diphenols gave polymers (161). A related type of polymer, involving bonds between a metal and unsaturated molecule, was obtained from nickel cycloctatetraene (162). Although no systematic study of thermal stability has been carried out, based on melting decomposition phenomena, this polymer was less stable than biscyclopentadienyl iron.

Ferrocene-containing polymers have been prepared from the self-condensation of the ferrocenyl Mannich base, N,N-dimethylaminomethylferrocene, in the presence of zinc chloride–HCl catalyst systems in the melt phase (163).

Such polymers were shown to be in the 4000–8000 molecular weight range.

Similarly, the acid-catalyzed polycondensation of ferrocenyl carbinols has been described (164). These polymers have also been prepared only in low molecular weight. Although no detailed study of the thermal stability of these polymers was made, based on melting point data, they appeared to decompose in the 300–400°C range (164).

1,1-Ferrocenedicarboxylic acid has been used in the preparation of polybenzimidazoles (165). The polymers which were soluble in dimethyl sulfoxide had inherent viscosities of 0.4 and decomposed extensively below 300°C (see Fig. 17). In addition, the polyesters (166,161), polyurethanes (166,161), and polyamides (166,161) prepared from 1,1-ferrocenedicarboxylic acid have melting points below 300°C and in most cases, these melting points also represented decomposition temperatures. Thus, it would appear that ferroceneodicarboxylic acid yields polymers with a lower level of thermal stability than those derived from aromatic dicarboxylic acids.

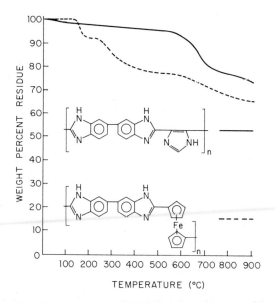

Fig. 17. TGA plot of 1,1-ferrocenedicarboxylic acid and 4,5-imidazolecarboxylic acid derivatives (heating rate, 6°C/min) (165).

References

1. Chemical Society, *Inorganic Polymers*, London, the Society, 1961.
2. F. G. A. Stone and G. Graham, *Inorganic Polymers*, Academic Press, New York, 1962.
3. M. F. Lappert and G. J. Leigh, *Developments in Inorganic Chemistry*, Elsevier, New York, 1962.
4. K. A. Andrianov, *Metalorganic Polymers*, Interscience, New York, 1965.
5. Ref. 3, p. 24.
6. H. S. Turner and R. J. Warne, *Proc. Chem. Soc.*, **1962**, 69.
7. Ref. 3, p. 26.
8. R. H. Wentorf, *J. Chem. Phys.*, **26**, 956 (1957).
9. E. C. Evers, W. O. Freitag, W. A. Kriner, and A. G. MacDirmid, *J. Am. Chem. Soc.*, **81**, 5106 (1959).
10. J. Goubeau and H. Grabner, *Chem. Ber.*, **93**, 1379 (1960).
11. M. F. Lappert and H. Pyszora, *Proc. Chem. Soc.*, **1960**, 350.
12. M. F. Lappert, *Proc. Chem. Soc.*, **1959**, 59.
13. H. J. Becker and S. Frick, *Z. Anorg. Chem.*, **295**, 83 (1958).
14. D. W. Aubrey, M. F. Lappert, and H. Pyszora, *J. Chem. Soc.*, **1961**, 1931.
15. W. Gerrard, Symposium on High Temperature Resistance and Thermal Degradation of Polymers, London, Sept., 1960; *SCI Monograph No. 13*, London, p. 328.
16. G. F. L. Ehlers, WADC-TR-61-622, Feb. 1962.
17. D. H. Campbell, T. C. Bissot, and R. W. Parry, *J. Am. Chem. Soc.*, **80**, 1549 (1958).
18. W. Gerrard, M. F. Lappert, and B. A. Mountfield, *J. Chem. Soc.*, **1959**, 1529.
19. K. Niedenzu and J. W. Dawson, *Angew. Chem.*, **72**, 920 (1960).
20. L. J. Schupp and C. A. Brown, 128th Meeting American Chemical Society, Minneapolis, Sept. 1955, p. 48A.
21. L. F. Hohnstedt and A. M. Pellicciotto, 137th Meeting American Chemical Society, Cleveland, April 1960, p. 70.
22. B. Rudner and J. J. Harris, 138th Meeting American Chemical Society, New York, Sep. 1960, p. 61P.
23. W. Gerrard, *J. Oil Colour Chemists' Assoc.*, **42**, 625 (1959).
24. D. Ulmschneider and J. Goubeau, *Ber.*, **90**, 2733 (1957).
25. R. L. Letsinger and S. B. Hamilton, *J. Am. Chem. Soc.*, **80, 1958**, 5411.
26. M. J. S. Dewar, V. P. Kubba, and R. Pettit, *J. Chem. Soc.*, 3076 (1958).
27. R. W. Upson, U.S. Pat. 2,517,944 (August 1950).
28. W. L. Ruigh, C. E. Erickson, F. C. Gunderloy, and M. Sedlak, WADC-TR-55-26, Part II, May 1955.
29. J. R. Thomas and O. L. Harle, U.S. Pat. 2,795,548 (June 1957).
30. J. Solms and H. Deuel, *Chimia*, **11**, 311 (1957).
31. F. J. Binda, U.S. Pat. 2,544,850 (March 1951).
32. M. Hyman and C. D. West, U.S. Pat. 2,445,579 (June 1948).
33. C. S. Marvel and C. E. Dennoon, *J. Am. Chem. Soc.*, **60**, 1048 (1938).
34. R. E. Rippere and V. K. LaMer, *J. Phys. Chem.*, **47**, 204 (1943).
35. H. C. Browne and E. A. Fletcher, *J. Am. Chem. Soc.*, **73**, 2808 (1951).
36. J. A. Blau, W. Gerrard, M. F. Lappert, B. A. Mountfield, and H. Pyszora, *J. Chem. Soc.*, **1960**, 380.
37. W. A. Banford and S. Fordham, Ref. 15, p. 320.

38. O. C. Musgrave, *Chem. Ind., London*, **1957**, 1152.
39. O. C. Musgrave and T. O. Park, *Chem. Ind., London*, **1955**, 1552.
40. R. J. Brotherton, Conference on High Temperature Polymer & Fluid Research, 1962, ASD-TDR-62-372, August, 1962, p. 389.
41. J. Cazes, *Compt. Rend.*, **247**, 2019 (1958).
42. R. L. Letsinger and S. B. Hamilton, *J. Am. Chem. Soc.*, **81**, 3009 (1959).
43. W. J. Lennarz and H. R. Snyder, *J. Am. Chem. Soc.*, **82**, 2169 (1960).
44. A. K. Hoffman and W. M. Thomas, *J. Am. Chem. Soc.*, **81**, 580 (1959).
45. A. K. Hoffman and W. M. Thomas, U.S. Pat. 2,934,526 (April 1960).
46. A. K. Hoffman and W. M. Thomas, U.S. Pat. 2,931,788 (April 1960).
47. H. R. Arnold, U.S. Pat. 2,402,590 (June 1946).
48. V. A. Sazanova and N. Y. Kronrod, *Zh. Obshch. Khim.*, **26**, 1876 (1956).
49. T. D. Parsons, M. B. Silverman, and D. M. Ritter, *J. Am. Chem. Soc.*, **79**, 5091 (1957).
50. H. Normant and J. Braun, *Compt. Rend.*, **248**, 828 (1959).
51. R. I. Wagner, R. M. Washburn, and K. R. Eilar, Ref. 40, p. 425.
52. A. B. Burg, Conference on High Temperature Polymer and Fluid Research, 1959, WADC Tech. Report 57-657, p. 19.
53. A. B. Burg and P. J. Slota, *J. Am. Chem. Soc.*, **82**, 2148 (1960).
54. A. B. Burg, *J. Inorg. Nucl. Chem.*, **11**, 258 (1959).
55. N. L. Paddock, Ref. 3, p. 87.
56. H. N. Stokes, *J. Am. Chem. Soc.*, **17**, 275 (1885).
57. H. N. Stokes, *J. Am. Chem. Soc.*, **18**, 629 (1896).
58. L. F. Audrieth, 17th IUPAC Congress, Munich, 1959.
59. C. P. Haber, Ref. 15, p. 115.
60. R. Searle, *Proc. Chem. Soc.*, **1959**, 7.
61. W. Bilbo, *Z. Naturforsch.*, **156**, 330 (1960).
62. C. P. Haber, H. Herring, and R. Lawton, *J. Am. Chem. Soc.*, **80**, 2116 (1958).
63. J. Dumont, *Kunstoffe-Plastics*, **6**, 30 (1959).
64. W. Humiec and R. Bezman, *J. Am. Chem. Soc.*, **83**, 2210 (1961).
65. R. Shaw and R. Stratton, *Chem. Ind., London*, **52** (1959).
66. C. P. Haber, Ref. 15, p. 119.
67. M. Becke-Goehring and J. Schulze, *Chem. Ber.*, **91**, 1188 (1958).
68. (*a*) M. Goehring and K. Niedenzu, *Chem. Ber.*, **89**, 1771 (1956). (*b*) J. E. Malowan and F. R. Hurley, U.S. Pat. 2,596,935 (June 1952).
69. S. H. Rose and B. P. Block, *J. Polymer Sci. A-1*, **4**, 573 (1966).
70. H. W. Coover, R. L. McConnell, and N. H. Shearer, *Ind. Eng. Chem.*, **52**, 412 (1960).
71. M. Becke-Goehring, Ref. 3, p. 130.
72. H. Zenftman and H. R. Wright, *Brit. Plastics*, **25**, 374 (1952).
73. M. L. Nielsen, Ref. 40, p. 504.
74. C. M. Murphy, C. E. Saunders, and D. C. Smith, *Ind. Eng. Chem.*, **42**, 2462 (1950).
75. L. E. Scala, W. M. Hickam, and M. H. Loeffler, *J. Appl. Polymer Sci.*, **2**, 297 (1959).
76. C. D. Doyle, *J. Polymer Sci.*, **31**, 95 (1958).
77. C. W. Lewis, *J. Polymer Sci.*, **33**, 153 (1958).
78. C. W. Lewis, *J. Polymer Sci.*, **37**, 425 (1959).
79. Brit. Pat. 799,067 (1958).
80. A. R. Gilbert, U.S. Pat. 2,717,902 (May 1955).

81. H. R. Baker and C. R. Singleterry, *U.S. Govt. Res. Rept.* PB161449.
82. P. D. George, M. Prober, and J. R. Elliot, *Chem. Rev.*, **56**, 1065 (1956).
83. R. MacFarlane and E. S. Yankura, Contract No. DA-19-020-ORD-5507, Quart. Rept. #7.
84. (a) J. E. Curry and J. D. Byrd, *J. Appl. Polymer Sci.*, **9**, 295 (1965). (b) J. E. Curry, J. D. Byrd, W. R. Dunnavant, R. A. Markle, and P. B. Stickney, *J. Polymer Sci. 1-A*, **5**, 707 (1967).
85. K. A. Andrianov, A. A. Zhdanov, N. A. Kurasheva, and V. G. Dulova, *Dokl. Akad. Nauk SSSR*, **112**, 1050 (1957).
86. K. A. Andrianov, A. A. Zhdanov, and A. A. Kazakova, *Izv. Akad. Nauk SSSR, Otdel, Khim. Nauk*, **1959**, 466.
87. K. A. Andianov and A. A. Zhdanov, *J. Polymer Sci.*, **30**, 513 (1958).
88. K. A. Andrianov, A. A. Zhdanov, and S. A. Pavlov, *Dokl. Akad. Nauk SSSR*, **102**, 85 (1955).
89. K. A. Andrianov, A. A. Zhdanov, and T. N. Ganina, *Bull.* Mendeleev, All-Union Chem. Soc., **3**, 2 (1955).
90. K. A. Andrianov, A. A. Zhdanov, and E. Z. Asnovich, *Dokl. Akad. Nauk SSSR*, **118**, 1124 (1958).
91. K. A. Andrianov, A. A. Zhdanov, and E. Z. Asnovich, *Izv. Akad. Nauk SSSR Otdel. Khim. Nauk*, **1959**, 1760.
92. K. A. Andrianov, T. N. Ganina, and W. Khrustaleva, *Izv. Akad. Nauk SSSR Otdel. Khim. Nauk*, **1956**, 798.
93. (a) K. A. Andrianov, *Usp. Khim.*, **26**, 895 (1957); (b) *ibid.*, **27**, 1257 (1958).
94. T. R. Patterson, F. J. Pavlick, A. A. Baldoni, and R. L. Frank, *J. Am. Chem. Soc.*, **81**, 4213 (1959).
95. USSR Pat. 71,115 (1947).
96. I. K. Stavitskii and S. N. Borishov, *Vysokomolekul. Soedin.*, **1**, 1496 (1959).
97. K. A. Andrianov and A. A. Zhdanov, *Dokl. Akad. Nauk SSSR*, **102**, 85 (1955).
98. C. L. Segal, H. H. Takimoto, and J. B. Rust, 137th Meeting American Chemical Society, Cleveland, April 1960, p. 35.
99. C. L. Segal, H. H. Takimoto, and J. B. Rust, *J. Org. Chem.*, **26**, 2467 (1961).
100. V. A. Zeitler and C. A. Brown, *J. Am. Chem. Soc.*, **79**, 4616 (1957).
101. P. E. Koenig and J. H. Hutchinson, WADC-TR-58-44, ASTIA Doc. No. AD-151197, May 1958.
102. S. M. Atlas and H. F. Mark, *Angew. Chem.*, **72**, 249 (1960).
103. R. D. Crain and P. E. Koenig, Ref. 52, p. 72.
104. W. S. Tatlock and E. G. Rochow, *J. Org. Chem.*, **17**, 1555 (1952).
105. J. F. O'Brien, WADC-TR-57-502, ASTIA Doc. No. 142100, October 1957.
106. F. A. Henglein, R. Lang, and K. Scheinost, *Makromol. Chem.*, **15**, 177 (1955).
107. K. A. Andrianov and L. M. Vokova, *Izv. Akad. Nauk, SSSR, Otdel. Khim. Nauk.*, **1957**, 303.
108. M. G. Voronkov and V. N. Zgonnik, *Zh. Obshch. Khim.*, **27**, 1476 (1957).
109. (a) M. Wick, *Kunststoffe*, **50**, 433 (1960). (b) S. Papetti, B. B. Schaeffer, A. P. Gray, and T. L. Heying, *J. Polymer Sci. A-1*, **4**, 1623 (1966).
110. R. R. McGregor and E. L. Warrick, U.S. Pat. 2,431,878 (1947).
111. R. R. McGregor, *Silicones and Their Uses*, McGraw-Hill, New York, 1954, p. 186.
112. F. S. Martin, U.S. Pat. 2,609,201 (1952).
113. A. P. Kreshkov and D. A. Karateev, *Zh. Prikl., Khim.*, **32**, 369 (1959).
114. A. P. Kreshkov and D. A. Karateev, *Zh. Obshch. Khim.*, **27**, 2715 (1957).

115. A. P. Kreshkov and D. A. Karateev, *Zh. Prikl. Khim.*, **30**, 1416 (1957).
116. A. P. Kreshkov and D. A. Karateev, *Zh. Obschch. Khim.*, **29**, 4082 (1959).
117. M. G. Voronkov and V. N. Zgonnik, *Zh. Obschch. Khim.*, **27**, 1483 (1957).
118. M. Schmidt and H. Schmidbaur, *Chem. Ber.*, **93**, 878 (1960).
119. N. S. Lezvov, L. A. Sabrin, and K. A. Andrianov, *Zh. Obschch. Khim.*, **29**, 1270 (1959).
120. F. A. Henglein, R. Lang, and K. Scheinost, *Makromol. Chem.*, **18–19**, 102 (1956).
121. S. N. Borisov, I. K. Stavitskii, V. A. Ponomarenko, N. G. Sviridova, and G. Ya. Zueva, *Vysokomolekul. Soedin.*, **1**, 1502 (1959).
122. L. M. Khananashvili, Ref. 4, p. 317.
123. A. P. Kreshkov, Ref. 4, p. 317.
124. R. F. Lang, *Makromol. Chem.*, **78**, 1 (1964); **83**, 274 (1965).
125. H. Anderson, *J. Am. Chem. Soc.*, **75**, 814 (1953).
126. U.S. Pat. 2,484,508.
127. (*a*) P. Pheiffer, *Ber.*, **35**, 3305 (1902). (*b*) M. Frankel, D. Gertner, D. Wagner, and A. Zilkha, *J. Appl. Polymer Sci.*, **9**, 3383 (1965). (*c*) S. D. Bruck, *J. Polymer Sci. B*, **4**, 933 (1966).
128. J. G. Noltes and G. J. M. van der Kirk, Ref. 40, p. 615.
129. Ref. 4, p. 364.
130. W. C. Fernelius, WADC-TR-56-203, May 1956.
131. J. P. Wilkins and E. L. Wittbecker, U.S. Pat. 2,659,711 (June 1953).
132. C. N. Kenny, Ref. 3, p. 258.
133. R. W. Kluiber and J. W. Lewis, *J. Am. Chem. Soc.*, **82**, 5777 (1960).
134. (*a*) Ref. 3, p. 260. (*b*). M. E. B. Jones, D. A. Thornton, and R. F. Webb, *Makromol. Chem.*, **49**, 62 (1961); **49**, 69 (1961). (*c*) M. E. B. Jones, D. A. Thornton, R. F. Webb, and J. R. Urivin, *Makromol. Chem.*, **50**, 232 (1961).
135. K. V. Martin, *J. Am. Chem. Soc.*, **80**, 233 (1958).
136. J. C. Bailer, Ref. 52, p. 35.
137. E. Horowitz, M. Tryon, R. G. Christensen, and T. P. Perros, *J. Appl. Polymer Sci.*, **9**, 2321 (1965).
138. A. A. Berlin and N. G. Matveeva, *Vysokomolekul. Soedin.*, **1**, 1643 (1959).
139. C. S. Marvel and J. H. Rossweiler, *J. Am. Chem. Soc.*, **80**, 1197 (1958).
140. C. S. Marvel and M. M. Martin, *J. Am. Chem. Soc.*, **80**, 6600 (1958).
141. E. Epstein and B. S. Wild, *J. Chem. Phys.*, **32**, 2, 324 (1959).
142. A. A. Berlin, N. G. Matveeva and A. I. Sherle, *Izv. Akad. Nauk. SSSR*, **12**, 13 (1959).
143. C. S. Marvel and N. Tarkoy, *J. Am. Chem. Soc.*, **80**, 832 (1958).
144. R. C. Degeiso, L. G. Donaruma, and E. A. Tomic, *J. Appl. Polymer Sci.*, **9**, 411 (1965).
145. T. W. Campbell and E. A. Tomic, *J. Polymer Sci.*, **62**, 379 (1962).
146. T. W. Campbell, E. A. Tomic, and V. S. Foldi, *J. Polymer Sci.*, **62**, 387 (1962).
147. A. H. Frazer and F. T. Wallenberger, *J. Polymer Sci. A*, **2**, 1825 (1964).
148. A. A. Berlin, N. G. Matveeva, and A. I. Sherle, *Izv. Akad. Nauk SSSR, Otdel. Khim. Nauk*, **1959**, 2261.
149. C. S. Marvel and M. M. Martin, *J. Am. Chem. Soc.*, **80**, 619 (1958).
150. V. Kugler, *J. Polymer Sci.*, **29**, 637 (1958).
151. T. R. Patterson, F. J. Pavlik, A. A. Baldoni, and R. L. Frank, *J. Am. Chem. Soc.*, **81**, 4213 (1959).
152. J. C. Bailar, WADC-TR-57-657, June 1957.

153. J. A. Elvidge and A. B. P. Lever, *Proc. Chem. Soc.*, **1959**, 195.
154. R. D. Joyner and M. E. Kenney, *J. Am. Chem. Soc.*, **82**, 5790 (1960).
155. G. Wilkinson and F. A. Cotton, *Progress in Inorganic Chemistry*, Vol. 1, F. A. Cotton, Ed., Interscience, New York, 1959, p. 1.
156. P. L. Pauson, *Quart. Rev. (London)*, **9**, 391 (1955).
157. F. S. Arimoto and A. C. Haven, *J. Am. Chem. Soc.*, **77**, 6295 (1955).
158. L. E. Coleman and M. D. Rausch, *J. Polymer Sci.*, **28**, 207 (1958).
159. K. L. Rinehart, Ref. 52, p. 75.
160. R. L. Schaaf and P. T. Kan, Ref. 52, p. 105.
161. F. W. Knobloch and W. H. Rauscher, *J. Polymer Sci.*, **54**, 651 (1961).
162. G. Wilke, *Angew. Chem.*, **72**, 581 (1960).
163. E. W. Neuse and D. S. Trifan, *J. Am. Chem. Soc.*, **85**, 1952 (1963).
164. (a) E. W. Neuse and E. Quo, *J. Polymer Sci. A*, **3**, 1499 (1965). (b) E. W. Neuse and K. Koda, *J. Macromol. Chem.*, **1**, 595 (1966). (c) E. W. Neuse and E. Quo, *Bull. Chem. Soc. Japan*, **39**, 1508 (1966). (d) E. W. Neuse and K. Koda, *Bull. Chem. Soc. Japan*, **39**, 1502 (1966). (e) E. W. Neuse and R. K. Crossland, *J. Organomet. Chem.*, **7**, 344 (1967). (f) E. W. Neuse and K. Koda, *J. Polymer Sci. A-1*, **4**, 2145 (1966). (g) M. L. Hallensleben and G. Greber, *Makromol. Chem.*, **92**, 137 (1966).
165. L. Plummer and C. S. Marvel, *J. Polymer Sci. A*, **2**, 2559 (1964).
166. M. Okawara, Y. Takeomoto, H. Kitaoka, E. Haruki, and E. Imoto, *Kogyo Kagaku Zasshi*, **65**, 685 (1962).

VI. LADDER POLYMERS

Classical polymer theory has divided all polymers into two groups: linear polymers, which are soluble and/or thermoplastic, and randomly crosslinked network polymers which are gels. Most of the polymers discussed in this text are of the first type. In recent years, however, there has been growing evidence that a third type of polymer may be far less inaccessible than was once thought. These polymers which have molecular structures describable as nonrandomly crosslinked or ordered networks will form the basis for this chapter.

A network polymer may be defined as an extended polycyclic structure of bridged, fused, or spirolinked rings. Network polymers can be obtained by polymerizing monomers having a functionality greater than 2, i.e., monomers which can become attached to more than two monomeric units. They can also be obtained by crosslinking linear polymers. The known network polymers include materials having many different molecular sizes, shapes and structural forms. Some are ordered, others are disordered; some are soluble, others are insoluble; some are gels or microgels, others are crystalline lattices or sheets or linear multichain polymers or spherelike cages.

As is well known, the uncontrolled, nonstereoselective polymerization of a polyfunctional monomer normally produces an insoluble, irregular, randomly connected, three-dimensional network structure. If the polymerization is carried out at high dilution, one may obtain microgels instead. These are irregular, randomly connected, highly polydisperse polymers. On the other hand, structurally selective polymerizations of polyfunctional monomers will produce ordered network polymers. It can be shown that there are several dozen topologically distinct structural forms which regular network polymers, i.e., those in which all the monomer units are structurally equivalent, may conceivably assume.

These alternative structural arrangements may be divided into four categories according to whether the structural form possesses indefinite extension in zero, one, two, or three dimensions. Thus, there are four main types of ordered network polymers which are: finite cagelike molecules; linear molecules of either the double chain or cylindrical types; two-dimensional sheetlike structures, and three-dimensional crystalline lattices. The most important of these for the purposes of this

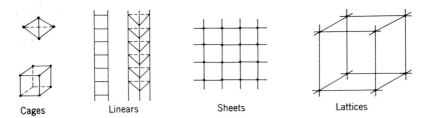

Cages   Linears   Sheets   Lattices

discussion are the double-strand or double-chain or ladder polymers. These ladder polymers are structures with two chains periodically bound together by chemical bonds. (See the diagram above.)

It is evident that the attaining of an ordered rather than a randomly crosslinked network is dependent upon the structural and stereochemical selectivity of the reaction used in its synthesis. The conditions for achieving stereoselective network polymerization do not seem to be particularly mysterious. They may be rationalized by noting that stereoselectivity in any chemical synthesis requires geometrical restrictions on the reaction system during the product-determining stage of the bond-forming process. These geometrical restrictions become most stringent in rigid sterically hindered cyclic or polycyclic systems. Thus, at one extreme it is to be expected that the network polymerization of a flexible monomer of relatively low functionality under irreversible conditions will be nonselective due to a low cyclization density and little rigidity in the product-determining stage of the reaction. All functional groups may be considered as equally likely to combine with each other, and a randomly connected gel should be produced. The behavior of the polymerization system should be describable quantitatively in terms of conventional nonlinear polymerization theory. At the opposite extreme, we must expect that homopolymerization of a rigid monomer of high functionality under reversible conditions, which implies a high cyclization density and much rigidity in the product-determining stage of the reaction, will be highly selective and produce one of the many types of ordered network polymers.

The interest in ladder polymers as polymers resistant to high temperatures lies not in the fact that these polymers are necessarily more thermodynamically stable than the single-strand or nonladder polymers, but rather that improved thermal stability will result because the polymer chain cannot be severed by a single bond-breaking reaction. If chain scission is considered to be an equilibrium process, then degradation of a single-strand polymer will occur with each break in the main chain since the fragmented ends can separate permanently (1). However, when a break occurs at any point along a double strand or ladder polymer, the second strand keeps the entire polymer system intact and maintains overall chain

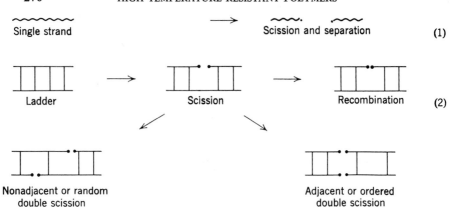

(1) Single strand → Scission and separation

(2) Ladder → Scission → Recombination / Nonadjacent or random double scission / Adjacent or ordered double scission

integrity (2). Moreover, the close proximity of the fragmented ends makes recombination probable. Statistically, the thermal attack, instead of cleaving one bond to sever a polymer chain, must now cleave at least two or more bonds in an ordered nonrandom cleavage. For random scission, four or more bonds must be cleaved to sever the polymer chain. Thus, given a single and a double strand polymer of comparable thermal stability based on the chemical structures involved, the double-strand or ladder polymer will be more thermally stable due to this effect.

A. PYROLYZED POLYACRYLONITRILE AND POLYMETHACRYLONITRILE

Probably the first ladder polymers recognized as such were those obtained from the pyrolysis of polyacrylonitrile (1–4) and polymethacrylonitrile (5). When polyacrylonitrile was pyrolyzed under a controlled oxygen atmosphere at temperatures of 160–300°C, a product with the structure **1** was obtained (1–4). Similarly when the pyrolysis of poly-

1

methacrylonitrile was carried out at a lower temperature (140°C), an analogous product was obtained (5).

Ladder polymers of this type showed outstanding resistance to short-time exposures at elevated temperatures. Black Orlon* or Fiber AF, a

---
* Orlon: Registered trademark for du Pont's acrylic fiber.

pyrolyzed polyacrylonitrile in fiber form, withstood exposure to temperatures of 700–800°C with an open flame with essentially no loss in properties (2). However, such fibrous materials did show a significant break in the thermogravimetry curve in air at temperatures as low as 360°C and significant losses in physical properties after exposures of less than 8 hr at this temperature (6).

B. POLYPHENYLSILSESQUIOXANES

The double chain polyphenylsilsesquioxanes prepared by elevated temperature (250°C) equilibrium of concentrated solutions of phenylsilsesquioxane polymers were shown to have the structure, **2** (7).

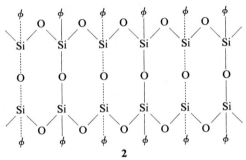

2

This *cis*-syndiotactic double chain polyphenylsilsesquioxane had properties consistent with that expected for a ladder polymer. A comparison of the properties of this ladder polymer with those of the single-strand analog showed marked differences. The double-strand polymer was more resistant to hydrolysis than the single-strand polyphenylsiloxane and heat-aged almost as well in steam as in air. It had approximately twice the tensile strength of a corresponding silicone resin having comparable composition and electrical properties which were retained at higher temperatures than the single-strand analog (7).

As might be expected, the thermal stability was significantly better. The first break in the thermogravimetry curve in air of the double-strand polymer occurred above 525°C, whereas the first break for the single-strand polyphenylsiloxane occurred at 350°C. This improved thermal stability was strikingly demonstrated by the fact that thin film strips were unaffected by exposure for short periods of time at temperatures of 650°C and instantaneous exposures to red heat (7).

C. CYCLIZED POLY-3,4-ISOPRENE

Cyclized poly-3,4-isoprene, the first example of the ladder polymer consisting only of carbocyclic rings, was prepared by the acid catalyzed cyclization of poly-3,4-isoprene (8,9). It was completely soluble in such solvents as hydrocarbons, tetrahydrofuran, and carbon disulfide, and based on physical and chemical properties, was shown to have the linear fused polycyclic structure **3**. The cyclized product had a softening tem-

3

perature of 140°C, a density of approximately 1 g/cc, and low residual unsaturation. Several x-ray powder diffraction spectra of this polymer did show very weak crystalline patterns which were totally absent in the amorphous poly-3,4-isoprene (9). Clear, stiff, flexible films with initial moduli of 250,000 psi, tensile strengths of 4200 psi, and elongations of 2% were prepared by conventional methods. The properties were significantly different from those of the elastomeric precursor poly-3,4-isoprene which had a density of less than 1 g/cc, was rubbery at 25°C, and whose films showed initial moduli and strengths of less than 500 psi with elongations of 500% (9).

Although no detailed study of the thermal stability of cyclized poly-3,4-isoprene has been carried out, these polymers were reported to be less susceptible to oxidative attack at elevated temperature than the un-cyclized poly-3,4-isoprene (10).

D. CYCLIZED POLYBUTADIENE, CYCLIZED POLYISOPRENE, AND CYCLIZED POLYCHLOROPRENE

Other polydiene ladder polymers have been prepared by polymerization of the parent diene with complex catalyst systems, followed by cyclization with concentrated sulfuric acid (10–12). Unlike the cyclized poly-3,4-isoprene, these polymers were not soluble in common polymer solvents. Based on chemical and spectral data, with the exception of the cyclized

4

polychloroprene, they have the linear ladder structure **4**. Based on the lability of the chlorine atoms, the cyclized polychloroprene appeared to have the spiro-ladder structure **5**. Like the cyclized 3,4-polyisoprene

TABLE I
Properties of Diene and Cyclopolymers (10–12)

Polymer	Decomp. pt. (°C)	Density, g/cc (25°C)
Cyclopolyisoprene ($C_6H_5MgBr/TiCl_4$ = 0.5)	370	1.082
Completely cyclized cyclopolyisoprene	388	—
Polybutadiene	—	0.90
Cyclized polybutadiene	—	1.07
Cyclopolybutadiene ($C_6H_5MgBr/TiCl_4$ = 0.5)	405	0.966
Completely cyclized cyclopolybutadiene	413	—
Cyclopolychloroprene ($C_6H_5MgBr/TiCl4_4$ = 0.5)	422	—

discussed in Section C, these ladder structures were also markedly different from the uncyclized parent polymer, both in decomposition temperature and in density (Table I). It is of interest to note that the cyclized

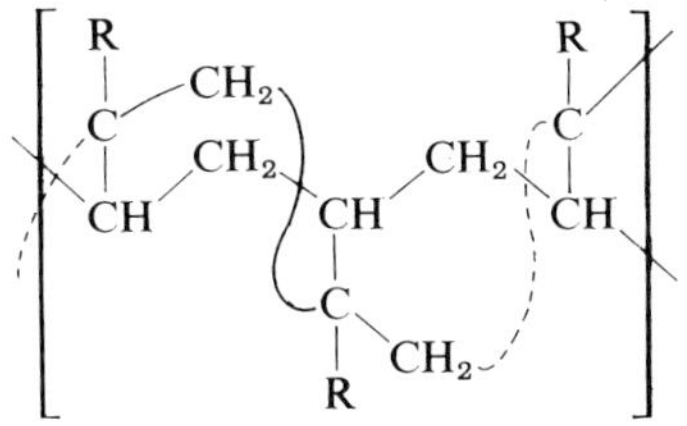

5

poly-1,4-isoprene had a much higher melting and decomposition point than the cyclized poly-3,4-isoprene.

E. LADDER POLYMERS FROM VINYL ISOCYANATE

These structures have been prepared (15) via the two-step synthetic routes outlined in reactions 3.

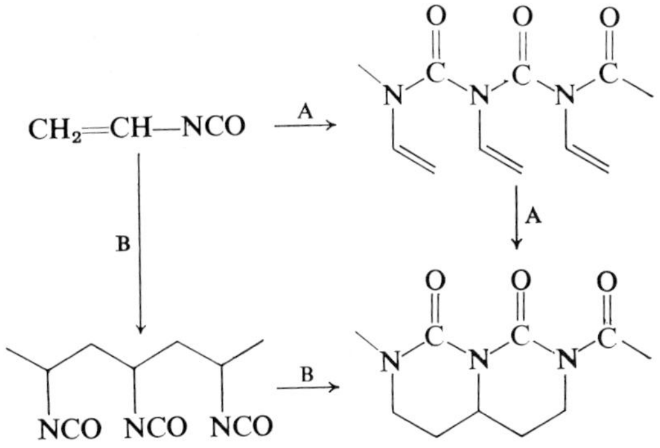

(3)

Since the isocyanate and vinyl groups can be polymerized under quite different conditions, the above procedures have been successfully used. The N-vinyl-1-nylon, prepared by anionic polymerization using sodium cyanide in N,N-dimethylformamide, was cyclized to the ladder polymer with azobisisobutyronitrile and ultraviolet light (Pathway A). The poly-(vinyl isocyanate), prepared by the uncatalyzed polymerization of vinyl isocyanate in dilute solutions, was cyclized by treatment with x-rays (Pathway B). The products from both routes were identical, and, based on chemical and spectral data, appeared to be ladder polymers of the structure shown above (13).

Although no detailed study of the thermal behavior of this ladder structure was carried out, the polymers had decomposition temperatures of 385–390°C (9).

F. POLYQUINONE ETHERS, POLYQUINONE THIOETHERS, AND POLYQUINONE IMINES

Ladder polymers of this type reportedly have been prepared via the reaction of chloranil with tetrahydroxy-1,4-benzoquinone at elevated temperatures in suitable solvents (14) (see reaction 4). The polymers as

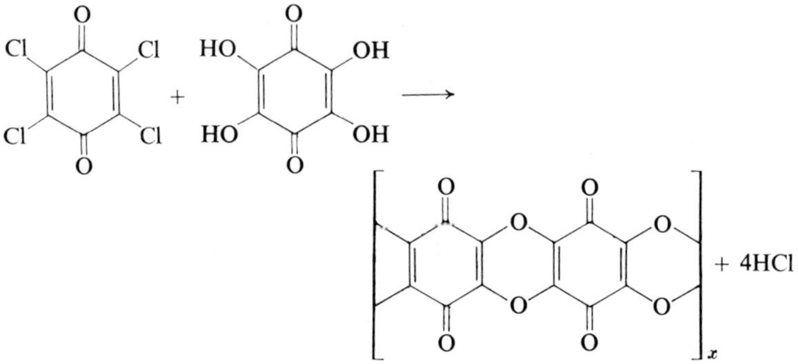

[chemical reaction scheme] + 4HCl   (4)

obtained were dark colored, of low molecular weight, but reportedly possessed appreciable thermal stability. No quantitative thermal data are available.

Similarly, chloranil was reported to react with sodium sulfide, hydrogen

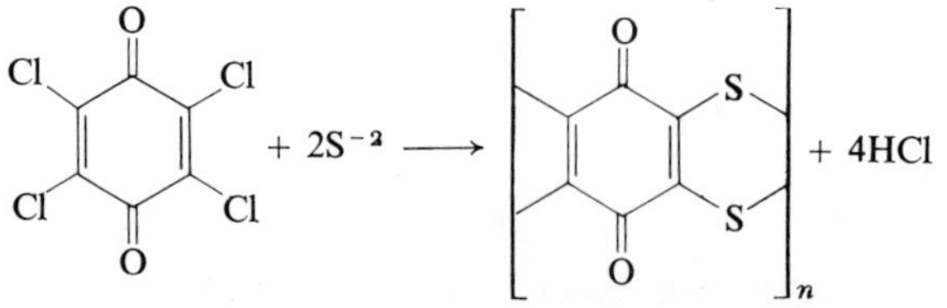

[chemical reaction scheme] + $2S^{-2}$ → [structure] + 4HCl   (5)

sulfide and/or potassium sulfide to produce polyquinone thioethers (15) (see reaction 5). From potassium sulfide a product with molecular weight of approximately 5500 was obtained. Here again, these polymers were deeply colored and did not decompose up to temperatures of 400°C. No additional data on the thermal stability of such polymers are available.

The chloranil reaction with ammonia or amides of Group 1 and 2 metals has been reported to yield polyquinoneimines (16) (see reaction 6).

$$\text{(chloranil)} + NH_3 \longrightarrow \text{(polyquinoneimine)} \quad (6)$$

These polymers have high melting points and decomposition temperatures, along with semiconductor properties. Here again, no quantitative data on thermal stability are available.

G. POLYQUINOXALINE LADDER POLYMERS

The quinoxaline ladder polymer has been reported via the reaction of 4-fluoro-3-nitroaniline at elevated temperatures (17). The initial condensation product, on heating at 270°C overnight, yielded the polyquinoxaline oxide which on further heating at this temperature yielded the ladder structure (reaction 7). The evidence for the formation of the polyquin-

$$\text{(4-fluoro-3-nitroaniline)} \longrightarrow \text{(poly-2-nitro-}p\text{-phenylene imine)} \longrightarrow$$

$$\text{(polyquinoxaline oxide)} \longrightarrow \text{(polyquinoxaline)} \quad (7)$$

oxaline structure from the poly-2-nitro-p-phenylene imine was the behavior of the heat-aged product in thermogravimetric analysis. The product was stable up to temperatures of 600°C and above under a nitrogen atmosphere (Fig. 1).

A second route to polymers of this type involved (1) the reaction of

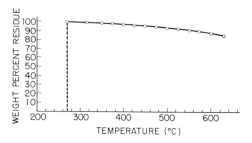

Fig. 1. TGA plot of polyquinoxaline in nitrogen (heating rate, 6°C/min) (17).

aromatic tetramines with aromatic tetrahydroxy, tetrachloro, and tetraphenoxy compounds, or (2) the self-condensation of 2,3-dihydroxy-7-8-diaminoquinoxaline in polyphosphoric acid (18) (see reaction 8). These

where R is = Cl, OH, Oϕ

(8)

polymers, based on thermogravimetric analysis, appeared to be stable up to temperatures of 600°C.

A third route to ladder polymers of polyquinoxaline has involved the reaction of ammonia with 3,6-diketo-4,5-dichlorocyclohexene followed by

(9)

oxidation (19a) (see reaction 9). Little experimental detail and practically no characterization of these products were reported.

A fourth route to ladder polymers of quinoxaline has involved the reaction of 1,2,6,7-tetraketopyrene with 1,2,4,5-tetraminobenzene tetrahydrochloride (19b) (see reaction 10). In thermogravimetric analysis,

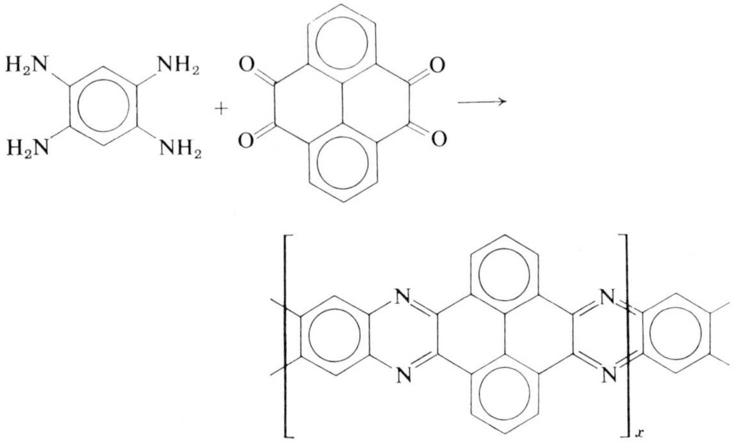

(10)

polymers of this type showed a major break at 460°C in air, while in nitrogen the break occurred at 683°C.

H. POLYDIOXINS

Polymers of this type reportedly have been prepared in low yield via the two-step reaction outlined below, involving the catalytic dehydrogenation of poly-2-hydroxy-6-methylphenylene oxide (20) (see reaction 11).

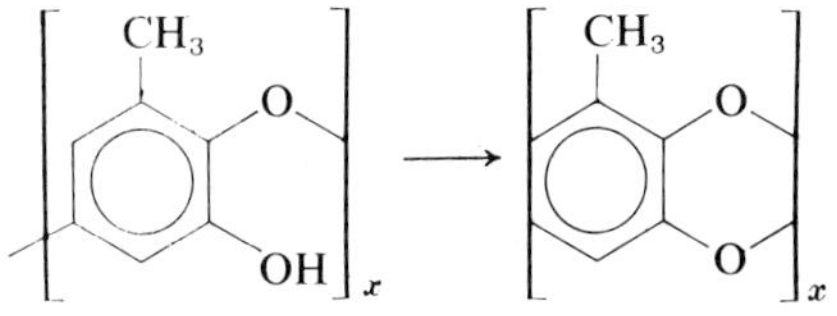

(11)

Thus far, only low molecular weight, incompletely cyclized products have been obtained; therefore, no measure of the thermal stability of ladder polymers of this structure can be ascertained.

I. LADDER POLYMERS FROM VINYLBUTADIENE WITH CYCLIC BISDIENOPHILES

The polymers of the above type have been prepared via the Diels-Alder reaction of 2-vinylbutadiene with (1) benzoquinone (21a) (reaction 12)

$$\text{(reaction scheme 12)} \tag{12}$$

and (2) with *trans,trans*-1,6,11,16-tetraoxa-3,13-cyclodocosadiene-2,5,12,-15-tetranone (21*b*) (reaction 13).

$$\text{(reaction scheme 13)} \tag{13}$$

The polymer from reaction 12 which precipitated softened at 350–400°C, was insoluble in all cold solvents, but was soluble in hot solvents. The soluble portion of the polymer, melting at a considerably lower temperature, 170–200°C, was found to have a molecular weight of 400. Although no direct determination of the molecular weight of the insoluble polymer was obtained, it was greater than 400 and could possibly have been as high as 2000. By thermogravimetric analysis this polymer showed only a 38.5 wt % loss up to 680°C.

The polymer from reaction 13 which was soluble in chloroform had a softening point of 143–150°C. By thermogravimetric analysis, this ladder polymer showed a $T_{1/2}$ (the temperature at which there was a 50 wt % loss) of 400°C whereas its linear polyester analog, poly(tetramethylene adipatic) had a $T_{1/2}$ of 300°C.

J. LADDER POLYMERS DERIVED FROM 4,4-DIMETHYL-1,6-HEPTADIENE-3,5-DIONE

The above monomer has been cyclopolymerized to the cyclic dione structure 6 (22). On monooximation, followed by dehydration at elevated

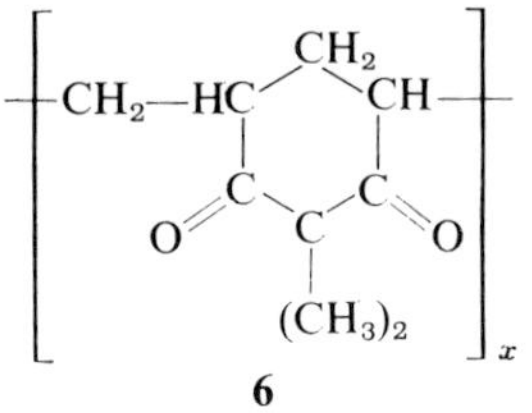

temperatures, a product was obtained which based on elemental analysis and infrared spectra had the structure 7.

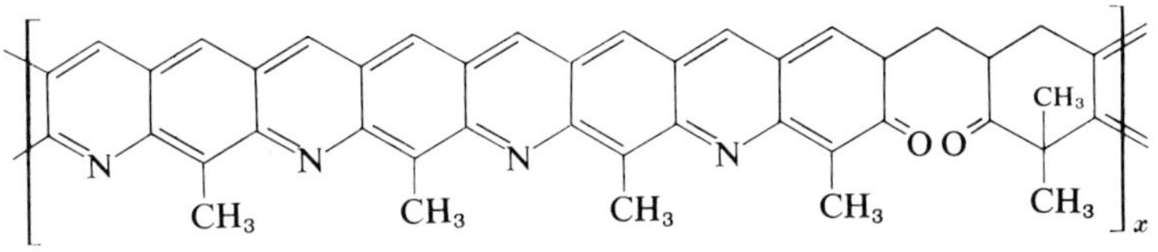

The thermogravimetry curve for this polymer supported the above structure in that a weight loss of 20% between 300 and 500°C was observed and then very little additional loss up to 900°C. This might well indicate that the unaromatized ring was broken out of the chain, leaving the ladder section of the polymer chain which was essentially unaltered by heating. The 20% weight loss corresponded rather closely to what would be expected if this type of decomposition took place. The stability of the ladder structure at 900°C again would be the behavior expected of such a structure.

K. POLY(IMIDAZOPYRROLONES)

By far this class of ladder polymers has been most extensively investigated (23–25). Schematically, the preparation of such polymers via the reaction of the dianhydrides with an aromatic tetramine can be represented as shown in reactions 14.

The initial condensation product, for the most part, was the amide amino acid, 8A, along with a small amount of the benzimidazole acid structure, 9. On initial heating, the major product was the aminoimide, 8. On further heating at elevated temperatures, both the aminoimide and the benzimidazole acid were converted to the imidazopyrrolone structure **10** (24).

Such polymers have been prepared in the completely ladder form when an aromatic tetramine such as 1,2,4,5-tetraminobenzene was used or in what has been referred to as a step- or pseudo-ladder form when an aromatic tetramine such as 3,3'-diaminobenzidine was used. Some of the structures which have been prepared are **11**, **12**, and **13**. These structures have been confirmed by chemical and spectral analysis, and the physical

11

12

13

and chemical properties of the polymers determined. Films showed outstanding resistance to high energy radiation (25).

Of particular interest is the thermal stability of these polymers. The differential thermal analysis of the step- or pseudo-ladder polymer from 3,3'-diaminobenzidine and pyromellitic dianhydride, **12**, showed no exothermic or endothermic activity below 600°C. Thermogravimetric analysis of polymer in air showed no significant weight loss until 550–600°C (Fig. 2) and a maximum rate of weight loss nearly identical to that for pyrolytic

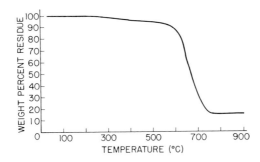

Fig. 2. TGA plot of **12** in dry air (heating rate, 6°C/min) (24).

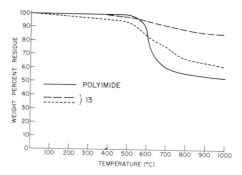

Fig. 3. TGA plot of **13** and polyimide in vacuum (heating rate, 6°C/min) (25).

graphite (25). A comparison of the thermogravimetric analysis in vacuum of the step- or pseudo-ladder polymer from 3,3',4,4'-tetraminobiphenyl ether and pyromellitic dianhydride, **13**, with a related polyimide (25) (Fig. 3) clearly showed the superior thermal stability of the stepladder. Similarly a comparison of these data with those of the polybenzimidazoles (see Chapter IV) showed these polymers to be more thermally stable. Moreover, the complete ladder polymer of this structure was reported to be even more stable (25), but no quantitative data are available.

L. POLY(BENZIMIDAZO-BENZOPHENANTHROLINES)

These polymers differ from the previously discussed poly(imidazopyrrolones) only in the presence of the fused six-membered cyclic diimide structures as shown in **14** and **15**. The preparation of such polymers has been effected via the reaction of 1,4,5,8-naphthalene tetracarboxylic acid

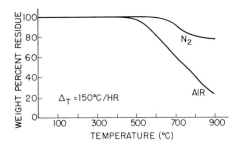

Fig. 4. TGA of **14**; 150°C/hr in nitrogen and air (26).

dianhydride with aromatic tetraamines in polyphosphoric acid at temperatures up to 220°C (26,27). As in the case of the poly(imidazopyrrolones), when 3,3',4,4'-tetraminophenyl was used a step- or pseudo-ladder polymer, **14**, was obtained, and when 1,2,4,5-tetraminobenzene was used, a ladder polymer, **15**, was obtained.

Thermogravimetric analysis of **14** and **15** (Figs. 4 and 5) in nitrogen showed no significant weight loss to over 600°C. In air, no significant weight loss was observed up to 500°C. The failure of the ladder polymer, **15**, to be significantly more thermally stable than the stepladder polymer, **14**, has been ascribed to incomplete cyclization or ladder formation in **15** (26).

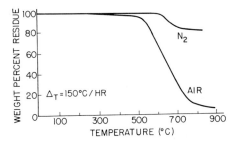

Fig. 5. TGA of **15**; 150°C/hr in nitrogen and air (26).

References

1. R. C. Houtz, *Textile Res. J.*, **20**, 786 (1950).
2. W. G. Vosburgh, *Textile Res. J.*, **30**, 882 (1960).
3. W. J. Burlant and J. L. Parsons, *J. Polymer Sci.*, **22**, 249 (1956).
4. A. V. Topchiev, M. A. Geiderikh, B. E. Davydov, V. A. Kargin, B. A. Krentsel, I. M. Kustanovich, and L. S. Polak, *Dokl. Akad., SSSR*, **128**, 312 (1959).
5. N. Grassie and I. C. McNeill, *J. Polymer Sci.*, **27**, 207 (1958).
6. G. F. L. Ehlers, WADS-TR-61-622, Feb. 1962.

7. J. F. Brown, *J. Polymer Sci. C*, **1**, 83 (1963).
8. S. Tocker, *J. Am. Chem. Soc.*, **85**, 640 (1963).
9. (*a*) R. J. Angelo, *J. Polymer Sci.*, in press; (*b*) R. J. Angelo, M. L. Wallach, and R. M. Ikeda, *Am. Chem. Soc., Div. Polymer Chem., Preprints*, **8**, No. 1, 221 (1967).
10. N. G. Gaylord, I. Kossler, M. Stolka, and J. Vodehnal, *J. Polymer Sci. A*, **2**, 3969 (1964).
11. N. G. Gaylord, I. Kossler, M. Stolka, and J. Vodehnal, *J. Am. Chem. Soc.*, **85**, 641 (1963).
12. (*a*) N. G. Gaylord, I. Kossler, M. Stolka, and J. Vodehnal, *J. Polymer Sci. A*, **2**, 3987 (1964); (*b*) N. G. Gaylord, I. Kossler, B. Matyska, and K. Mach, *Polymer Preprints*, **8**, No. 1, 174 (1967).
13. C. G. Overberger, S. Ozaki, and H. Mukamal, *J. Polymer Sci. B*, **2**, 627 (1964).
14. Ger. Pat. 1,179,716 (1964).
15. Ger. Pat. 1,179,715 (1964).
16. Ger. Pat. 1,179,228 (1965).
17. H. K. Reimschuessel and F. Boardman, AFML-TR-369, Nov. 1964.
18. (*a*) C. S. Marvel, ML-TDR-65-39, Part II, Feb. 1965; (*b*) C. S. Marvel and M. Okada, *Am. Chem. Soc., Div. Polymer Chem., Preprints*, **8**, No. 1, 229 (1967). (*c*) H. Jadamus, F. DeSchryver, W. DeWinter, and C. S. Marvel, *J. Polymer Sci. A-1*, **4**, 2831 (1966). (*d*) C. S. Marvel and F. DeSchryver, *J. Polymer Sci. A-1*, **5**, 542 (1967).
19. (*a*) H. Inone and E. Imoto, *Bull. Univ. Osaka, Prefect. Ser. A*, **10**, 61 (1961). (*b*) J. K. Stille and E. L. Mainen, *J. Polymer Sci. B*, **4**, 39 and 665 (1966); *Am. Chem. Soc., Div. Polymer Chem., Preprints*, **8**, No. 1, 244 (1967).
20. N. P. Loire, AFML-TR-64-338, Oct. 1964.
21. (*a*) W. J. Bailey, E. J. Fetter, and J. Economy, *J. Org. Chem.*, **27**, 3479 (1962). (*b*) W. J. Bailey and B. D. Feinberg, *Am. Chem. Soc., Div. Polymer Chem., Preprints*, **8**, No. 1, 165 (1967).
22. W. DeWinter and C. S. Marvel, *J. Polymer Sci. A*, **2**, 5123 (1964).
23. F. Dawans and C. S. Marvel, *J. Polymer Sci. A*, **3**, 3549 (1965).
24. J. G. Colson, R. H. Michel, and R. M. Paufler, *J. Polymer Sci. A-1*, **4**, 59 (1966).
25. V. L. Bell and G. F. Pezdirtz, *Am. Chem. Soc., Div. Polymer Chem., Preprints*, **6**, No. 2, 747 (1965); V. L. Bell and R. A. Jewell, *Am. Chem. Soc., Div. Polymer Chem., Preprints*, **8**, No. 1, 235 (1967), *J. Polymer Sci. A-1*, **5**, 5439 (1967).
26. R. L. Van Deusen, *J. Polymer Sci. B*, **4**, 211 (1966).
27. A. A. Berlin, B. I. Liogonikii, and R. M. Gitina, *Bull. Acad. Sci. USSR Div. Chem. Sci.* (*Engl. Transl.*), **1966**, 909.

VII. APPLICATIONS

As stated in the first chapter, since the ultimate test of thermal stability of any polymer is in a use or application, this final chapter will be devoted to the discussion of the end uses in which the previously discussed polymers are finding applications. Since most of the polymers are not in full commercial production, most of the available data has been generated on experimental or developmental quantities. It would be expected that the products derived from these relatively new polymers are capable of being improved, and this improvement will be reflected in higher levels of properties. Thus the comparative data cited must be considered in that light.

A. ADHESIVES

Of the previously discussed polymers the polybenzimidazoles (1–3), polybenzoxazoles (6,7), polybenzothiazoles (6,7), polyimides (3–7), poly-1,3,4-oxadiazoles (8), poly-1,2,4-oxadiazoles (8), polyphenylene sulfides (8), and polyquinoxalines (6,7) have been studied as high temperature adhesives. The polybenzimidazoles and the polyimides have been most extensively investigated and therefore will be the subject of most of this discussion.

Polyimide adhesives exhibited the excellent thermal stability of the parent polymer, but thus far the adhesive bonds which have been prepared (see Table I) had shear strengths lower than the comparable commercial epoxy-phenolic (3). However, the polyimide adhesive bonds retained over 50% original strength after exposure at 300°C for 1000 hr, whereas the commercial epoxy-phenolic had lost all strength after only 100 hr (3). Similarly, the polyimide adhesives retained in excess of 50% original bond strength after 24 hr at temperatures of 360°C while the epoxy-phenolic bond had completely deteriorated after 1 hr at this temperature (3). In addition, the polyimide adhesive exhibited outstanding resistance to water, salt spray, jet fuel, and hydraulic oil (3,5). The polybenzimidazole adhesives (see Table I) exhibited extremely high initial lap shear strengths, comparable to that of the commercial epoxy-phenolics (1–3). Like the polyimides, the polybenzimidazole bonds were found to be superior to those of the epoxy-phenolics on exposure to elevated temperatures (1–3).

TABLE I
Lap Shear Strengths of Bonds from Various Adhesives (1-5)

Test temp. (°C)	Aging conditions	Tensile shear strengths (psi)			
		Polyimide	Polyimide formulated	Polybenz-imidazole formulated	Commercial epoxy-phenolic
25	None	2500	2800	4000	>4000
300	1 hr at 300°C	1800	1600	3500	2100
300	100 hr at 300°C	1800	1900	2500	0
300	200 hr at 300°C	1900	1400	1100	0
300	500 hr at 300°C	1600	1300	700	—
300	1000 hr at 300°C	1400	1400	0	—
360	1 hr at 360°C	1400	1300	2000	900
360	10 hr at 360°C	1300	1100	2000	0
360	24 hr at 360°C	1300	1200	900	0
360	60 hr at 360°C	0	1100	0	—
77	30 days/25°C water	2200	2100	1900	2600
77	30 days/45°C = 100% R.H.	2300	1900	1900	2900
77	30 days/40°C = 5% salt spray	2200	2000	1900	3200
77	7 days in JP-4 jet fuel	2300	2300	3900	>4000
77	7 days in hydraulic oil	2200	2300	3900	>4000
77	−80°C	2800	3600	4000	>4000

However, these bond strengths were not retained nearly as well as those of the polyimides on long-term exposure at elevated temperatures. For example, after 100 hr at 300°C one-half of the initial bond strength was lost, and after 200 hr at this temperature, bond strengths for the polybenzimidazoles were less than those of the polyimides. Similar results were obtained at higher temperatures (360°C) (1-3). Surprisingly, the polybenzimidizoles did not exhibit the outstanding resistance to hydrolytic attack that might be expected based on the performance of the polymer in bulk hydrolytic tests (5). Thus, it would appear that both the polyimide and the polybenzimidizoles are far superior to the commercial epoxy-phenolic adhesives now in use as a high temperature adhesive. The polybenzimidazoles yielded bonds of greater strength than the polyimides, but the polyimide adhesives exhibited greater thermal stability.

The polybenzoxazoles, the polybenzothiazoles, and the polyquinoxalines have undergone only limited evaluation as adhesive materials. In all cases, as summarized in Table II, these polymers yielded adhesive bonds of only marginal acceptability, i.e., tensile strength in excess of 1000 psi, but

VII. APPLICATIONS

TABLE II
Lap Shear Strengths of Adhesive Bonds from Other Aromatic Heterocyclic Polymers (6)

Test temp. (°C)	Aging conditions	Tensile shear strength (psi)		
		Poly-quinoxalines	Poly-benzothiazoles	Poly-benzoxazoles
25	None	2300	2165	1950
326	None	2120	1810	1730
360	1 hr/360°C	1415	1547	1200
360	10 hr/360°C	830	573	410
425	None	1230	1100	1000
425	1 hr/425°C	930	520	430
535	None	800	601	520
535	1 hr/535°C	225	350	225

did exhibit stability to extremely high temperatures, i.e., temperatures of 400°C or greater (6,7). No data on long-term thermal stability at lower temperatures (300–360°C) is available, but, based on the behavior of the polybenzimidizoles and the similarity of these polymers in their thermal behavior to the polybenzimidazoles, it would be expected that these would give adhesive bond strengths similar to those of the polybenzimidazoles.

The poly-1,3,4-oxadiazoles, the poly-1,2,4-oxadiazoles, and the polyphenylene sulfides have also received cursory examination as adhesive candidates (8). Unfortunately, for all of these polymers, sufficient bond strengths have not as yet been developed. Therefore, no assessment of their utility as high temperature adhesives can be made.

B. COATINGS

Of all the previously discussed polymers the polyimides have been most extensively studied as coatings both as wire enamels and as coatings for papers. This latter class will be discussed in the section on papers. The polyimide coatings, in addition to their outstanding thermal stability, have moisture resistance, and insensitivity to solvents, greases, lubricants, exotic fluids, and strong acids (4). Moreover, these coatings retained properties over a wide temperature range, from −250 to over 300°C. As suggested by their high absorptivity (0.92–0.93) in the visible and near infrared, and their emissivity of about 0.9 from 40 to over 350°C the coatings have excellent heat transfer properties.

In high temperature electrical tests, 1 mil of aluminum foil coated with

TABLE III
Comparison of the Initial Properties of Wire Coated with Polyimide Wire Enamel and Several Other Commercial Enamels
(Applied on #18 Copper Wire)

Test	Polyimide	Teflon	Typical Formvar	Lecton
Thermal classification	220°C	180°C or higher	105°C	125°C
Build (diam increase)	0.0028 in.	0.0030 in.	0.0030 in.	0.0030 in.
Dry dielectric strength (1)	3400 V/mil	1900 V/mil	2000 V/mil	2700 V/mil
Wet dielectric strength (2)	1900 V/mil	1200 V/mil	600 V/mil	1000 V/mil
Insulation resistance	Infinite	Infinite	Infinite	Infinite
Dielectric constant				
(100 cycles)	3.6	2.0	3.71	3.67
(1000 cycles)	3.6	2.0	3.64	3.48
Dissipation factor				
(100 cycles)	0.004	0.0002	0.0275	0.0390
(1000 cycles)	0.004	0.0002	0.0220	0.0370
G.E. scrape abrasion	15–34 scrapes	1–4 scrapes	60–80	35–50
NEMA unidirectional scrape	1.3–1.5 kg	Very low	1.5–1.7 kg	1.3–1.5 kg
Cut through temperature	Above 500°C	325°C	240°C	300°C
Extractables	Less than 1%	Less than 1%	3.0–6.0%	0.5–1.0%
Heat shock	Passes to 425°C	Passes to 215°C	Passes to 124°C	Passes to 175°C
Solderability	N.G.	N.G.	Fair	Good

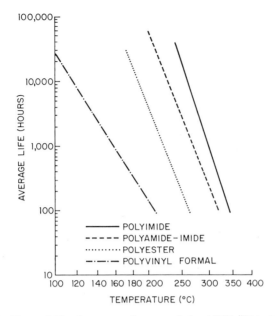

Fig. 1. Thermal life of magnet wire enamels by AIEE #57 test (3,10).

½ mil of polimide withstood passage of 200 amps at 3 V and a surface temperature of 350°C for extended periods of time (4). Uncoated foil subjected to a current of 120 amps at 2 V burned at surface temperatures of 260–315°C (4).

A comparison of the initial properties of wire coated with polyimide vs. those of other commercial enamels clearly showed superiority of the polyimide in this application (4) (Table III). Similarly the comparative thermal lives of coated magnet wire (Fig. 1) indicated the excellent durability of the polyimide coating (4,10). Similar data on the copolyamide-imides as wire enamels showed that reduced thermal life resulted from the introduction of the amide group and demonstrated once again the greater stability of the aromatic heterocyclic structure (4,10).

C. COMPOSITE STRUCTURE (GLASS-REINFORCED LAMINATES)

As in the case of the previously discussed polymers for adhesive applications, the polybenzimidazoles (1–3), the polybenzothiazoles (6,7), the polybenzoxazoles (6,7), the polyimides (3–7), the poly-1,3,4-oxadiazoles (8), and the copolyamide-imides (9,10) have been studied as binders or laminating resins in composite structures. The polybenzimidazoles and polyimides have been most extensively investigated.

TABLE IV
Flexural Strength of Fiberglas Laminates with Various Binders (1–7,9,10)

Test temp. (°C)	Aging conditions	Flexural strength (psi × 10⁻³)				
		polyimide	Copoly-amide-imide	Epoxy	Silicone	Polybenz-imidazole
25	None	43.5	45.0	70	39	113.3
25	2 hr in boiling water	42.1	—	—	—	74.4
300	½ hr at 300°C	33.0	22.0	22	20	74.6
300	100 hr at 300°C	31.5	22.0	21	19	74.0
300	200 hr at 300°C	27.9	21.0	20	19	50.1
300	500 hr at 300°C	21.5	18.0	19	17	15.1
300	1000 hr at 300°C	16.6	8.0	—	—	8.0
300	1500 hr at 300°C	12.2	0.0	—	—	—
300	2000 hr at 300°C	13.8	0.0	—	—	—
320	½ hr at 320°C	26.0	18.5	—	—	112.1
320	100 hr at 320°C	28.7	22.0	—	—	47.0
320	200 hr at 320°C	23.7	18.0	—	—	12.1
320	500 hr at 320°C	13.7	10.0	—	—	0.0
320	1000 hr at 320°C	2.9	0.0	—	—	—
360	½ hr at 360°C	15.6	—	—	—	89.2
360	16 hr at 360°C	18.6	—	—	—	12.1
360	34 hr at 360°C	13.0	—	—	—	6.3
425	½ hr at 425°C	15.6	—	—	—	66.8

Fiberglas laminates made with polyimide binders exhibited the expected high temperature resistance characteristic of the polyimides themselves (Table IV). These had high flexural strength, superior to that of the silicones, comparable to that of the copolyamide-imides, and inferior to that of the epoxys and polybenzimidazoles. However, at temperature of 300–320°C the flexural strength of the polyimide and copolyamide-imide impregnated laminates was superior to that of both the epoxy and the silicone. These retained their strength even after long-term exposure at these temperatures, whereas the epoxy and the silicone composites did not, showing no strength at 320°C. At temperatures above 300°C, the flexural strength of composite structures made with the copolyamide-imide as laminating resin did not exhibit the long term thermal stability of the polyimide.

The effect of the presence of the amide grouping in the copolyamide-imide on electrical properties at elevated temperatures was best demonstrated by the temperature-power factor relationship (Fig. 2) (9). The

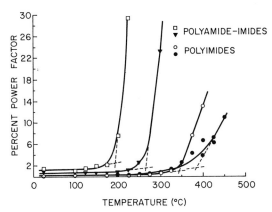

Fig. 2. Effect of temperature on power factor (100 tan δ) for several aromatic polyimides at 1 kc determination of T_g. Intersection of dotted lines approximates T_g (9).

curve for the copolyamide-imide composite showed a marked change in slope approximately 200°C lower than that of the polyimide.

The polybenzimidazole laminates exhibited behavior similar to that found in adhesive applications. These laminates initially had extremely high flexural strength at room and elevated temperatures (Table IV) but showed rapid losses in strength on exposure for long periods of time, to elevated temperatures (1–3). This behavior, in addition to being due to the oxidative attack on the polybenzimidazole resins, also reflected the extremely low density of these laminates due to the high loss of volatiles. Current research aimed at obtaining more dense laminates has yielded experimental composites which have improved retention of strength on aging at elevated temperatures.

In filament wound composite structures, the polyimides exhibited excellent thermal stability and retention of properties at elevated testing temperatures (3). In hoop tensile strength (Table V) the polyimide

TABLE V
Hoop Tensile Strength for Filament Wound Composites (1–7)

Resin	Hoop tensile composite strength (psi)	
	30°C, no age	300°C, ½ hr
Polyimide	185,000	171,700
Polybenzimidazole	160,000	120,000
Epoxy	171,300	111,500
Silicone	139,800	89,600

TABLE VI
Horizontal Shear Strength for Filament Wound Composites (1–7)

Resin	Resin content (%)	Horizontal shear strength (psi)			
		30°C	130°C, ½ hr	300°C, ½ hr	360°C, ⅙ hr
Polyimide	16.0	4600	3700	1990	980
Polybenzimidazole	13.6	6100			1333
Epoxy	19.0	7800	5460	1040	160
Silicone	24.2	3100	1400	460	375

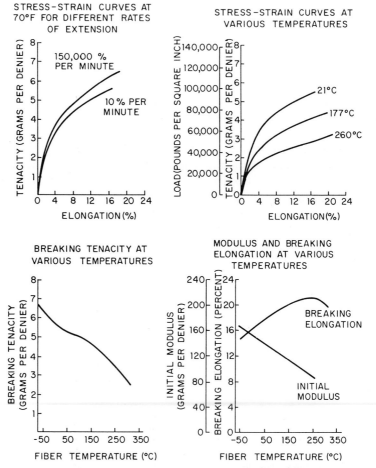

Fig. 3. Stress–strain data for filament yarns of Nomex high temperature resistant nylon. Specimens tested at elevated temperatures were conditioned by heating for 5 min in dry air (12).

composites had superior strength both at room temperature and at 300°C. In horizontal shear strength (Table VI), they showed acceptable strength levels, with better retention of these levels at temperatures of 300°C and above than the polybenzimidazoles or the commercially available resins.

The polybenzoxazoles, polybenzothiazoles, and polyquinoxalines have received only preliminary evaluation in this particular application (6,7). For the most part, the flexural strengths of the composites have been too low for meaningful evaluation. However, in all cases, these composites showed a rapid falling off in strength on high temperature ageing.

D. FIBERS

By far, the most extensive application for the high temperature resistant polymers has been in fiber uses. In addition to the polyaromatic heterocyclics, polyphenylene amides (11–16), polyphenylene hydrazides (17), polyphenylene sulfonamides (18), poly-N-alkylterephthalamides (19), and copolyphenylene amides (20) have been evaluated in this application. By far the most thoroughly investigated class has been the polyphenylene amides.

The aromatic polyamides, as exemplified by Nomex nylon fiber, a commercial product, have an excellent balance of physical and chemical properties, and should have broad utility in industrial and military applications involving exposure to temperatures up to 250–300°C (Fig. 3 and Table VII). This fiber is approximately 50% as strong at 285°C as it is at

TABLE VII

Typical Properties of Nomex High Temperature Resistant Nylon at 70°F (21°C) (12)

	Straight test	Loop test
Breaking strength, lb (for 200 den yarn)		
Conditioned at 65% RH	2.4	2.3
Wet	1.8	—
Breaking tenacity, g/den		
Conditioned at 65% RH	5.5	5.2
Wet	4.1	—
Breaking elongation, %		
Conditioned at 65% RH	17	14
Wet	14	—
Specific gravity	1.38	
Specific heat, cal/g-°C	0.29	
Coefficient of linear expansion (after heat-setting at 500°F)		
in./in.-°F between 70 and 400°F	1.12×10^{-5}	
Thermal conductivity, BTU/hr-ft²-°F for 1 in. thickness	0.9	
Finish on fiber and yarn, %	1.3	

TABLE VIII
Chemical Resistance of Nomex High Temperature Resistant Nylon (12)

Chemical	Concentration (%)	Temperature (°F)	Temperature (°C)	Time (hr)	Effect on breaking strength				
					None[a]	Slight[b]	Moderate[c]	Appreciable[d]	Degraded[e]
Acids									
Benzoic	100	70	21	10	X				
Formic	90	70	21	10	X				
Hydrochloric	10	203	95	8				X	
Hydrochloric	35	70	21	10	X				
Hydrochloric	35	70	21	100				X	
Nitric	10	70	21	100	X				
Nitric	70	70	21	100				X	
Peracetic	100	70	21	10	X				
Sulfuric	10	70	21	100	X				
Sulfuric	10	140	60	1000			X		
Sulfuric	60	140	60	100			X		
Sulfuric	70	70	21	100	X				
Sulfuric	70	203	95	8				X	
Alkalis									
Ammonium hydroxide	28	70	21	100	X				
Sodium hydroxide	10	70	21	100	X				
Sodium hydroxide	10	140	60	100		X			
Sodium hydroxide	10	203	95	8				X	
Sodium hydroxide	40	70	21	10	X				
Sodium hydroxide	50	140	60	100					X

VII. APPLICATIONS

Miscellaneous Chemicals				
N,N-Dimethylformamide	100	158	70	168	X
Ethylene glycol	100	158	70	168	X
Jet fuel	100	158	70	168	X
Perchloroethylene	100	158	70	168	X
Phenol	100	70	21	10	X
Sodium chlorite	0.5	70	21	10	X
Sodium chlorite	0.5	140	60	100	X
Sodium hypochlorite	0.4	70	21	10	X
Stoddard solvent	100	70	21	10	X
Xylene	100	158	70	168	X
Sealed-Tube Exposures				
Air + 5% water + 5% sulfur dioxide	—	347	175	100	X
Freon-22 refrigerant	—	356	180	1000	X
Sulfur hexafluoride	—	356	180	1000	X
Steam	—	311	155	100	X

[a] None: 0–9% strength loss.
[b] Slight: 10–24% strength loss.
[c] Moderate: 25–44% strength loss.
[d] Appreciable: 45–79% strength loss.
[e] Degraded: 80–100% strength loss.

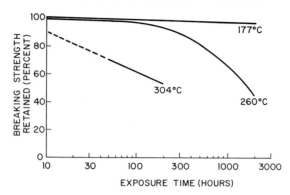

Fig. 4. Strength retained by Nomex nylon after exposure to hot, dry air. Specimens tested at 70°F, 65% RH (12).

room temperature, and shows a similar retention of modulus at this temperature (11). In addition, it has excellent resistance to oxidation at elevated temperatures, retaining over half its original strength even after exposure to over 1000 hr at 260°C and over 200 hr at 300°C. At lower temperatures ($\sim$200°C) this fiber has withstood several thousand hours of exposure with essentially no change (Fig. 4) (12). As might be expected, it

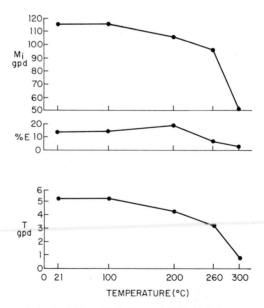

Fig. 5. Fiber properties (initial modulus, M_i; elongation, E; tenacity, T) of **1** at elevated testing temperatures (17).

VII. APPLICATIONS

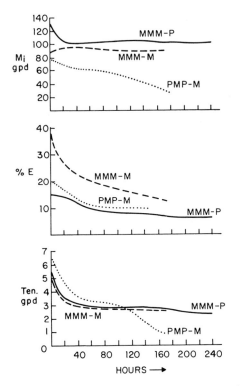

Fig. 6. Thermal durability of selected ordered copolyamide fibers in air at 300°C (20).

TABLE IX
Fiber Properties of Aromatic Polyhydrazides (17)

Properties	1[a]	2[b]
Tenacity, g/den	5.2	6.0
Elongation, %	14	8.0
Initial modulus, g/den	115	151
Knot tenacity, g/den	2.1	1.0
Knot elongation, %	7.0	4.2
Knot initial modulus, g/den	50	90
Fiber stick temperature, °C	320	335
Moisture regain	4.5	4.3
Density, g/cc	1.443	1.452

[a] 1: Fiber from poly(isophthalic hydrazide).
[b] 2: Fiber from alternating copolyhydrazide of isophthalic and terephthalic acids.

does not melt as most synthetic fibers do and is ignited only with the greatest of difficulty by direct exposure to a flame. The flame is rapidly self-extinguishing as soon as the ignition source is removed. In addition, this polyamide fiber possesses a useful balance of other properties including strength, coverage, dimensional stability, and resistance to chemical and solvent attack (12). This resistance to solvent attack is amply demonstrated in Table VIII.

The other phenylene-R polymers which have been evaluated in fiber form have been studied only under experimental or development programs.

TABLE X-A
Fiber Properties of Wholly Aromatic Ordered Copolyamides (20)

Properties	MMM-M[a]	MMM-P[b]	MMM-B[c]	MMM-N[d]	PMP-P[e]
Tenacity, g/den	5.1	6.0	5.9	6.3	6.2
Elongation, %	3.7	23	16	19	20
Initial modulus, g/den	89	101	93	94	90
Moisture regain, %	—	5.2	—	4.5	—
Density, g/cc	1.35	1.36	—	1.35	1.36

[a] MMM-M

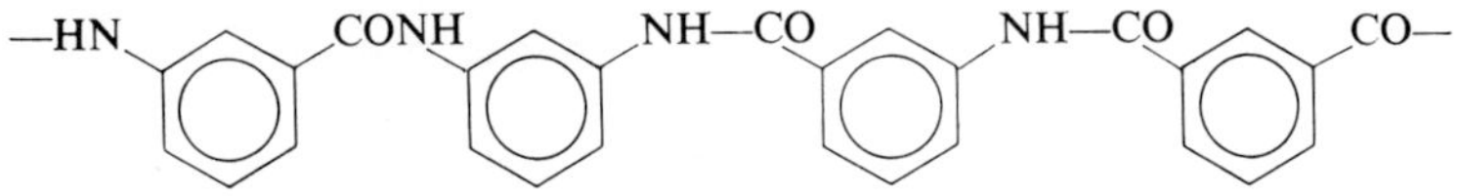

—HN—[C₆H₄]—CONH—[C₆H₄]—NH—CO—[C₆H₄]—NH—CO—[C₆H₄]—CO—

[b] MMM-P

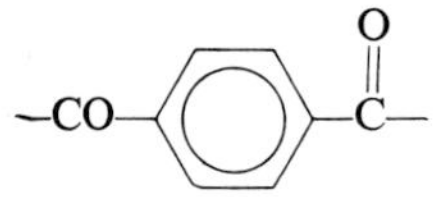

—CO—[C₆H₄]—CO—

[c] MMM-B

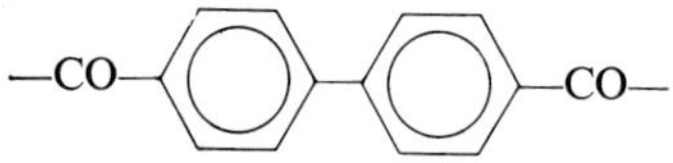

—CO—[C₆H₄]—[C₆H₄]—CO—

[d] MMM-N

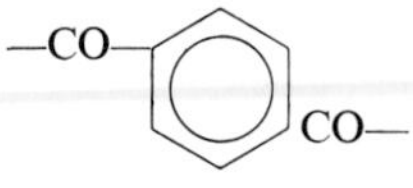

—CO—[C₆H₄]—CO—

[e] PMP-M

—H—N—[C₆H₄]—CO—NH—[C₆H₄]—NH—CO—[C₆H₄]—NH—CO—[C₆H₄]—CO

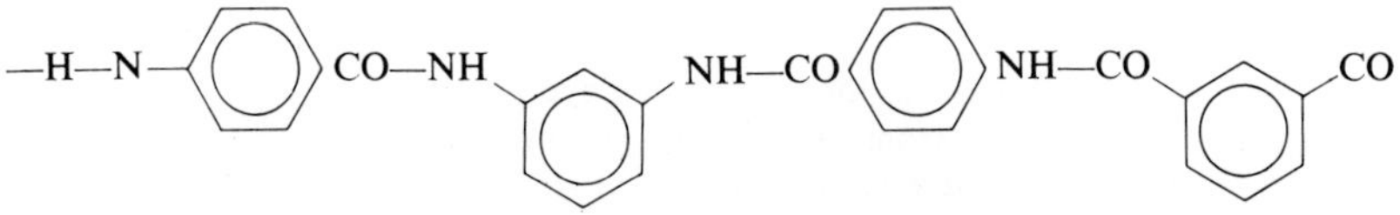

TABLE X-B
Tensile Properties of Fibers from Wholly Aromatic Ordered Copolymers at Elevated Temperatures (20)

Temp. (°C)	T/E/Mi[a]				
	MMM-M[b]	MMM-P[b]	MMM-B[b]	MMM-N[b]	PMP-P[b]
100	4.6/24/85	—	3.4/13/58	5.6/—/—	—
150	—	4.0/12/107	—	—	4.9/20/6
200	3.0/29/68	—	—	—	—
250	—	—	1.6/15/58	—	3.3/21/46
300	1.4/21/11	2.0/11/50	—	2.0/19/—	2.3/19/36
350	0.5/15/7	1.1/11/24	0.67/13/6	1.2/24/—	1.0/14/14
400	—	0.8/11/16	0.6/13/6	0.86/10/—	—
450	—	0.6/5/—	0.2/7/—	—	0.4/5/13
ZST (°C)[c]	>400	485	>450	455	470

[a] T/E/Mi- tenacity, gpd/elongation, %/initial modulus, gpd.
[b] See Table X-A for code.
[c] ZST (°C): zero strength temperature (°C).

The polyphenylene hydrazides exhibited properties similar to the aromatic polyamides (17) (Table IX and Fig. 5). However, on exposure to temperatures in excess of 260°C, these fibers were converted to the more stable poly-1,3,4-oxadiazole fibers (17). The polyphenylene sulfonamides (18) and the poly-N-alkylterephthalamides (19) fibers had very high softening temperatures but did not exhibit the same measure of thermal stability as the aromatic polyamides. Recently the properties of fibers from ordered copolyamides have been reported (20). These properties, in general, were quite similar to those of aromatic polyamides (Fig. 6 and Table X). The overall stability of these fibers appeared to be comparable to those derived from the polyamides.

All of the investigations of the aromatic heterocyclic polymers are in the experimental or developmental stage. The polybenzimidizoles have been fabricated into fibers and showed properties similar to that of the parent polymer (21,22) (Table XI). These fibers had lower thermal durability in air up to 300°C than the aromatic polyamides. At temperatures above 300°C the overall stability of the polybenzimidazole fibers to long-term exposure in air was poorer than that of the other polyaromatic heterocyclic fibers and was only marginally better than that of the aromatic polyamide fibers. The one outstanding property of the polybenzimidazole fibers appeared to be their retention of strength at elevated testing temperatures. For certain applications, such as decelerators, this particular property may be of great importance (21,22).

TABLE XI
Properties of Fibers from Polyaromatic Heterocyclics (20–26)

Polymer[a]	Tensile properties (30°C)						Thermal properties		
	Straight			Loop			$\dfrac{\text{Tenacity at 300°C}}{\text{Tenacity at 30°C}} \times 100$	Tenacity half-life at 283°C[b]	Tenacity half-life at 400°C[b]
	Tenacity (gpd)	Elong. (%)	Modulus (gpd)	Tenacity (gpd)	Elong. (%)				
Polyimide	6.9	13	78	84	75		43	750 hr	6 hr
Polybenzimidazole	4.9	21	115	80	30		75	<250	3
Poly-1,3,4-oxadiazole	5.3	10	216	70	50		60	>1000	24
Poly-1,3,4-thiadiazole	3.5	14	78	—	—		60	—	24
Poly-N-phenyltriazole	3.1	8	86	—	—		32	<100	<50
Polyoxadiazoleamide I	6.7	11	114	—	—		37	~168	—
Polyoxadiazoleamide II	6.5	7	126	—	—		34	>168	>6
Polythiazoleamide	3.4	13	88	—	—		47	96	—
Polythiazoleamide	7.8	7	168	—	—		40	—	—

VII. APPLICATIONS   301

a Polyoxadiazoleamide I

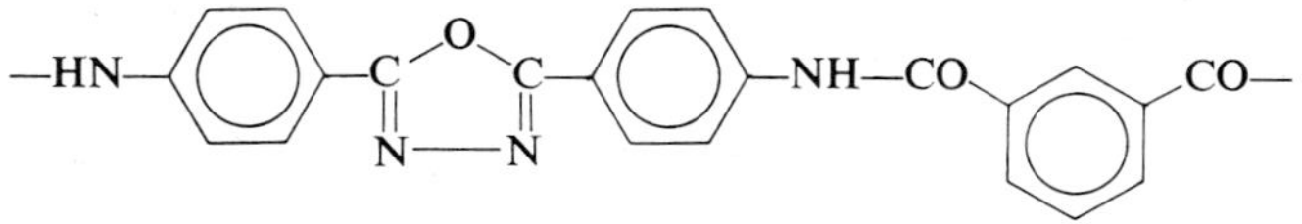

Polyoxadiazoleamide II

Polythiadiazoleamide

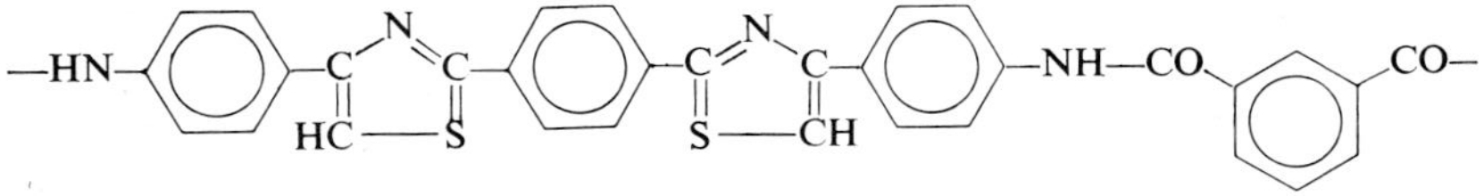

Polybithiazoleamide

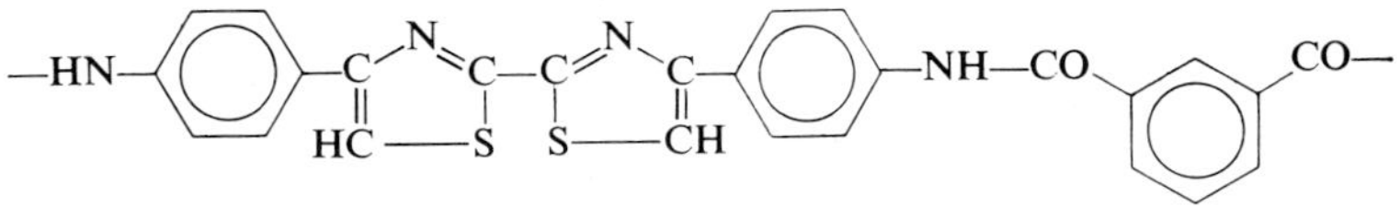

b Tenacity half-life: time required in hours to reach one-half original tenacity.

The polyimide fibers have been prepared with excellent mechanical properties, i.e., high strength, and have exhibited good long-term stability under air aging conditions at temperatures in excess of 300°C (23) (Table XI). This, coupled with their oustanding chemical resistance, make them extremely attractive as high temperature resistant fibers for applications requiring stability above that of the aromatic polyamide fibers (23).

The poly-1,3,4-oxadiazoles and the poly-1,3,4-thiadiazoles have also been fabricated in fiber form and have been evaluated as high temperature resistant fibers (24,25) (Table XI). Both of these polymers exhibited an extremely high level of durability in air aging tests. In the case of the poly-1,3,4-oxadiazoles, fibers have been obtained with high strength (24), but poly-1,3,4-thiadiazoles have been obtained with only marginally acceptable strength levels (25). Fibers from these polymers were more stable in thermal air aging tests than any of the previously discussed polyaromatic heterocyclic fibers and thus may find some utility on the basis of this particular property.

The poly-N-phenyltriazoles have also been evaluated as high temperature resistant fibers (26). The physical properties of these fibers were somewhat similar to those of the properties poly-1,3,4-thiadiazole fibers, but they unfortunately had rather disappointing thermal stability (Table XI).

Fibers from ordered aromatic heterocyclic-amide copolymers have been prepared and evaluated as high temperature resistant fibers (20) (Table XI). These fibers possessed excellent mechanical properties and thermal stabilities superior to wholly aromatic polyamide fibers but inferior to the wholly aromatic heterocyclic fibers containing the same structural units.

The one example of a ladder polymer in fiber form is Black Orlon acrylic fiber, pyrolyzed polyacrylonitrile fiber (27,28). Although having marginally acceptable tensile properties, it showed no change in properties after instantaneous exposure to an open flame, 5 min initial exposure at 500°C, and 10 sec exposures at 760°C.

E. FILMS

Although most of the previously discussed polymers have been prepared in film form for characterization purposes, only one, the polyimides, have been extensively studied and developed as a self-supporting film. From this work a commercial product, Kapton polyimide film, has emerged.

This polyimide film has excellent mechanical properties throughout a wide temperature range, from liquid helium temperatures to as high as 450°C, with high tensile and impact strength, and high resistance to tear initiation (29). In general, its room temperature properties are comparably to those of Mylar polyester films (Table XII). The film exhibits moderately

VII. APPLICATIONS

TABLE XII
Mechanical and Physical Properties of Kapton Polyimide Film and Mylar Polyester Film (29)

			Value	
Property	ASTM test	Temp. (°C)	Kapton	Mylar
Tensile strength, psi	D-882	25	25,000	23,000
		200	17,000	7,000
Yield point, psi	—	25	14,000	12,000
		200	9,000	1,000
Stress to product 5% elongation, psi	D-882	25	13,000	13,000
Ultimate elongation, %	D-882	25	70	100
		200	90	Large
Tensile modulus, psi	D-882	25	430,000	550,000
		200	260,000	50,000
Bursting strength, psi	D-774	25	75	30
Density, g/cc	—	25	1.42	1.40
Coefficient of friction, kinetic, film-to-film	D-1505	25	0.42	0.45
Area factor, sq ft/lb-mil	—	25	135	140

high dielectric constant, low dissipation factor, high volume and surface resistivity, and high dielectric strength (Table XIII). Although the electrical properties are affected by temperature and frequency, these effects are considerably less than those for Mylar polyester film. The dissipation

TABLE XIII
Electrical Properties of Kapton Polyimide film and Mylar Polyester Film (30)

			Value	
Property	ASTM test	Temp. (°C)	Kapton	Mylar
Dielectric strength at 60 cps, 1 mil thick, V/mil	D-149	25	7000	7000
		150	6000	5000
Dielectric constant at 10^3 cps	D-150	25	3.5	3.1
		200	3.0	
Dissipation factor at 10^3 cps	D-150	25	0.003	0.0047
		200	0.002	0.01
Volume resistivity, ohm-cm	D-257	25	10^{18}	10^{18}
		200	10^{14}	5×10^{11}
Surface resistivity at 1 kV, 50% RH, ohms	D-257	25	10^{16}	10^{16}

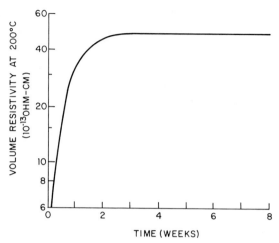

Fig. 7. Effect of thermal aging at 300°C of 1-mil Kapton polyimide film on volume resistivity. Resistivity was measured at 200°C (29).

factor is relatively constant through 50–200°C, and dielectric constant is essentially unchanged with frequency variations at all temperatures. The dissipation factor increases with frequency at room temperature and becomes constant at higher temperature.

The zero strength of the polyimide film (20 psi load, 5 sec to failure) is 800°C, cut-through temperature, 435°C. The film has no melting point and only chars when exposed to direct flame (30).

Prolonged thermal aging (8 weeks at 300°C) has no effect on volume resistivity after the first 2 weeks (Fig. 7). In the same test, dielectric strength after 8 weeks is 90% of the original value. Thermal aging to embrittlement (extrapolated), would require 10 years at 260°C.

F. PAPERS

Although some work has been done on a number of the previously discussed polymer systems in paper applications, only one such evaluation has been successful enough that a commercial product has been realized. This is from the aromatic polyamides, and the product is the Nomex nylon papers.

These polyamide papers, like the polyamide fibers, do not melt, are consumed by fire, but do not burn independently. All the interesting electrical, thermal, mechanical, and chemical properties characteristic of the polyamide fibers are found in the papers (11).

The mechanical properties of this aromatic polyamide paper are illustrated in Table XIV and Figs. 8 and 9. The tensile strength of the paper

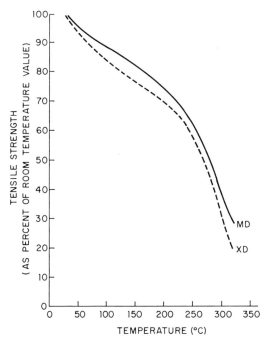

Fig. 8. Tensile strength vs. temperature. 10-mil Nomex nylon paper (31).

TABLE XIV
Physical Properties of Typical Nomex Nylon Papers

Thickness, mils		2	3	5	7	10	15	20	30	ASTM test no.
Tensile strength, lb/in.	MD[a]	22	37	59	120	170	300	440	550	D-828-60
	XD[b]	13	23	35	68	100	180	250	380	
Elongation, %	MD	7	10	10	15	17	22	19	19	D-828-60
	XD	6	9	9	12	13	15	15	15	
Elmendorf tear, g	MD	95	150	290	380	550	850	1100	2000	D-689
	XD	140	250	470	620	900	1400	2100	2700	
Finch edge tear, lb/in.	MD	13	23	43	77	100	110	150	190	D-827-47
	XD	6	10	18	22	32	43	54	76	
Shrinkage at 285°C, %	MD	1.6	1.2	0.9	0.5	0.3	0.3	0.3	0.2	—
	XD	1.9	1.6	1.3	1.0	0.4	0.4	0.4	0.4	
Thermal conductivity, BTU in./hr-ft^2-°F		—	—	—	—	0.76	—	—	—	C-177-45
Basis weight, oz/yd^2		1.2	1.9	3.2	5.1	7.3	11	15	23	

[a] MD = machine direction.
[b] XD = cross direction.

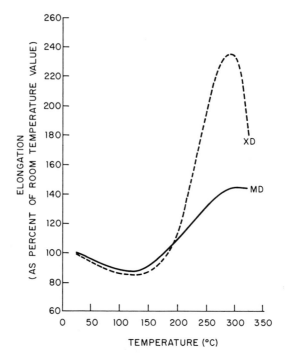

Fig. 9. Elongation vs. temperature. 10-mil Nomex nylon paper (31).

decreases only gradually with temperature and still maintains 68% of its room temperature value at 225°C.

Elongation is essentially constant up to 200°C (31). With extremely thick papers, dimensional stability is quite outstanding, showing less than 1% shrinkage at temperatures up to 285°C, and as thickness decreases, shrinkage increases up to 2%. The dimensions of the paper are unaffected

TABLE XV
Dimensional Stability of Nomex Nylon Papers to Changes in Relative Humidity (31)

Relative humidity (%)	3 Mil Nomex paper			10 Mil Nomex paper		
	Regain (%)	Expansion (%)		Regain (%)	Expansion (%)	
		MD	XD		MD	XD
Oven dry	0	0	0	0	0	0
50	2.9	0.4	0.5	3.5	0.4	0.5
65	4.9	0.6	0.5	5.1	0.6	0.9
95	7.7	0.9	1.6	8.4	1.1	1.8

TABLE XVI
Electrical Properties of Typical Nomex Nylon Papers (31)

Thickness (mils)	2	3	5	7	10	15	20	30
Dielectric strength (V/mil)[a]								
ac rapid rise	450	500	600	750	750	800	700	600
ac 1-min hold	—	—	—	—	600	—	—	—
ac 1-hr hold	—	—	—	—	500	—	—	—
dc rapid rise	—	—	—	—	1200	—	—	—
Dielectric constant (10^3 cps)[b]	2.2	2.3	2.6	2.8	2.9	3.1	3.4	3.3
Dissipation factor (10^3 cps)[b]	0.008	0.010	0.011	0.013	0.014	0.015	0.016	0.016

[a] ASTM D-149, using 2 in. diam electrodes.
[b] ASTM D-150, 202, using 1 in. diam electrodes under 20 psi pressure.

by changes in humidity up to 65% and even at 95% there is less than 2% linear expansion (Table XV) (31).

The electrical properties of these papers are shown in Table XVI. The volume resistivity of the paper at room temperatures and ambient humidity is about 1.3×10^{16} ohm-cm, regardless of thickness. Resistivity decreases somewhat with increasing temperature but still exceeds 10^{11} ohm-cm at 250°C (Fig. 10). The dielectric constant of the paper is maintained at low levels over a wide range of temperatures and frequencies (Fig. 11), and the dissipation factor reaches a minimum at 180°C, in contrast to other insulating materials (Fig. 12). The dielectric strength of these papers is

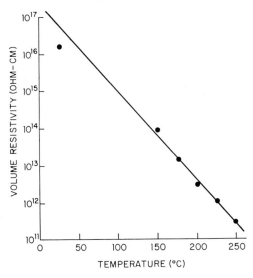

Fig. 10. Volume resistivity vs. temperature. 10-mil Nomex nylon paper (31).

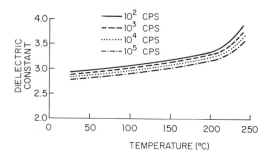

Fig. 11. Dielectric constant vs. temperature. 10-mil Nomex nylon paper (31).

unaffected by variations in temperature up to 200°C, and about 95% of its room temperature value is maintained at 250°C (31).

The resistance to high temperature of this aromatic polyamide paper reflects again the stability of the parent polymer, and this resistance permits the retention of electrical and mechanical properties over long periods of continuous exposure at high temperatures (Figs. 13–15). These data suggest that the polyamide papers could be expected to maintain a dielectric strength of at least 300 V/mil, as well as approximately 50% initial tensile strength and break elongation after 10 years of continuous service at 220°C. The long-term resistance of the paper considerably exceeds the minimum requirements of Class H materials (32).

These papers, like the aromatic polyamide fiber, show outstanding stability and resistance to attack in the presence of other chemical agents (Table XVII).

The properties of polyimide coated polyamide paper have been reported (33). This product offers a unique combination of polyamide–polyimide polymer properties in this particular application. In Tables XVIII and XIX are tabulated the mechanical and electrical properties at

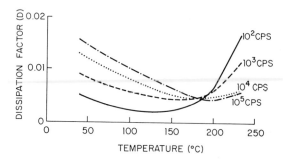

Fig. 12. Dissipation factor vs. temperature. 10-mil Nomex nylon paper (31).

VII. APPLICATIONS 309

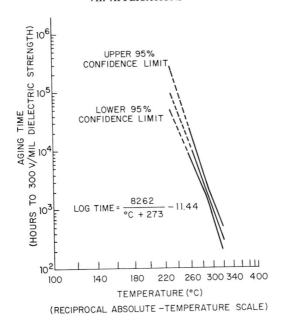

Fig. 13. Useful life vs. temperature. 10-mil Nomex nylon paper (31).

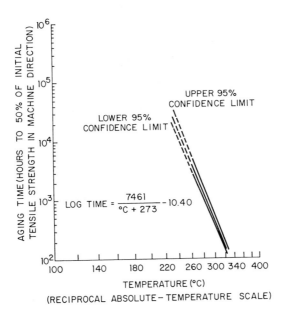

Fig. 14. Useful life vs. temperature. 10-mil Nomex nylon paper (31).

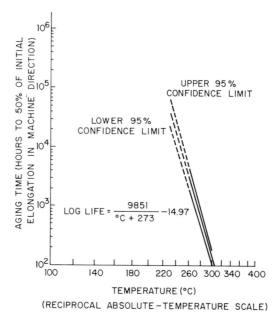

Fig. 15. Useful life vs. temperature. 10-mil Nomex nylon paper (31).

room temperature of the polyimide-coated polyamide papers along with those of the uncoated polyamide papers. The coated papers had a slightly higher cut-through temperature and a somewhat reduced moisture absorption than the uncoated papers. The major difference was that the

TABLE XVII
Chemical Resistance of Nomex Nylon Papers (31)

Chemical	Temp. (°C)	Exposure time	Tensile strength retained (%)
70% Sulfuric acid	21	100 hr	100
70% Sulfuric acid	95	8 hr	50
70% Nitric acid	21	100 hr	50
10% Sodium hydroxide	21	100 hr	100
10% Sodium hydroxide	95	8 hr	50
Perclene	70	7 days	99
Xylene	70	7 days	97
Ethylene glycol	70	7 days	93
N,N-Dimethylformamide	70	7 days	93
Freon F-11, F-12, F-22	25	150 days	100
Freon F-11	150	150 days	95
Freon F-12	150	150 days	100
Freon F-22	150	150 days	80

TABLE XVIII
Mechanical Properties of Polyimide-Coated Polyamide Papers at Room Temperature (33)[a]

	Polyamide and polyimide-coated polyamide papers									Polyimide coated glass fabric			Polyimide film	Remarks
	2 and 2.5–3.2 mil			5 and 5.1–5.6 mil			10 and 10.5–11.0 mil			4 mil	7 mil	10 mil		
	Uncoated	B-staged	Cured	Uncoated	B-staged	Cured	Uncoated	B-staged	Cured					
Tensile strength, psi				12,000	12,000	12,000				22,000	28,000	32,000	25,000	ASTM D-828
Tensile modulus, psi					2.8×10^4	4.6×10^4	3.0×10^4				10^6	10^6	10^6	43.0×10^4	ASTM D-828
Elongation, % (MD)				10	9	8				3	3	2	70	ASTM D-828
Elmendorf tear, g/mil (MD)	48	20	33	58	53	45	55	27	43	31	57	160	8	ASTM D-689
Specific gravity	0.73	0.95	0.93	0.87	0.90	0.90	0.96	0.98	0.96	1.9	1.7	1.7	1.42	
Cut-through temp., °C (°F)			400° (752°)	390° (730°)	390° (734°)	430° (802°)							525° (977°)	New value of Film Dept. (See note)
Moisture absorption, %	6	14	2	7	6	4	7	7	6	2	2	2	2	48 hr at 73°F and 95% RH
Basis weight, oz/sq yd	1.2	2.3	1.7	3.2	3.7	3.5	7.3	8.1	7.9	2.9	7.0	10.2	5.3 (5 mil)	
Coating weight, oz/sq yd	—	1.1	0.5	—	0.5	0.3	—	0.8	0.6	1.5	3.8	4.8	—	

[a] Equipment specified in ASTM D-876-61. A 1/16 in. diameter steel ball is pressed into sheet material under a 1 kg load. Equipment is located in an oven and a temperature rise of 0.5°C/min is used. Cut-through temperature is reported as temperature at which 110 volt circuit is completed between the ball and the steel backing plate.

TABLE XIX
Electrical Properties of Polyimide-Coated Polyamide Papers at Room Temperature (33)

	Polyamide and polyimide-coated polyamide papers									Polyimide coated glass fabric			Polyimide film	Remarks
	2 and 2.5–3.2 mil			5 and 5.1–5.6 mil			10 and 10.5–11.0 mil							
	Uncoated	B-staged	Cured	Uncoated	B-staged	Cured	Uncoated	B-staged	Cured	4 mil	7 mil	10 mil		
Dielectric strength V/mil 2 in. flat electrode	425	850	700	600	650	700	750	850	950	500	975	600	7000	ASTM D-149-59
Curved	600	850	1000	650	800	750	850	950	900					ASTM 149
Crease	425	350	550	500	550	550	650	600	550					Paper creased, cured, and then measured
Dissipation factor[a]	0.022	0.028	0.009	0.013	0.017	0.017	0.020	0.030	0.020	0.005	0.005	0.005	0.003	10^3 cps ASTM D-150
Dielectric constant[a]	4.4	6.6	5.8	5.9	4.8	4.1	5.1	5.1	5.3	3.5	3.3	3.3	3.5	10^3 cps ASTM D-150
Volume resistivity, -cm	9×10^{16}	3×10^{16}	5×10^{17}	6×10^{16}	2.5×10^{16}	6×10^{17}	1.4×10^{16}	1.2×10^{16}	1.8×10^{17}	10^{13}–10^{14}	10^{13}–10^{14}	10^{13}–10^{14}	10^{18}	ASTM D-257-58

[a] Values determined according to ASTM D-150 using silver paint electrodes.

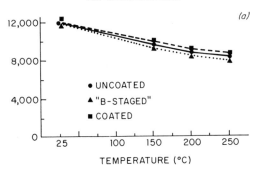

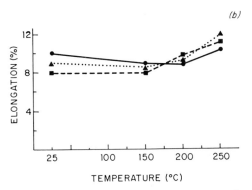

Fig. 16. (a) Tensile strength vs. temperature. 5-mil Nomex nylon paper (33). (b) Elongation vs. temperature. 5-mil Nomex nylon paper (33).

electrical properties of the coated paper were improved not only because of the improved barrier properties of the film coating, but also because of the better electrical properties of the polyimides (33).

At elevated temperatures, the tensile properties of the coated and uncoated polyamide papers were practically identical, showing only a slight

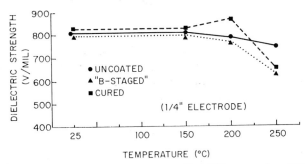

Fig. 17. Dielectric strength vs. temperature. 10-mil Nomex nylon paper (33).

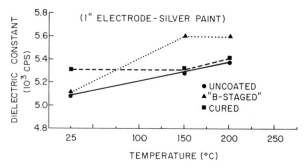

Fig. 18. Dielectric constant vs. temperature. 10-mil Nomex nylon paper (33).

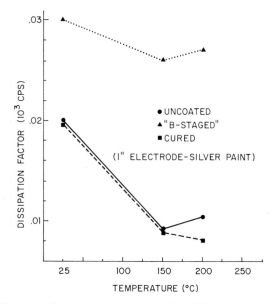

Fig. 19. Dissipation factor vs. temperature. 10-mil Nomex nylon paper (33).

decrease in tensile strength and increase in elongation with increasing temperature Fig. 16 (*a*), (*b*). The electrical properties of the coated and uncoated papers at elevated temperatures were quite similar (Figs. 17–19).

The major advantage of these coated structures is that these papers can be used in heat sealable applications.

G. RESINS

The use of the previously discussed polymers as moldable resins has received little attention primarily because these intractable materials are exceedingly difficult to fabricate into useful forms. The one exception to this has been the polyimides which have been utilized in resin applications (34,35).

The polyimide resins are strong, rigid materials which have excellent resistance to heat, abrasion, and nuclear radiation. In addition, they have good electrical properties along with other properties characteristic of the other polyimide-based products which have been discussed previously (34).

Fabricated polyimide products have been used continuously in air at temperatures of 260°C and in an inert atmosphere as high as 325°C. Even intermittent or short-time exposures at operating temperatures as high as 450°C were possible. In addition, laboratory and field tests of various component parts in liquid nitrogen system, have suggested potential cryogenic applications.

The polyimide resins have high retention of tensile strength and stiffness at elevated temperatures (Table XX) along with excellent creep resistance. In addition, because of the infusibility of the polyimide, these resins are useful in those applications in which wear and friction are encountered. The rubbing surface reaches temperatures above 350°C before a rapid increase in wear rate signals that the PV limit of the material had been reached. Short-term PV values in excess of 100,000 are possible without

TABLE XX

Mechanical and Thermal Properties of a Polyimide Resin (34)

Property	ASTM test	Temp. (°F)	Value
Tensile strength, psi	D-708	73	13,500
		302	9,700
		482	7,700
		600	5,000
		752	3,500
Elongation, %	D-708	77	6–8
Shear strength, psi	D-732	73	11,900
Compressive strength, psi	D-695	73	24,400
Flexural modulus, 1000 psi	D-790	−310	515
		73	450
		392	225
		482	210
		752	90
Coef, of linear expan., in./in.-deg F	—	73–750	28–35
Thermal conductivity, BTU-in./sq ft-hr-deg F	—	—	2.20
Specific heat, BTU/lb-deg F	D-648	—	0.27
Heat distortion temp., 264 lb (F)	D-648	—	>473

TABLE XXI
Friction and Wear of Polyimide Resin Against Carbon Steel (34)

Environment	Pressure, P (psi)	Velocity, V (fpm)	Coefficient of friction[a,b]	Wear rate (in./1000 hr)
		$PV = 25,000$		
Air	30, 60	834,417	0.17	0.3
Nitrogen	30, 60	834,417	0.10	0.0005
		$PV = 100,000$		
Nitrogen	240	417	0.05	0.002

[a] Initial surface finish, 8–10 in. rms.
[b] After break-in.

lubrication, but the wear rates are usually the controlling factor (34,35) (Table XXI).

The electrical properties of the polyimide resins are similar to those of the other polyimide products, having a good combination of high dielectric strength, low dielectric constant and low dissipation factor (Table XXII). Similarly, the chemical resistance of the polyimide resins is similar to that of the other polyimide products showing excellent resistance to attack by organic solvents and acids.

TABLE XXII
Comparison of Electrical Properties of Some Plastics (34)

Property	ASTM test	Material			
		Polyimide	DAP[a]	Phenolic[b]	TFE
Dielectric strength[c] ⅛ in. thick (V/mil)	D-149	400	450	300–425	400
Dielectric constant at 10^6 cps	D-150	3.4	3.4	4–7	2.1
Dissipation factor at 10^6 cps	D-150	0.003	0.01	0.03–0.07	0.0002
Volume resistivity, ohm-cm	D-257	10^{17}	10^{16}	10^{11}	10^{18}
Arc resistance, sec	D-495	230	118	5	Will not track

[a] DAP = unfilled diallyl phthalate; TFE = fluorocarbon resin.
[b] General purpose grade, cellulose filled.
[c] Short time.

References

1. H. H. Levine, M. B. Smith, M. B. Sheratee, K. J. Kjoller, and E. C. Janis, ML-TDR-64-8, Vol. I, Nov., 1963.
2. H. H. Levine, J. R. O'Neal, K. J. Kjoller, V. H. Noto, W. Wrasidlo, and M. B. Sheratte, AFML-TR-64-365, Pt. I, Vol. 1., Dec., 1963.
3. J. R. Courtright and C. K. Ikeda, Paper presented at SPE Regional Meeting, San Diego, Calif., 1964.
4. *Machine Design*, **36**, 234 (1964).
5. New Product Technical Information, PI-3301 FGR-Polyimide Adhesive and Binder, E. I. du Pont de Nemours & Co., Inc. (1964).
6. P. M. Hergenrother, J. L. Kukmeyer, and H. H. Levine, High Temperature Structural Adhesives, Contract NOW-64-0524-C, April, 1964.
7. P. M. Hergenrother, J. L. Kukmeyer, and H. H. Levine, High Temperature Structural Resins, Contract NOW-64-0524-C, May, 1965.
8. H. H. Levine, N. P. Loire, W. Wrasidlo, and V. E. Sanderson, ML-TDR-64-8, Vol. II, Nov., 1963.
9. J. H. Freeman, E. J. Traynor, J. Miglarese, and R. H. Lunn, *SPE Trans.*, **2**, No. 3, 216 (1962).
10. J. H. Freeman, L. W. Frost, G. M. Bower, and E. J. Traynor, *SPE Trans.*, **5**, No. 2, 75 (1965).
11. L. K. McCune, *Textile Res. J.*, **32**, 762 (1962).
12. New Product Technical Bulletin, NP-33, Properties of "Nomex" High Temperature Resistant Nylon Fiber, E. I. du Pont de Nemours & Co., Inc. (1964).
13. E. Mikolajewski and V. E. Swallow, Tech. Note CPM. 57, Jan., 1964.
14. H. Romeyn, ML-TDR-64-275, Aug., 1964.
15. H. Bockman, ML-TDR-64-208, Sept., 1964.
16. J. H. Ross and C. O. Littke, RTD-TDR-63-4031, Oct. 1964.
17. A. H. Frazer and F. T. Wallenberger, *J. Polymer Sci. A*, **2**, 1147 (1964).
18. S. A. Sundet, W. A. Murphey, and S. B. Speck, *J. Polymer Sci.*, **40**, 389 (1959).
19. V. E. Shashoua and W. M. Eareckson, *J. Polymer Sci.*, **40**, 343 (1959).
20. (a) J. Preston and F. Dobinson, *J. Polymer Sci. B*, **2**, 1171 (1964). (b) J. Preston and W. B. Black, *ibid.*, **3**, 845 (1965). (c) J. Preston and W. B. Black, *ibid.*, **4**, 267 (1966). (d) J. Preston and R. W. Smith, *ibid.*, **4**, 1033 (1966). (e) J. Preston, *J. Polymer Sci. A-1*, **4**, 529 (1966). (f) J. Preston and F. Dobinson, *ibid.* **4**, 2093 (1966). (g) J. Preston and W. B. Black, *ibid.*, *J. Polymer Sci. C*, **19**, 17 (1967). (h) J. O. Weiss, H. S. Morgan, and M. R. Lilyquist, *ibid.*, **19**, 29 (1967). (i) J. Preston, R. W. Smith, and C. J. Stehman, *ibid.*, **19**, 7 (1967).
21. R. W. Singleton, H. D. Noether, and J. F. Tracy, IR-8-163A (I), August, 1965.
22. R. W. Singleton, H. D. Noether, and J. F. Tracy, *J. Polymer Sci. C*, **19**, 65 (1967).
23. R. S. Irwin and W. Sweeny, *J. Polymer Sci. C*, **19**, 41 (1967).
24. A. H. Frazer, W. P. Fitzgerald, and T. A. Reed, ML-TDR-64-285, Aug. 1964.
25. A. H. Frazer, W. P. Fitzgerald, T. A. Reed, and L. W. Wilson, AFML-TR-221, Pt. 1, July, 1965.
26. J. R. Holsten and M. R. Lilyquist, *J. Polymer Sci. C*, **19**, 77 (1967).
27. R. C. Houtz, *Textile Res. J*, **20**, 786 (1950).
28. W. G. Vosburgh, *Textile Res. J.*, **30**, 882 (1960).
29. *Machine Design*, **36**, 232 (1964).
30. Technical Information Bulletin, H-1, "Kapton" Polyimide Film Summary of Properties, E. I. du Pont de Nemours & Co., Inc. 1965.

31. New Product Technical Bulletin, NP-31, Properties and Processing of "Nomex" High Temperature Resistant Nylon Paper, E. I. du Pont de Nemours & Co., Inc. (1964).
32. W. R. Clay, EIEET-137-12, Elect. Insul. Conf., Washington, February, 1962.
33. New Product Technical Properties and Processing of Heat-Sealable "Pyre-ML" Polyimide-Coated "Nomex" Nylon Paper, E. I. du Pont de Nemours & Co., Inc. (1964).
34. *Machine Design*, **36**, 230 (1964).
35. N. W. Todd, Paper presented at SPE Regional Meeting, Milwaukee, Wis., 1964.

AUTHOR INDEX

Numbers in parentheses are reference numbers and indicate that the author's work is referred to although his name is not mentioned in the text. Numbers in *italics* show the pages on which the complete references are listed.

A

Abe, Y., 6(51), *35*
Abramo, S. V., 31(107), *37*, 129(184), *137*, 160–164(44), 166–168(44), *211*
Abshire, E. J., 178(69), 184(69), 204(69), 205(69), *212*
Achhammer, B. D., 20(73), *35*
Adams, C. H., 11(68b), 13–16(68b), *35*
Adicoff, A., 26(87), *36*
Adkins, H., 40(24), *76*
Aelony, D., 107(93), *135*
Aftergut, S., 116(160), *136*
Akiyama, M., 178(68), *212*
Akiyoshi, S., 78(1), 79, *132*
Alexander, P., 85(32), *133*
Aliferis, J., 27(92), *36*
Alishoeva, A. B., 100(63), *134*
Aloiso, C. J., 5(30), *34*
Amborski, L. E., 30(106), *37*, 129(80), *137*, 160(46), *211*
American Society for Testing and Materials, 24(78a), *36*
Anderson, H., 243(125), *266*
Andrianov, K. A., 214(4), 229(4), 230(4), 234(85–91), 236(88–93,97), 237(92), 238(91,93), 239(107), 240(119), 247(129), 248(129), *263*, *265*, *266*
Angelo, R. J., 160(55), 173–176(64), *211*, *212*, 272(9), 274(9), *284*
Anyos, T., 152–154(25), *210*
Arimoto, F. S., 261(157), *267*
Arnold, F. E., 192–196(89), *212*
Arnold, H. R., 221(47), *264*
Arnold, R. G., 132(188,189), *137*
Asnovich, E. Z., 234(90,91), 236(90,91), 238(91), *265*
Atlas, S. M., 90–92(44), *134*, 238(102), *265*
Aubrey, D. W., 217(14), *263*
Audrieth, L. F., 223(58), *264*

Auspos, L. A., 53(55,56), 57–60(56), 61, 62(56), 63(56), *76*

B

Baccaredda, M., 5(41), *35*
Bailer, J. C., 250(136), 252(136), 253(136), 259(152), *266*
Bailey, W. J., 277(21a), 278(21b), *284*
Baker, H. R., 230(81), *265*
Baker, W. O., 18(71), *35*, 107(128), 120(164), *136*, *137*
Balckinton, R. J., 116(160), *136*
Baldoni, A. A., 236(94), 259(151), *265*, *266*
Ballou, J. W., 59, *77*
Bamba, Y., 208(107), *213*
Banford, W. A., 220(37), *263*
Bartholomew, R., 1(1), *34*
Beachell, H. C., 132(198), *137*
Beaman, R. G., 3, *34*, 78(13), 79(13), 80, 81, 83, 87, 88, 92, 94, 98, 102, 109, 111, 112, 117, 128, *133*, 143, 153, 157, 165, 177, 184, 190, 193, 198, 201, 228, 233, 248, 250
Becke-Goehring, M., 225(67,68), 226(71), *264*
Becker, H. J., 217(13), *263*
Bekkendahl, N., 5(31), *34*
Bell, V. L., 279(25), 281(25), *284*
Bennett, A. R., 4(25), *34*
Bennett, M., 38(3), *75*
Berchet, G. F., 78(28), 79(28), *133*
Berger, L., 49, *76*
Berlin, A. A., 39(10–12), 44(10,11), *75*, 253(138), 255(142), 258(148), *266*, 283(27), *284*
Berlin, A. M., 188(82,83), 189–191(84), *212*
Berr, C. E., 31(107), *37*, 129(184), *137*, 160–162(44), 163(44,62), 164(44), 166–168(44), 170, 173, 174(62), *211*

Bertram, J. L., 189–191(85), *212*
Beste, L. F., 60
Bezman, R., 224(64), *264*
Bickelhaupt, F., 39(15), *75*
Bilbo, A. J., 38(8), *75*
Bilbo, W., 223(61), 224(61), *264*
Biletch, H., 114(135), *136*
Billmeyer, F. W., Jr., 4(16), 8, *34*, *35*
Binda, F. J., 220(31), *263*
Bischoff, C. A., 101(71), *134*
Bissot, T. C., 218(17), *263*
Black, W. B., 90(51), 92–94(51), *134*, 182 (76), 184(76), *212*, 293(20), 297–300 (20), 302(20), *317*
Blake, E. F., 114(134), *136*
Blanchard, H. S., 116(153,155,156), *136*
Blau, J. A., 220(36), *263*
Block, B. P., 225(69), *264*
Bloomfield, P. R., 28, *36*
Bloomstrom, D. C., 176–178(68a), *212*
Boardman, F., 275(17), 276(17), *284*
Bock, H., 85(34), *133*
Bock, L. H., 107(94), *135*
Bockman, H., 293(15), *317*
Bogert, H., 151(22), 152(22), 155(29,30), 159, *210*, *211*
Bogert, T. M., 159, *211*
Bogomalnyy, V. Y., 39(16), *75*
Borishov, S. N., 236(96), 240(121), *265*, *266*
Bower, G. M., 30(101), *36*, 129(182), *137*, 160(50), 164(50), 165(50), 167(50), 173 (63), 175(63), 176(73), *211*, 289(10), 290(10), *317*
Boyer, R. F., 2(4), 4, 5(32), 7(4), *34*
Braithwaite, D., 39(17), *75*
Brauer, G. M., 4(23–25), 5(23,44,45), *34*, *35*
Braun, J., 221(50), *264*
Braz, G. I., 153–155(28), *210*
Brinker, K. C., 139, 151, *210*
Brotherton, R. J., 220(40), 221(40), 241 (40), *264*
Brown, C. A., 219(20), 237(100), *263*, *265*
Brown, C. J., 55, 56(63), 57(63), 60(63), *77*
Brown, G. P., 116, *136*, *137*
Brown, J. F., 40(33), *76*, 271(7), *284*
Browne, H. C., 220(35), *263*
Bruck, S. D., 163(59–61), 170, 171(59), 172(61), 173(61), *211*, 243(127), *266*

Brumfield, R., 155(32), *211*
Brysin, Yu. P., 78(8), 90(8), *133*
Buchdahl, R., 4(20), *34*
Buckner, W., 188(80), *212*
Bunn, C. W., 107(123), 108(123), 114 (136), *136*
Burck, S. D., 2(4), 7(4), *34*
Burg, A. B , 222(52–54), *264*
Burlant, W. J., 270(3), *283*
Burnham, C. W., 53(56), 56–63(56), *76*
Busch, M., 39(13), *75*
Butta, E., 5(41), *35*
Butte, W. A., 115(149,151), *136*
Byrd, J. D., 230–233(84), *265*

C

Cachia, P. M., 107(127), *136*
Caldwell, J. R., 106(78–80), 113(80), *134*
Cameron, D. D., 151(24), *210*
Campbell, D. H., 218(17), *263*
Campbell, H. N., 70, 77, 107(124), *136*
Campbell, T. W., 256(145,146), *266*
Carpenter, A. S., 107(119), 108(119), *135*
Carpenter, D. R., 6(46), *35*
Carrington, W. K., 124(170), 127(170), *137*
Carsivell, T. S., 11(68b), 13–16(68b), *35*
Cassidy, P. E., 40(28), 43(28), 45(28), *76*
Cazes, J., 242(41), *264*
Chemical Age (*London*), 38(9), 45(9), *75*
Chemical Society, 214(1), *263*
Chiang, Y., 26(82), 28(82), *36*
Chistyakova, M. V., 69(77), *77*
Chopey, N. P., 100(64), *134*
Christensen, R. G., 252(137), 253(137), *266*
Chu, N. S., 115(150), *136*
Chujo, R., 6(51), *35*
Clar, E., 38(2), *75*
Clay, W. R., 30(104), 31(104), *37*, 308(32), *318*
Clendinning, R. A., 131(187), *137*
Coffman, D. D., 78(28), 79(28), *133*
Cole, C., 152–154(25), *210*
Cole, T. B., 201–204(101), *213*
Coleman, L. E., 261(158), *267*
Colson, J. G., 279–281(24), *284*
Conix, A., 97, 98(52,53), 99(53), 107 (102), 109(102), 128(175), *134*, *135*, *137*
Cooke, Troughton and Simms, Inc., 27 (90), *36*

Cool, L. G., 2(4), 7(4), *34*
Coover, H. W., 226(70), *264*
Corley, R. S., 53(54), 60(54), 61(54), 68 (54), *76*
Cotter, J. L., 187(78), *212*
Cotton, F. A., 261(155), *267*
Coulson, C. A., 56, *77*
Courtright, J. R., 285(3), 286(3), 289–292 (3), *317*
Cox, J. M., 116(163), 118–122(163), *137*
Cox, W. P., 5(28), *34*
Craig, D. P., 56(70), *77*
Crain, R. D., 238(103), *265*
Cram, D. J., 55(62), *77*
Cramer, F. B., 80, 81, 83, 87, 88, 92, 94, 98, 102, 109, 111, 112, 117, 128, 143, 153, 157, 165, 177, 184, 190 193, 198, 201, 228, 233, 248, 250
Craven, J. M., 200, 201(99), 202(99), *213*
Crossland, R. K., 261(164), 262(164), *267*
Cudby, M. E. A., 130(186), *137*
Curry, J. E., 230–233(84), *265*
Curtus, R., 188(81), *212*

D

Dains, D., 151(21), *210*
Dakin, T. W., 20(75), *36*
Dale, J. W., 138(3), *210*, 17(70), 25(70), 35, 38(1), *75*
Damm, P., 40, *75*
Dammont, F. R., 53(48), *76*
Daniels, E. A., 115(137), *136*
Dannis, M. L., 5, 6(37), *35*
Daudel, E., 60(74), *77*
Davis, A., 104–106(77), *134*
Davydov, B. E., 270(4), *283*
Dawans, F., 40, 41(34), 44(34), *76*, 202(103), *213*, 279 (23), *284*
Dawson, J. W., 218(19), *263*
Dech, W., 151(21), *210*
De Gandemaris, G., 192(90), 193(90), *212*
Degeiso, R. C., 256(144), 257(144), *266*
Delano, C. B., 141(16), 144(16), *210*
De Montmollin, R., 197(92), *212*
Denisenko, Ya. I., 40(26), *76*
Dennoon, C. E., 220(33), *263*
DeSchryver, F., 276(18), *284*
Deuel, H., 220(30), *263*
Dewar, M. J. S., 115, *136*, 219(26), *263*

DeWinter, W., 276(18), 279(22), *284*
Diatkina, R., 56(71), *77*
Dine-Hart, R. A., 90(45), 91(45), 94–96 (45), 129(45), *134*
Doak, K. W., 107(124), *136*
Dobinson, F., 90(51), 92–94(51), *134*, 293 (20), 297–300(20), 302(20), *317*
Dolgoplosk, B. A., 39(16), *75*
Donaruma, L. G., 256(144), 257(144), *266*
Doyle, C. D., 25(81), 26(81), 28, 30(98), 36, 229(76), *264*
Drake, G. L., 26(82), 28(82), *36*
Drewitt, J. G. N., 107(99), *135*
Dudek, T. J., 2(4), 7(4), *34*
Dulova, V. G., 234(85), *265*
Dumont, J., 224(63), *264*
Dunnavant, W. R., 230–233(84), *265*
Du Pont de Nemours, E. I., & Co., Inc. 285(5), 286(5), 289–292(5,12), 293(12), 294(12), 296(12), 298(12), 303(30), 304 (30), 305–307(33), 308(33), 309(31), 310(31), 311–314(33), *317–318*
Duval, C., 27(91), *36*
Dye, W. T., Jr., 90(46), 92(46), *134*

E

Eareckson, W. M., 78(15,20), 79,80(15), 107(101), 109(101), 110(101), *133*, *135* 293(19), 299(19), *317*
Eberlin, E. C., 5, *35*
Economy, J., 277(21a), *284*
Edgar, O. B., 5(27), 8, *34*, *35*, 78(3), 79, *132*
Edwards, G. A., 38(6,7), *75*
Edwards, W. M., 31(107), *37*, 129(184), *137*, 159, 160, 161(42–44), 162(42–44), 163(44), 164(44), 166–168(44), *211*
Egorova, Yu. V., 107(103), *135*
Ehlers, G. F. L., 25(80), 27(88,93), 28, 36, 69(78), 77, 99(30,31), 114(30,31), 123, 128, 129(30,31,179), 132,*134*, *137*, 187, 188, *212*, 217, 219(16), 222(16), 224, 226(16), 229(16), 235–239(16), 241(16), 242(16), 247(16), 250(16), 252 (16), *263*, 271(6), *283*
Eilar, K. R., 221(51), 222(51), *264*
Eisenberg, A., 2(4), 7(4), *34*
Ellery, E., 8, *35*

Elliot, J. R., 230(82), *265*
Elvidge, J. A., 261(153), *267*
Endres, G. F., 116(153,155,156,158,159), 117(159), *136*
Endrey, A. L., 31(107), *37*, 129(184), *137*, 160(44,48,53), 161(44), 162(44,48,53), 163(44,53), 164(44), 166–168(44), *211*
Epstein, E., 255(141), *266*
Erickson, C. E., 220(28), 221(28), *263*
Erofeev, B. V., 40(23,29,30), *75*, *76*
Errede, L. A., 53(47), 54, 57(73), *76*, *77*
Eustance, J. W., 116(153),156), *136*
Evans, R. D., 107(100), *135*
Evans, W. V., 39(17), *75*
Eventova, M. S., 40(26), *76*
Evers, E. C., 216(9), 229(9), 230(9), *263*
Ewing, G. W., 138(2), *210*
Eykamp, G. R., 1(1), *34*

F

Farb, H., 2(4), 7(4), *34*
Farnham, A. G., 131(187), *137*
Farthing, A. C., 55, 56(63), 57(63), 60(63), *77*
Faucher, J. A., 4(26), 5(26), *34*
Feasey, R. G., 130(186), *137*
Fedorova, L. S., 74(86), *77*
Fedotova, O. Ya., 78(8), 90(8), *133*
Feinberg, B. D., 278(21b), *284*
Fernelius, W. C., 249(130), *266*
Ferry, J. D., 5(33), *34*
Fetter, E. J., 277(21a), *284*
Finkbeiner, H. L., 116(155,156), *136*
Fischer, T. M., 200(99), 201(99), 202(99), *213*
Fisher, J. W., 78(4,5), 85(4,5), 107(96), *132*, *133*, *135*
Fitzgerald, W. P., 197(93), 198(93), 199(93,94), *212*, 300(24,25), 302(24,25), *317*
Fletcher, E. A., 220(35), *263*
Florin, R. E., 115(145), *136*
Flory, P. J., 4(18), 5(18), 34(6), *133*
Foldi, V. S., 256(146), *266*
Fordham, S., 220(37), *263*
Foster, R. T., 141(9), 143(9), 146(9), *210*
Fox, D. W., 101(74), *134*
Fox, T. G., 4(18), 5(18), *34*
Frank, R. L., 236(94), 259(151), *265*, *266*
Frankel, M., 243(127), *266*

Frazer, A. H., 86(36), 87(36,38–40), 88(41), 89(41), *133*, 149(19), 180(73,74), 181(74), 183, 184(73), 185–187(74), 197(93), 198(93), 199(93,94), *210*, *212*, 258(147), *266*, 293(17), 296(17), 297(17), 299(17), 300(24,25), 302(24,25), *317*
Freeman, J. H., 30(102), *36*, 173(63), 175(63), 176(63), *211*, 289(9,10), 290(9,10), 291(9), *317*
Freitag, W. O., 216(9), 229(9), 230(9), *263*
Frey, D. A., 40(35), *76*
Frick, S., 217(13), *263*
Friedlander, P., 208(106), *213*
Frosch, C. J., 107(97), *135*
Frost, L. W., 30(101), *36*, 129(182), *137*, 160(50,54), 164(50,54), 165(50,54), 167(50), 173(63), 175(63), 176(63), *211*, 289(10), 290(10), *317*
Fry, J. S., 100(66), *134*
Fujimoto, F., 178(68), *212*
Fukami, A., 163(58), 165(58), 168, *211*
Fuller, C. S., 107(125,128), *136*
Fulsami, A., 129(183), *137*
Fushchillo, N., 4(10), *34*

G

Ganina, T. N., 234(89,92), 236(89,92), 237(92), *265*
Gabriel, E., 199(95), *212*
Gaylord, N. G., 272(10–12), 273(10–12), *284*
Geiderikh, M. A., 270(4), *283*
George, P. D., 230(82), *265*
Gerrard, W., 217(15), 218(15,18), 219(23), 220(18,36), *263*
Gertner, D., 243(127), *266*
Gibbs, W. E., 1(1), *34*
Gilbert, A. R., 230(80), *264*
Gilch, H. G., 55(67), *77*
Gilkey, R., 106(79), *134*
Gillam, A. E., 44(40), *76*
Gillham, J. K., 26(83–87), *36*
Gitina, R. M., 100(63), *134*, 153–155(28), 157(28), 158(28), *210*, 283(27), *284*
Gladkovaskii, G. A., 51(44,45), 52(45), *76*
Glukhov, N. A., 74, *77*
Goldberg, E. P., 55(66), 70–72(66), *77*

Golden, J. H., 104(77), 105(77), 106, 115, 117, 118(144), *134*, *136*
Goldfinger, G., 38, *75*
Goldman, A., 116(161,162), *136*, *137*
Gorham, W. F., 55(67), *77*
Gorman, N. H., 138(2), *210*
Gormby, J. J., 2(4), 7(4), *34*
Goubeau, J., 216(10), 219(24), *263*
Grabner, H., 216(10), *263*
Graham, G., 214(2), *263*
Grassie, N., 19, *35*, 270(5), *283*
Gray, A. P., 239(109), *265*
Gray, D. N., 141(6a,b), 146(6), *210*
Greber, G., 261(164), 262(164), *267*
Gregorian, R. S., 54(60), *77*
Gunderloy, F. C., 220(28), 221(28), *263*

H

Haas, H. C., 49, 50, 53(54), 60(54), 61(54), 68(54), *76*
Haber, C. P., 223(59), 224(62,66), *264*
Hale, W. F., 131(187), *137*
Hall, L. A. R., 53(55,56), 55(67), 56–60(56), 61(55,56), 62(56), 63(56), *76*, *77*
Hallensleben, M. L., 261(164), 262(164), *267*
Hamilton, S. B., 219(25), 221(42), *263*, *264*
Hammant, B. L., 104–106(77), *134*
Handlovits, C. E., 72(81), 73(81), *77*, 124(171,172), 125(172), 127(171), *137*
Hara, S., 182(75), 184(75), 187(75), *212*
Harle, O. L., 220(29), *263*
Harris, J. J., 219(22), *263*
Hartzell, G. E., 40, 43(28), 45(28), *76*
Haruki, E., 262(166), *267*
Hasegawa, H., 78(9), 90(9), *133*
Hasegawa, M., 40(35), *76*
Hashimoto, S., 78(1), 79, *132*
Haven, A. C., 261(157), *267*
Hay, A. S., 116, 117(154,159), *136*
Hazell, E. A., 104–106(77), *134*
Heacock, J. F., 163(62), 170, 173, 174(62), *211*
Hedenstroem, A., 101(71), *134*
Hellmannn, M., 38, *75*
Henglein, F. A., 99(61), *134*, 239(106), 240(120), *265*, *266*
Hergenrother, P. M., 155(35,36), 156(35,36), 160(57), 164(57), *211*, 285(6,7), 287(6,7), 289–293(6,7), *317*
Herring, H., 224(62), *264*
Hey, D. H., 44(40), *76*
Heying, T. L., 239(109), *265*
Hickam, W. M., 229(75), *264*
Hilditch, T. P., 124, *137*
Hill, H. W., 90(42), 91(42), *133*
Hill, R., 8, *35*, 78(3,7), 79, 90(7), 107(118,129), 108(118,129), *132*, *135*, *136*
Hoeg, D. F., 55(66), 70, 71(66), 72(66), *77*
Hoffman, A. K., 221(44–46), *264*
Hoffman, F., 40, *75*
Hoffman, R. M., 60, *77*
Hofmann, K., 138(1), *210*
Hohnstedt, L. F., 219(21), *263*
Holsten, J. R., 205(104), 206(104), *213*, 300(26), 302(26), *317*
Holub, F. F., 107(106,110–113,117), 108(106), 110(111–113,117), *135*
Hopwood, S. L., 54(58), *76*
Horowitz, E., 252(137), 253(137), *266*
Hout, J. M., 54(59,60), *76*, *77*
Houtz, R. C., 40(24), *76*, 269(1), 270(1), *283*, 302(27), *317*
Hsu, L. C., 42(39), 45(39), 48(39), *76*
Hubbard, J. K., 53(55,56), 56–60(56), 61(55,56), 62(56), 63(56), *76*
Huffman, W. A. H., 90(46), 92(46), *134*
Huggins, C. M., 6(46), *35*
Hughes, R. B., 173(63), 175(63), 176(63), *211*
Hughes, R. E., 115(151), *136*
Huisgen, R., 176(67), 178(67), *212*
Humiec, W., 224(64), *264*
Hunter, I., 155(31), *210*
Hunter, W. H., 115, *136*
Hurley, F. R., 225(68), *264*
Husted, W., 155(30), *210*
Hutchinson, J. H., 238(101), *265*
Hyman, M., 220(32), *263*

I

Ikeda, C. K., 285(3), 286(3), 289–292(3), *317*
Ikeda, K., 208(107,108), *213*
Ikeda, R., 160–164(45), 167(45), *211*
Ikeda, R. M., 272(9), 274(9), *284*

AUTHOR INDEX

Imai, Y., 141(11–13), 143(11,12), 149(11–13), 150(12,13), 153(27,28), 154(27,28), 155(27,28), 157(27,28), 158(27,28), 173(63), 175(63), 176(63), 178(63), 182(76), 184(76), *210–212*
Imoto, E., 262(166), *267*, 277(19a), *284*
Imoto, M., 74(88), 75(89), *77*
Ingold, C. K., 49, *76*
Ingold, E. H., 49, *76*
Inone, H., 277(19a), *284*
Inone, S., 141(13), 149(13), 150(13), *210*
Ioba, M., 74(84), *77*
Irwin, R. S., 300(23), 302(23), *317*
Iskenderov, M. A., 107(105,108,116), 109(105,108), *135*
Ito, K., 6(51), *35*
Iwakura, Y., 141, 143(11,12), 149(11–13), 150(12,13), 153, 154(27,28), 155, 157(27,28), 158, 173(63), 175(63), 176(63), 178(68), 182(75,76), 184(75,76), 187, *210–212*
Izard, E. F., 107(120–122), 108(120–122), *135*, *136*

J

Jadamus, H., 276(18), *284*
Jaffe, L. D., 1(2), *34*
James, A. N., 115, *136*
Janis, E. C., 285(1), 286(1), 288–292(1), *317*
Jenckel, E., 4(19), 7(56), *34*, *35*
Jenkins, L. T., 26(82), 28(82), *36*
Jennings, B. E., 130(186), *137*
Jensen, P. W., 5,9(67), *35*
Jewell, R. A., 279(25), 281(25), 282(25), *284*
John, J. B., 17(70), 25(70), *35*, 38(1), *75*
Johns, I. B., 138(3), *210*
Johnson, R. N., 131(187), *137*
Jones, H. O., 124, *137*
Jones, J. I., 160(41), 163(41), 166(41), *211*
Jones, J. T., 129(181), *137*
Jones, M. E. B., 130(186), *137*, 250(134), 251(134), 260(134), *266*
Jones, R., 155(31), *210*
Jones, W. D., 132(192), *137*
Joyce, F. E., 115(139), *136*
Joyner, R. D., 261(154), *267*
Jung, S. L., 78(21), *133*

K

Kalinsnikov, G. S., 74(83), *77*
Kan, P. T., 261(160), *267*
Kane, J. J., 149(19), *210*
Kane, M. W., 53(54), 60(54), 61(54), 68(54), *76*
Kantor, S. W., 107(106,110–113,117), 108(106), 110(111–113,117), *135*
Karateev, D. A., 240(113–116), *265*, *266*
Kargin, V. A., 270(4), *283*
Kass, P., 107(98), *135*
Kato, Y., 4(8), *34*
Katz, M., 78(14,17), 81(14), *133*
Kaufman, M. H., 53(53), 56–60(53), *76*
Kazakova, A. A., 234(86), *265*
Ke, B., 26(82), 28(82), *36*
Kearney, J. J., 5, *35*
Keattch, C. J., 26(82), 28(82), *36*
Kenney, C. N., 249(132), 250(132), *266*
Kenney, M. E., 261(154), *267*
Kesse, I., 160(54), 164(54), 165(54), *211*
Ketley, A. D., 49(43), 50(43), *76*
Khananashvili, L. M., 240(122), *266*
Kharlamov, V. V., 107(108,116), 109(108), *135*
Khrustaleva, W., 236(92), 237(92), *265*
Killay, F. R., 128(174), *137*
Kiprianov, V., 155(34), *211*
Kirk, G. J. M. van der, 243–247(128), *266*
Kirk, W., 53(55,56), 56–60(56), 61(55,56), 62(56), 63(56), *76*
Kitaoka, H., 262(166), *267*
Kjoller, K. J., 141(16), 144(16), *210*, 285(1,2), 286(1,2), 288–292(1,2), *317*
Kline, G. M., 20(73), *35*
Kluiber, R. W., 250(133), *266*
Knight, G. J., 187(78), *212*
Knobloch, F. W., 261(161), 262(161), *267*
Knoor, R., 188(79), *212*
Koch, F. W., 42(38), 45(38), *76*
Hopper, R. J., 42(39), 45(39), 48(39), *76*
Koda, K., 261(164), 262(164), *267*
Koenig, P. E., 238(101,103), *265*
Koike, M., 6(47), *35*
Koike, Y., 4(8), *34*
Kolb, H. J., 107(122), 108(122), *136*
Kolesnikov, G. S , 74, *77*
Koller, C. R., 78(13), 79(13), *133*
Koontz, F. H., 78(18), *133*

AUTHOR INDEX

Korshak, V. V., 69, 74, 77, 109(84), 107 (103–105,107–110,115,116), 109(105, 107–110,115), 110(106), *135*, 153(28), 154(28), 178(68), 188, 189 190(84), 191 (84), *210*, *212*
Kossler, I., 272(10–12), 273(10–12), *284*
Koton, M., 56(72), 77
Kovacic, J. P., 43(39), 45(39), 48(39), 76
Kovacic, P., 42, 43(37), 45(36–39), 48 (39), 76
Kozlov, P. V., 100(63), *134*
Kraskovyak, M. G., 55(64), 70, 77
Krause, S., 2(4), 7(4), *34*
Krentsel, B. A., 270(4), *283*
Kreshkov, A. P., 240(113–116,123), 241(123), *265*, *266*
Kreuchunas, A., 130(185), *137*
Kriner, W. A., 216(9), 229(9), 230(9), *263*
Kromov, S. I., 40(26), 76
Kronganz, E. S., 153–155(28), 157(28), 158(28), 178(68), 188(82–84), 189–191 (84), *210*, *212*
Kronrod, N. Y., 221(48), *264*
Kruger, P., 176(65), *212*
Kubba, V. P., 219(26), *263*
Kubota, T., 152, 153–155(26), *210*
Kugler, V., 259(150), *266*
Kukami, A., 90–92(44), *134*
Kukhareva, L. V., 55(64), 77
Kukmeyer, J. L., 285(6,7), 287(6,7), 289–293(6,7), *317*
Kurashev, M. V., 40(31), 76
Kurasheva, N. A., 234(85), *265*
Kurian, C. J., 115(148), 117(148), *136*
Kurihara, M., 208(107,108), *213*
Kustanovich, I. M., 270(4), *283*
Kwiatek, J., 116(158), *136*
Kwolek, S. L., 78(11,12,22,23,25–27), 79 (27), 81(25–27), 90(42,43), 91(42,43), 128(27), *133*
Kyriakis, A., 42(36), 43(37), 45(36,37), 76

L

Ladenburg, A., 151(20), *210*
LaMer, V. K., 220(34), *263*
Lancaster, J. M., 116(163), 118–122(163), *137*
Landrum, B. F., 54(57), 76
Lang, R., 239(106), 240(120), *265*, *266*
Lang, R. F., 242(124), *266*
Lange, R. M., 42(39), 45(39), 48(39), 76
Lappert, M. F., 214(3), 215(7), 216(11, 12), 217(14), 218(18), 220(18,36),229 (3,5,7), 230(3,5,7), 250(134), 251(134), 260(134), *263*, *266*
Lasidon, U., 128(175), *137*
Lawton, R., 224(62), *264*
Lebedev, N. N., 107(103), *135*
Lee, J. T., 28(95), *36*
Lee, L., 28(95), *36*, 103(76), 104, *134*
Lefebvre, G., 40(34), 41(34), 44(34), 76
Leigh, G. J., 214(3), 215(7), 229(3,5,7), 230(3,5,7), 250(134), 251(134), 260 (134), *263*, *266*
Lennarz, W. J., 221(43), *264*
Lenz, R. W., 72(81), 73(81), 77, 124, 125 (172), 127, *137*
Letsinger, R. L., 219(25), 221(42), *263*, *264*
Lever, A. B. P., 261(153), *267*
Lever, A. E., 24(77), *36*
Levi, D. W., 28(95), *36*
Levine, H. H., 30(99), *36*, 141, 144(16), 155, 156, 160(57), 164(57), *210*, *211*, 285(1,2,6–8), 286(1,2), 287(6–8), 288 (1,2), 289(1,2,6–8), 290–292(1,2,6,7), 293(6,7), *317*
Levkoey, J. J., 100(63), *134*
Lewis, A. F., 26(83–87), *36*
Lewis, C. W., 229(77,78), *264*
Lewis, J. W., 250(133), *266*
Lezvov, N. S., 240(119), *266*
Lilyquist, M. R., 205(104), 206(104), *213*, 293(20), 297–300(20), 300(26), 302 (20,26), *317*
Lincoln, J., 107(99), *135*
Liogonikii, B. I., 283(27), *284*
Littke, C. O., 293(16), *317*
Littlefield, J. B., 155(33), *211*
Livingstone, D. I., 49(43), 50(43), 53(54), 60(54), 61(54), 68(54), 76
Loeffler, M. H., 229(75), *264*
Lohr, J. J., 2(4), 7(4), *34*
Loire, N. P., 277(20), *284*, 285(8), 287 (8), 289(8), *317*
Loncrini, D. F., 173(63), 175(63), 176 (63), *211*
Longone, D. T., 200–202(100), *213*
Losev, I. P., 78(8), 90(8), *133*

M

Lunn, R. H., 30(102), *36*, 173(63), 175 (63), 176(63), *211*, 289–291(9), *317*
Lusk, D. I., 55(66), 70–72(66), *77*
Lyman, D. J., 78(21), *133*

M

Macallum, A. D., 124, *137*
Maccoli, A., 56(70), *77*
McConnell, R. L., 226(70), *264*
McCune, L. K., 30(103), 31(103), *36*, 293 (11), 296(11), 304(11), *317*
MacDirmid, A. G., 216(9), 229(9),230(9), *263*
McDonald, R. N., 70, *77*
McElhill, E. A., 17(70), 25(70), *35*, 38(1), *75*, 138(3), *210*
MacFarlane, R., 230(83), *265*
McFarlane, S. B., 85(35), 132(192), *133*, *137*
McGregor, R. R., 239(110,111), *265*
Mach, K., 272(12), 273(12), *284*
Machine Design, 285–287(4), 289–292 (4), 302–304(29), 315(34), 316(34), *317*, *318*
McIntyre, E., 5(40), *35*
McNeill, I. C., 270(5), *283*
Madorsky, J. L., 61, *77*
Magat, E. E., 78(13), 79(13), *133*
Maimans, J., 151(22), 152(22), *210*
Mainen, E. L., 277(19b), *284*
Maksimova, T. P., 40(30), *76*
Malowan, J. E., 225(68), *264*
Mandelkern, L., 5(34), *34*
Mandell, E. R., 4(12,13), *34*
Manufacturing Chemists Association, Inc., 9–15(68a), *35*
Marchionna, V. J., 42(39), 45(39), 48(39), *76*
Mark, H. F., 8(63), *35*, 53(53), 56–60 (53), *76*, 90–92(44), *134*, 238(102), *265*
Markle, R. A., 230–233(84), *265*
Marlies, C. A., 6(54), 7(54), 17(54), 21 (76), *35*, *36*
Martin, F. S., 239(112), *265*
Martin, G. M., 5(34), *34*
Martin, K. V., 250(135), *266*
Martin, M. M., 254(140), 258(149), *266*
Marvel, C. S., 40, 43(28), 45(28), 52(46), 53(46), 76, 139, 141, 142(5,7,10), 143 (7–9,15), 145(5,7), 146, 147(7), 154(7), 178(69), 184(69), 200, 201(98), 202(98, 103), 204(69), 205(69), *210*, *212*, *213*, 220(33), 254(139,140), 256(143), 258 (149), 262(165), *263*, *266*, *267*, 276 (18), 279(22,23), *284*
Material Design Engineering, 30(100), *36*
Matlack, J. D., 26(82), 28(82), *36*
Matreyek, K. W., 18(71), *35*
Matsurka, S., 5(30), *34*
Matveeva, N. G., 253(138), 255(142), 258 (148), *266*
Matyska, B., 272(12), 273(12), *284*
Maxwell, B., 5(30), *34*
Mecum, W. D., 155(33), *211*
Melnikova, E. P., 51(44,45), 52(45), 55 (64), *76*, *77*
Merriam, C. N., 131(187), *137*
Merz, E. H., 4(20), *34*
Mesrobian, R. B., 53(53), 56–60(53), *76*
Metzger, A. P., 26(82), 28(82), *36*
Michel, R. H., 279–281(24), *284*
Miglarese, J., 30(102), *36*, 173(63), 175 (63), 176(63), *211*, 289–291(9), *317*
Mikolajewski, E., 293(13), *317*
Miller, A. L., 85(35), *133*
Miller, R. L., 4(7), *34*
Mitchell, J., 28(95), *36*
Mitin, Yu. V., 74, *77*
Mitsuhashi, K., 143(15), *210*
Miyake, A., 97(55), 98(55), *134*
Modern Plastics Encyclopedia, 24(78b), *36*
Moldenhauer, O., 85(34), *133*
Moore, B. J. C., 90(45), 91(45), 94–96 (45), 129(45), *134*
Morgan, H. S., 293(20), 297–300(20), 302 (20), *317*
Morgan, P. W., 78, 79(13,24,27), 81(25–27), 90(42,43), 91(42,43), 128(27), *133*
Morton, A., 155(33), *211*
Mountfield, B. A., 218(18), 220(18,36), *263*
Moyer, W. W., 100(66), *134*, 152, 153 (25), 154, *210*
Muenster, A., 7, *35*
Mukamal, H., 274(13), *284*
Mulvaney, J. E., 200, 201(98), 202(98), *213*
Murphey, W. A., 78(19), 128(19), *133*, 293(18), 299(18), *317*

AUTHOR INDEX

Murphy, C. M., 226(74), 229(74), *264*
Musgrave, O. C., 220(38,39), *264*
Muchkalo, K., 155(34), *211*

N

Nakanishi, R., 152, 153(26), 154(26), 155(26), 208(107,108), *210, 213*
Namiot, S., 56(71), *77*
Naraba, T., 4(8), *34*
Nason, H. K., 11(68b), 13–16(68b), *35*
Naumova, S. F., 40(22,23,29,30), *75, 76*
Nelson, J. A., 132(189), *137*
Nems, S., 5(42), *35*
Neuse, E. W., 261(163,164), 262(164), *267*
Newkirk, A. E., 27(92), *36*
Newman, S., 5(28), *34*
Nichold, F. C., 138(2), *210*
Nicolesen, L., 74(84), *77*
Niedenzu, K., 218(19), 225(68), *263, 264*
Nielsen, L. E., 4(20), 5(40), *34, 35*
Nielsen, M. L., 226–228(73), *264*
Nishioka, A., 4(8), *34*
Nishizaki, S., 90–92(44), 129(183), *134, 137*, 163(58), 165(58), 168, *211*
Noether, H. D., 299(21,22), 300(21,22), *317*
Noltes, J. G., 243–247(128), *266*
Normant, H., 221(50), *264*
Noto, V. H., 285(2), 286(2), 288–292(2), *317*

O

O'Brien, J. F., 238(105), *265*
Ochynski, F. W., 129(181), *137*, 160(41), 163(41), 166(41), *211*
Odajima, A., 6(47,51), *35*
Ogata, N., 90–92(44), *134*
Okada, M., 276(18), *284*
Okawara, M., 262(166), *267*
Oliver, K. L., 31(107), *37*, 129(184), *137*, 160–164(44), 166–168(44), *211*
Olson, A. O., 115(137), *136*
O'Neal, J. R., 285(2), 286(2), 288–292(2), *317*
Overberger, C. G., 86–88(37), *133*, 178(68), 179(72), 184(72), *212*, 274(13), *284*
Overhults, W. C., 49(43), 50(43), *76*
Owaki, M., 4(8), *34*
Ozaki, S., 274(13), *284*

P

Paddock, N. L., 223(55), *264*
Papetti, S., 239(109), *265*
Parini, V. P., 39(10,11), 44(10,11), *75*
Park, T. O., 220(39), *264*
Parks, G. S., 5(33), *34*
Parry, R. W., 218(17), *263*
Parsons, J. L., 270(3), *283*
Parsons, T. D., 221(49), *264*
Patterson, T. R., 236(94), 259(151), *265, 266*
Paufler, R. M., 279–281(24), *284*
Pauling, L., 16(69), *35*, 107(126), 113(126), *136*
Paushkiw, Y. M., 40(31), *76*
Pauson, P. L., 261(156), *267*
Pavilk, F. J., 236(94), 259(151), *265, 266*
Pavlov, S. A., 234(88), 236(88), *265*
Pearson, R., 39(17), *75*
Pellicciotto, A. M., 219(21), *263*
Perkins, R. M., 26(82), 28(82), *36*
Perros, T. P., 252(137), 253(137), *266*
Peterson, W. R., 78(28), 79(28), *133*
Pettit, R., 219(26), *263*
Pezdirtz, G. F., 279(25), 281(25), 282(25), *284*
Pheiffer, P., 243(127), *266*
Phillips, R., 147, 148(17), 149(17), *210*
Pickesimer, L. G., 201–204(101), *213*
Plorin, R. E., 20(74), *35*
Plummer, L., 141(8), 143(8), 146(8), *210*, 262(165), *267*
Polak, L. S., 40(31), *76*, 270(4), *283*
Pollard, R. E., 5(40), *35*
Ponomarenko, V. A., 240(121), *266*
Pope, N. R., 18(71), *35*
Potts, K. T., 202(102), 204(102), *213*
Powles, J. G., 4(9), 6(48), *34, 35*
Preston, J., 90(50,51), 92–94(50,51), 95, *134*, 182(76), 184(76), *212*, 293(20), 297–300(20), 302(20), *317*
Preve, J., 192(90), 193(90), *212*
Price, C. C., 115, 117(148), 124, 126, *136, 137*
Prober, M., 230(82), *265*
Pullman, A., 56(70), *77*
Pummer, W. J., 115(145), *136*
Pyszora, H., 216(11), 217(14), 220(36), *263*

AUTHOR INDEX

Q

Quinn, F. A., Jr., 5(34), *34*
Quo, E., 261(164), 262(164), *267*

R

Rackley, F. A., 129(181), *137*, 160(41), 163(41), 166(41), *211*
Rathmann, F. H., 115(143), *136*
Rausch, M. D., 261(158), *267*
Rauscher, W. H., 261(161), 264(161), *267*
Ray, S., 40(28), 43(28), 45(28), *76*
Reed, T. A., 197(93), 198(93), 199(93,94), *212*, 300(24,25), 302(24,25), *317*
Reeves, W. A., 26(82), 28(82), *36*
Reich, L., 26(82), 28(82,95), *36*
Reichel, B., 202(103), *213*
Reimschuessel, H. K., 275(17), 276(17), *284*
Renfrew, M. M., 107(93), *135*
Renshaw, R. R., 159, *211*
Rhys, J., 24(77), *36*
Rinehart, K. L., 261(159), *267*
Rippere, R. E., 220(34), *263*
Ritter, D. M., 221(49), *264*
Robinson, I. M., 139, 151(24), 159, *210, 211*
Robinson, R., 199(96), *212*
Rochow, E. G., 238(104), *265*
Rolthoff, I. M., 28(95), *36*
Roman, N., 2(4), 7(4), *34*
Romeyn, H., 293(14), *317*
Rose, J. B., 130(186), *137*
Rose, S. H., 225(69), *264*
Ross, J. H., 1(3), *34*, 293(16), *317*
Rossweiler, J. H., 254(139), *266*
Rouch, L. L., 141(6b), 146(6), *210*
Rudner, B., 219(22), *263*
Rudner, J., 155(32), *211*
Ruigh, W. L., 220(28), 221(28), *263*
Rusenov, A. L., 153–155(28), 157(28), 158(28), 178(68), *210, 212*
Russkova, E. F., 100(63), *134*
Rust, J. B., 236(98,99), *265*
Rybnikar, F., 5(39), *35*

S

Sabrin, L. A., 240(119), *266*
Salazkin, S. N., 107(104), 109(84), *135*
Sanderson, V. E., 285(8), 287(8), 289(8), *317*
Santarovic, P. S., 40(20,21), *75*
Santome, K., 182(76), 184(76), *212*
Sarasohn, I. M., 44(40), *76*, 88(41), 89(41), *133*, 180(74), 181(74), 183, 185–187(74), *212*
Sato, K., 182(76), 184(76), *212*
Satok, S., 6(51), *35*
Sauer, J., 176(67), 178(67), *212*
Sauer, J. A., 4(10), 6(49,51), *34, 35*
Saunders, C. E., 226(74), 229(74), *264*
Saunders, M., 49(43), 50(43), *76*
Sazanova, V. A., 221(48), *264*
Scala, L. E., 229(75), *264*
Schaaf, R. L., 261(160), *267*
Schaefer, J. P., 189–191(85), *212*
Schaeffer, B. B., 239(109), *265*
Schaefgen, J. R., 53(55,56), 56, 57(56), 58(56,69), 60(56), 61(54,56), 62(56), 63, *76*, 77, 78(18), *133*
Scheidler, A. L., 30(105), *37*
Scheinost, K., 239(106), 240(120), *265, 266*
Schmidbaur, H., 240(118), *266*
Schmidt, A. X., 6(54), 7(54), 17(54), 21(76), *35, 36*
Schmidt, M., 240(118), *266*
Schnell, H., 100(65,70), 102(65), *134*
Schoenborn, E. M., 27(89), *36*
Schulze, J., 225(67), *264*
Schupp, L. J., 219(20), *263*
Scott, N. D., 5(52), 6(52), 9, *35*
Searle, R., 224(60), *264*
Sedlak, M., 220(28), 221(28), *263*
Segal, C. L., 236(98,99), *265*
Seyfried, L. M., 115(141), *136*
Shashoua, V. E., 78(15), 79, 80(15), *133*, 293(19), 299(19), *317*
Shaw, A., 53(52), *76*
Shaw, R., 224(65), *264*
Shearer, N. H., 226(70), *264*
Sheehan, W. C., 201–204(101), *213*
Sheratee, M. B., 285(1,2), 286(1,2), 288–292(1,2), *317*
Sherle, A. I., 255(142), 258(148), *266*
Shetter, J. A., 2(4), 7(4), *34*
Shimanouchi, T., 6(51), *35*
Shirashi, S., 178(68), *212*
Shirokova, N. I., 100(63), *134*
Shriner, R. L., 49, *76*
Shulman, G. P., 141(6a), 146(6), *210*

Sieffert, L. E., 27(89), *36*
Sillion, B., 192(90), 193(90), *212*
Silverman, M. B., 221(49), *264*
Simons, E. L., 27(92), *36*
Simpson, J. C. E., 192(86), *212*
Simroth, O., 197(92), *212*
Singleterry, C. R., 230(81), *265*
Singleton, R. W., 299(21,22), 300(21,22), *317*
Sinnot, K. M., 4(11), *34*
Slade, P. E., Jr., 26(82), 28(82), *36*
Slichter, W. P., 4(12,13), 6(50), *34, 35*
Slota, P. J., 222(53), *264*
Slyapnikova, I. A., 40(20,21), *75*
Smirnova, T. V., 74(86,87), *77*
Smith, D. C., 226(74), 229(74), *264*
Smith, D. D., 107(95), *135*
Smith, H. A., 124(172), 125(172), *137*
Smith, J. C., 59, *77*
Smith, J. O., 17(70), 25(70), *35*, 38(1), *75*, 138(3), 155(32), *210, 211*
Smith, M. B., 285(1), 286(1), 288–292(1), *317*
Smith, R. W., 90(46,51), 92(46,51), 93(51), 94(51), *134*, 293(20), 297–300(20), 302(20), *317*
Smothers, W. J., 26(82), 28(82), *36*
Snyder, H. R., 221(43), *264*
Sohma, J., 6(47), *35*
Solms, J., 220(30), *263*
Son, C. P. N., 132(198), *137*
Sorenson, W. R., 90(43), 91(43), *133*
Sosin, S. L., 69(77), *77*
Spanagel, E. W., 78(28), 79(28), *133*
Speck, S. B., 53(55,56), 56–60(56), 61(55,56), 62(56), 63(56), *76*, 78(19), 128(19), *133*, 293(18), 299(18), *317*
Spencer, R. S., 4, 5(32), *34*
Sroog, C. E., 31(107), *37*, 129(184), *137*, 160(44,45), 161(44,45), 162, 163(44, 45), 164(44,45), 166(44), 167(44,45), 168(44), *211*
Staffin, G., 115, *136*
Starr, L., 90–92(44), *134*
Statton, W. O., 79(29), *133*
Stavitskii, I. K., 236(96), 240(121), *265, 266*
Steck, E. A., 138(2), *210*
Stehman, C. J., 293(20), 297–300(20), 302(20), *317*

Steinberg, H., 55(62), *77*
Stephens, C. W., 78(16), 82, 83(16), 90(47), 92–94(47), *133, 134*
Stewart, H. C., 30(107), *37*
Stickings, R., 151(23), *210*
Stickney, P. B., 230–233(84), *265*
Still, R. H., 26(82), 28(82), *36*
Stille, J. K., 52(46), 53(46), *76*, 192(87–89), 193(87–89), 194(88,89), 195(89), 196(89), *212*, 277(19b), *284*
Stokes, H. N., 223(56,57), *264*
Stolka, M., 272(10–12), 273(10–12), *284*
Stolle, R., 176(66), 197(91), *212*
Stone, F. G. A., 214(2), *263*
Stratton, R., 224(65), *264*
Straus, S., 61, *77*
Strauss, E. L., 141(6b), 146(6), *210*
Strum, H. J., 176(67), 178(67), *212*
Stuart, H. A., 2(4), 4(26), 5(26,29), 7(4, 55), 8(58,59,61), 9(65,66), *34, 35*
Stuchlen, H., 40(19), *75*
Stull, R., 155(29), *210*
Sundet, S. A., 78(19), 128(19), *133*, 293(18), 299(18), *317*
Sunshine, N. B., 38(3,4), 44(4), *75*
Sviridova, N. G., 240(121), *266*
Swallow, V. E., 293(13), *317*
Sweeny, W., 87(40), *133*, 180(73), 183(73), 184(73), *212*, 300(23), 302(23), *317*
Syrkin, Y. A., 56(71), *77*
Szwarc, M., 53, 55, 56(51), 57(73), *76, 77*

T

Takami, K., 78(1), 79, *132*
Takeomoto, Y., 262(166), *267*
Takimoto, H. H., 236(98,99), *265*
Takse, Y., 182(76), 184(76), *212*
Taoka, I., 153(27,28), 154(27,28), 155(27, 28), 157(27,28), 158(27,28), *210*
Tarkoy, N., 256(143), *266*
Tarrasch, H., 99(61), *134*
Tasker, H. S., 124, *137*
Tasumi, M., 6(51), *35*
Tatlock, W. S., 238(104), *265*
Taylor, E. J., 173(63), 175(63), 176(63), *211*
Temin, S. C., 2(4), 7(4), *34*
Thayer, H., 40(19), *75*
Thomas, A. M., 79
Thomas, J. R., 220(29), *263*

Thomes, W. M., 221(44–46), *264*
Thomson, D. W., 128, 129(179), *137*
Thornton, D. A., 250(134), 251(134), 260 (134), *266*
Tiemann, F., 176(65), *212*
Tietz, R. F., 78(18), *133*
Tocker, S., 272(8), *284*
Todd, N. W., 315(35), 316(35), *318*
Tohyama, S., 208(107,108), *213*
Tomic, E. A., 256(144–146), 257(144), *266*
Topchiev, A. V., 40(31), 76, 270(4), *283*
Tracy, J. F., 299(21,22), 300(21,22), *317*
Travnikova, A. P., 188(83), *212*
Traynor, E. J., 30(102), *36*, 173(63), 175 (63), 176(63), *211*, 289(9,10), 290(9,10), 291(9), *317*
Trifan, D. S., 261(163), *267*
Truffault, R., 40(25), *76*
Tryon, M., 252(137), 253(137), *266*
Tseitlin, G. M., 153–155(28), 157(28), 158 (28), *210*
Tsunawaki, S., 124, 126, *137*
Tsykalo, L. G., 40(22,23,29), *75,76*
Turner, H. S., 215(6), 229(6), 230(6), *263*
Tversakaya, L. S., 40(31), *76*
Tyron, M., 20(73), *35*

U

Ueberreiter, K., 3(5), 4(19,21,22), 5(42), *34, 35*
Ulmschneider, D., 219(24), *263*
Un, H. H., 200–202(100), *213*
Unishi, T., 173(63), 175(63), 176(63), *211*
Uno, K., 141(11–13), 143(11,12), 149 (11–13), 150(12,13), 153(27,28), 154 (27,28), 155(27,28), 157(27,28), 158 (27,28), 173(63), 175(63), 176(63), 178 (68), 182(75,76), 184(75,76), 187(75), *210–212*
Upson, R. W., 220(27), *263*
Urivin, J. R., 250(134), 251(134), 260 (134), *266*

V

Valetskii, R. M., 107(107), 109(107), *135*
Vandenberg, E. J., 86–88(37), *133*, 179 (72), 184(72), *212*
Van Deusen, R. L., 283(26), *284*

Vansheidt, A. A., 51, 52(45), 55(64), 70, 76, 77
Verbanc, J. J., 132(188,189), *137*
Vinogradova, S. V., 107(104,105,107,108, 110,115,116), 109(84,105,107–110,115), 110(106), *135*
Vodehnal, J., 272(10–12), 273(10–12), *284*
Vogel, H. A., 130(186), *137*, 139, 141(5, 7,10), 142(5,7,10), 143(7), 145(5,7), 146 (5,7), 147(7), 154(7), *210*
Vokova, L. M., 239(107), *265*
Voronkov, M. G., 239(108), 240(117), *265, 266*
Vosburgh, W. G., 270(2), 271(2), *283*, 302(28), *317*

W

Wagner, D., 243(127), *266*
Wagner, R. I., 221(51), 222(51), *264*
Walker, E. E., 107(118), 108(118), *135*
Wall, L. A., 20(74), *35*, 115, *136*
Wall, R. A., 4(10), *34*
Wallach, M. L., 160(45), 161(45), 162, 163(45), 164(45), 167(45), *211*, 272(9), 274(9), *284*
Wallenberger, F. T., 86(36), 87(36,38–40), *133*, 149(19), 180(73), 183(73), 184(73), *210, 212*, 258(147), *266*, 293(17), 296 (17), 297(17), 299(17), *317*
Wallsgrove, E. R., 128(174), *137*
Walton, W. B., 173(63), 175(63), 176(63), *211*
Warne, R. J., 215(6), 229(6), 230(6), *263*
Warrick, E. L., 239(110), *265*
Washburn, R. M., 221(51), 222(51), *264*
Watanabe, W. H., 2(4), 7(4), *34*
Webb, R. F., 250(134), 251(134), 260 (134), *266*
Weber, W., 39(13), *75*
Weiss, J. O., 293(20), 297–300(20), 302 (20), *317*
Weissberger, A., 28(95), *36*
Wendlant, W. W., 26(82), 28(82), *36*
Wentorf, R. H., 216(8), 229(8), 230(8), *263*
West, C. D., 220(32), *263*
Whatley, E. W., 78(5), 85(5), *133*
Wheelwright, W. L., 55(67), *77*
Whencel, C. S., 85(32), *133*
White, D. M., 40(33), *76*

Whitman, L. C., 30(105), *37*
Whitney, R. B., 115(142), *136*
Wick, M., 239(109), *265*
Wielichi, E. A., 107(100), *135*
Wild, B. S., 255(141), *266*
Wiley, F. E., 5(35), *34*
Wiley, R. H., 4(23–25), 5(23,43–45), *34, 35*, 199(97), 201(97), *213*
Wilfong, R. E., 108, 109(81), 111(81), 135
Wilke, G., 261(162), *267*
Wilkins, J. P., 249(131), *266*
Wilkinson, G., 261(155), *267*
Williams, W. H., 39(14), *75*
Williamson, J. R., 192(87–89), 193(87–89), 194(88,89), 195(89), 196(89), *212*
Willis, P., 40(19), *75*
Wilson, C. L., 132(195), *137*
Wilson, L. W., 199(94), *212*, 300(25), 302(25), *317*
Winslow, F. H., 18, *35*, 120(164), *137*
Wirsing, W., 188(81), *212*
Wittbecker, E. L., 78, 79(13), *133*, 249(131), *266*
Wittig, G., 39(15), *75*
Wleugel, S., 208(106), *213*
Wolfe, J. K., 114(133), *136*
Wollett, G. H., 115(138,140), *136*
Woods, G. F., 38(3,4), 44(4), *75*
Woodward, A. E., 4(10,14), 6(49,51), *34, 35*
Wrasidlo, W., 155(35,36), 156(35,36), 160(57), 164(57), *211*, 285(2,8), 286(2), 287(8), 288(2), 289(2,8), 290–292(2), *317*

Wright, B. A., 116(163), 118–122(163), *137*
Wright, B. H., 116(163), 118–122(163), *137*
Wright, H. R , 226(72), *264*
Wright, W. W., 90(45), 91(45), 94, 95(45), 96(45), 116, 118, 119–122(163), 129(45), *134, 137*, 147, 148, 149(17), *210*
Wynstra, J., 100(66), *134*

Y

Yager, W. A., 120(164), *137*
Yamashita, Y., 6(51), *35*
Yankura, E. S., 230(83), *265*
Yoda, N., 97, 98(54–60), 99(54), 100(60), *134*, 208(107,108), *213*
Yukelson, A. A., 26(87), *36*

Z

Zeitler, V. A., 237(100), *265*
Zelinskii, N. D., 40(26), *76*
Zenftman, H., 226(72), *264*
Zgonnik, V. N., 239(108), 240(117), *265, 266*
Zhdanov, A. A., 234(85–91), 236(88–91, 97), 238(91), *265*
Ziegler, K., 40(27), *76*
Zilkha, A., 243(127), *266*
Zueva, G. Ya., 240(121), *266*

SUBJECT INDEX

A

Adhesives, 30, 285–287
Aliphatic–aromatic polyamides,
  differential thermal analysis, 95
  melting point, 81, 82
  polymer melt temperature, 81, 82
  preparation, 79
  properties, 81
  thermal stability, 84
  thermogravimetric analysis, 84
Aluminum–nitrogen polymers,
  preparation, 242
  thermal stability, 242
Aluminum–oxygen polymers,
  preparation, 241
  thermal stability, 242
  thermogravimetric analysis, 242
Aminophenols, 151–153
Aminothiophenols, 155, 156
Applications, 28–30, 285–316
Aromatic copolyamides, fibers, 292–299
  ordered, 92, 93
  polymer melt temperature, 92, 93
  preparation, 91, 95
  properties, 91, 95
  random, 91, 92
  thermal stability, 95
  thermogravimetric analysis, 95
Aromatic dicarboxylic acids, 86, 90–95,
  106–108, 139–141, 151–153,
  155, 156, 176, 179
Aromatic disulfonic acids, 128
Aromatic polyamide fibers, solvent
  resistance, 294, 295
  tensile properties, 292–297
  thermal stability, 292–297
Aromatic polyamides, differential
  thermal analysis, 95
  fibers, 292–299
  papers, 304–314
  polymer melt temperature, 91
  preparation, 90
  properties, 91, 92, 94–97
  thermal stability, 94–97
  thermogravimetric analysis, 94–96
Aromatic polyamide papers, chemical
  resistance, 308
  electrical properties, 307–314
  mechanical properties, 305–307
  physical properties, 305–310
  thermal stability, 305–314
Aromatic polycarbonates, differential
  thermal analysis, 103, 104
  melting point, 101, 102
  polymer melt temperature, 101, 102
  preparation, 100, 101
  properties, 101, 102
  thermal stability, 103–106
  thermogravimetric analysis, 104
Aromatic polyesters, differential thermal
  analysis, 113, 114
  polymer melt temperatures, 109–112
  preparation, 106–108
  properties, 108–114
  thermal stability, 113, 114
  thermogravimetric analysis, 114
Aromatic polyethers, polymer melt
  temperature, 117
  preparation, 115, 116
  properties, 117
  thermal stability, 117–123
  thermogravimetric analysis, 118–123
Aromatic polysulfides, differential
  thermal analysis, 127
  melting point, 124
  preparation, 124
  properties, 124
  thermal stability, 124–127
  thermogravimetric analysis, 124–126
Aromatic polysulfonates, polymer melt
  temperature, 128
  preparation, 128
  properties, 128
  thermal stability, 128, 129
  thermogravimetric analysis, 129
Aromatic polysulfones, differential
  thermal analysis, 130
  preparation, 130

SUBJECT INDEX

thermal stability, 130–132
thermogravimetric analysis, 130–132
Aromatic sulfonyl chlorides, 130

B

Bischloroformates, 100, 101
Bisphenols, 100, 101
Boron-containing polymers, 214–223
Boron–nitrogen polymers, 216–220
Boron–oxygen–carbon polymers, 220, 221
Boron–phosphorus polymers, 221–223
Borozen polymers, 215
Borozin polymers, 215, 216

C

Catalyst, for benzene polymerization, 42
  for polybenzyl formation, 49
  for polycarbonate formation, 101
  for polyether formation, 116
Coatings, 30, 287–289
Composites, 30, 289–293
Copolyimides, thermal stability, 174–176
  thermogravimetric analysis, 174–176
Copolymers of p-xylylene, 54
Cyclization, of polyacylamideoximes, 178
  of polyamic acids, 160–162
  of polydithiahydrazides, 197
  of polyhydrazides, 179–182
  of polyhydrazones, 188–190
  of polyhydroxyamides, 151–153
  of polyoxathiahydrazides, 197
Cyclized poly-3,4-isoprene, preparation, 271, 272
  properties, 272
  thermal stability, 272
Cyclized polydienes, 273

D

Diacid chlorides, aliphatic, 88
  aromatic, 86, 90–95, 106–108, 139–141, 151–153, 155, 156, 176, 179
  heterocyclic, 87
  sulfonyl aromatic, 82
Diamines, aliphatic, 79–85, 160
  aromatic, 90–94, 160, 161
Diazooxides, 115
p-Dichlorobenzene, 124
Differential thermal analysis, of aliphatic–aromatic polyamides, 85

of aliphatic–aromatic polyhydrazides, 88–90
of aliphatic polyhydrazides, 88, 89
of aromatic copolyamides, 95
of aromatic polyamides, 95
of aromatic polycarbonates, 103, 104
of aromatic polyesters, 113, 114
of aromatic polyhydrazides, 88–90
of aromatic polysulfides, 127
description, 25
measurement, 26
of polyimides, 168, 169
of poly-1,3,4-oxadiazoles, 186, 187
of poly-p-phenylenes, 45, 47, 48
of polypyrazoles, 192
of poly-p-xylyenes, 65–68
N,N-Dimethylacetamide, 86, 90–93, 151–153, 155, 156, 160–162, 176, 193, 200

F

Fibers, 30, 292–302
Films, 30, 302–304

G

Germanium–oxygen polymers, 243
Glass transition temperature, definition, 3, 4
  measurement, 5–7

H

Hydrazides, aliphatic, 86
  aromatic, 86
  heterocyclic, 86

I

Inorganic polymers, 214–264
Interfacial polymerization, 79, 85, 86, 90, 100, 107, 108, 116, 128
Irreversible thermal changes, 14–22
Isoteniscope, 25

K

Kinetics of thermal degradation, of polybenzimidazoles, 140, 141
  of polyimides, 169–176

L

Ladder polymers, definition, 269
  degradation, 270

Ladder polymers from Diels-Alder reactions, preparation, 279, 280
  thermal stability, 280–282
  thermogravimetric analysis, 281–282
Ladder polyquinoxalines, preparation, 275–277
  thermal stability, 276, 277
  thermogravimetric analysis, 276
Lap shear strength, 286, 287
Low temperature polymerization, 78, 79, 82, 86, 90, 100, 107, 108, 116, 128

M

Macallum polymerization, 124
Mechanism, of polybenzimidazole preparation, 140, 141
  of polybenzothiazole preparation, 156
  of poly-p-xylylene preparation, 55, 56
Melting point, of aliphatic–aromatic polyamides, 81, 82
  of aromatic polysulfides, 124
  definition, 7, 8
  measurement, 9
  of polyanhydrides, 98
  of polycarbonates, 101, 102
  of poly(methylene phenylenes), 50–52
  of sulfonyl aromatic polyamides, 83
Metal acetylacetonate chelate polymers, preparation, 249, 250
  thermal stability, 250–252
  thermogravimetric analysis, 252
Metal halogenophenoxides, 115, 116, 124

O

Ordered network polymers, 269–271
Organoarsenic polymers, preparation, 247
  thermal stability, 248, 249
  thermogravimetric analysis, 249
Organogermanium polymers, preparation, 243
  properties, 244, 245
  thermal stability, 246, 247
  thermogravimetric analysis, 246, 247
Organotin polymers, preparation, 243
  properties, 244, 245
  thermal stability, 246
  thermogravimetric analysis, 246, 247

P

Papers, 30, 304–314
Poly-O-acylamideoximes, 176, 177
Poly(alkylene phenylenes), 69
Polyamic acids, 160–162
Polyanhydrides, aromatic, 97
  aromatic–heterocyclic, 97
  heterocyclic, 97
  melting point, 98
  polymer melt temperature, 98
  preparation, 97
  properties, 98
  thermal stability, 99
Polybenzimidazoles, adhesives, 286
  composites, 290–293
  fibers, 299–302
  polymer melt temperature, 141–144
  preparation, 139–141
  properties, 141–146
  thermal stability, 145–151
  thermogravimetric analysis, 146–151
Polybenzimidazole adhesives, 286
Polybenzimidazole composites, electric strength, 290
  flexural strength, 291
  hoop tensile strength, 291, 293
  shear strength, 290
  thermal stability, 290–293
Polybenzimidazole fibers, 299–302
Poly(benzimidazo-benzophenanthrolines), 283
Poly(benzoxazinones), preparation, 208
  thermal stability, 209
  thermogravimetric analysis, 209
Polybenzoxazoles, adhesives, 287
  polymer melt temperature, 153
  preparation, 151–153
  properties, 153
  solubility, 154
  thermal stability, 154, 155, 158
  thermogravimetric analysis, 154, 155, 158
Polybenzothiazoles, adhesives, 287
  polymer melt temperature, 157
  preparation, 155, 156
  properties, 157
  thermal stability, 154, 155, 158
  thermogravimetric analysis, 154, 155, 158
Polybenzyls, 49, 50

Polycyanoterephthalylidenes, infrared
  spectra, 73
  preparation, 72
  properties, 72
  thermal stability, 72, 73
Poly-1,3-cyclohexadiene, 40, 41
Polydioxins, 278
Polyhydrazide fibers, 296, 297
Polyhydrazides, differential thermal
    analysis, 88, 89
  polymer melt temperature, 86–88
  preparation, 86
  properties, 86–88, 178–182
  thermal stability, 88, 90
  thermogravimetric analysis, 88, 90
Polyhydroxamic acids, 151–153
Poly(imidazopyrrolone), preparation,
    279, 280
  thermal stability, 280–282
  thermogravimetric analysis, 281, 282
Polyimides, adhesives, 286
  coatings, 287–289
  composites, 289–293
  differential thermal analysis, 168, 169
  fibers, 299–302
  films, 302–304
  kinetics of thermal degradation, 169–176
  polymer melt temperature, 164, 165
  preparation, 159–162
  properties, 163–165
  resins, 314–316
  thermal stability, 163–176
  thermogravimetric analysis, 163–168
Polyimide coatings, 287–289
Polyimide composites, 290–293
Polyimide fibers, 299–302
Polyimide films, 302, 303
Polyimide resins, 314, 316
Poly(methylene phenylenes), from aromatic hydrocarbons, 51, 52
  melting point, 50–52
  from poly(methylene cyclohexanes), 52, 53
  properties, 50–52
Poly-1,3,4-oxadiazole fibers, 299–302
Poly-1,2,4-oxadiazoles, polymer melt
    temperature, 177, 178
  preparation, 176

properties, 177
thermal stability, 178
Poly-1,3,4-oxadiazoles, differential
    thermal analysis, 186, 187
  fibers, 299–302
  polymer melt temperature, 184
  preparation, 178–182
  properties, 182, 184
  solubility, 182
  thermal stability, 183–188
  thermogravimetric analysis, 183–186
Poly-p-phenylenes, from benzene, 42, 43
  from bisdiazonium salts, 39
  differential thermal analysis, 45, 47, 48
  from halogenated benzenes, 38, 39
  from poly-1,3-cyclohexadiene, 40, 41
  infrared spectra, 44
  properties, 43–49
  structure, 43, 44
  thermal stability, 44–49
  thermogravimetric analysis, 45–47
  ultraviolet absorption spectra, 44
  x-ray diffraction, 44
Polyphenylsilsesquioxanes, 271
Poly-N-phenyl-1,2,4-triazoles, differential
    thermal analysis, 206
  fibers, 299–302
  preparation, 204, 205
  properties, 205
  thermal stability, 206, 207
  thermogravimetric analysis, 206
Polyphthalocyanine chelates, prepartion,
    253–255
  thermal stability, 254, 255
Polypyrazoles, differential thermal
    analysis, 192
  polymer melt temperature, 190
  preparation, 188–190
  properties, 190
  thermal stability, 191, 192
  thermogravimetric analysis, 191
Polyquinoxalines, adhesives, 287
  polymer melt temperature, 193
  preparation, 192, 193
  properties, 193
  thermal stability, 194–197
  thermogravimetric analysis, 194–197
Poly(quinazolinediones), preparation, 207
  thermal stability, 209
  thermogravimetric analysis, 209

Polyquinone ethers, 274
Polyquinone imines, 274
Polyquinone thioethers, 275
Polysiloxanes, thermal stability, 229–234
  thermogravimetric analysis, 231
Polysulfides, 130
Polysulfonamides, 128
Polytetraazopyrenes, polymer melt temperature, 202
  preparation, 202, 203
  properties, 202
  thermal stability, 203, 204
  thermogravimetric analysis, 203, 204
Poly-1,3,4-thiadiazoles, fibers, 299–302
  polymer melt temperature, 198
  preparation, 197
  properties, 197, 198
  thermal stability, 198, 199
  thermogravimetric analysis, 198, 199
Polythiazoles, 200, 201
Poly-p-xylylenes, α and β forms, 57
  differential thermal analysis, 65–68
  from halogenated hydrocarbons, 55
  from p-methylbenzyltrimethyl ammonium halides, 55
  from p-xylene, 53, 54
  from p-xylylene, 55
  from p-xylylidene, 55
  properties, 56–69
  thermal stability, 61–69
  thermogravimetric analysis, 63–65
  x-ray diffraction, 56–57
Poly-p-xylylidenes, 70, 71
Polymer melt temperature, of aliphatic–aromatic polyamides, 80, 81
  of aromatic–aliphatic polyhydrazides, 88
  of aromatic copolyamides (ordered), 92
  of aromatic copolyamides (random), 91, 92
  of aromatic copolyesters, 110, 111
  of aromatic–heterocyclic polyhydrazides, 87
  of aromatic polyamides, 91
  of aromatic polyanhydrides, 98
  of aromatic polyesters, 109, 111, 112
  of aromatic polyethers, 117
  of aromatic polyhydrazides, 87
  definition, 80
  of polybenzimidazoles, 141–144
  of polybenzoxazoles, 153
  of polybenzothiazoles, 157
  of polycarbonates, 101, 102
  of polyimides, 164, 165
  of poly-1,2,4-oxadiazoles, 177, 178
  of poly-1,3,4-oxadiazoles, 184
  of polypyrazoles, 190
  of polyquinoxalines, 193
  of polysulfonamides, 128
  of polysulfonates, 128
  of polytetraazopyrenes, 202
  of poly-1,3,4-thiadiazoles, 198
  of polythiazoles, 201
Polymeric boron complexes, 216
Polymerization of poly-1,3-cyclohexadiene, 40
Polyphosphoric acid, 141, 154, 156, 160, 176, 182, 194, 205

R

Resins, 314–316
Reversible thermal changes, 2, 9–14

S

Silicon–containing polymers, 228–241
Silicon–oxygen–aluminum polymers, 234, 235
Silicon–oxygen–antimony polymers, 240
Silicon–oxygen–arsenic polymers, 238, 239
Silicon–oxygen–boron polymers, 239
Silicon–oxygen–germanium polymers, 240, 241
Silicon–oxygen–lead polymers, 241
Silicon–oxygen–phosphorus polymers, 240
Silicon–oxygen–sulfur polymers, 240
Silicon–oxygen–tin polymers, 237–239
Silicon–oxygen–titanium polymers, 236–237
Softening behavior, 23–25
Solution polymerization, 79, 85, 86, 90, 100, 107, 108, 116, 128, 141, 152–154, 155, 156, 160, 177–179, 182, 189, 192, 193, 200, 205–208, 271, 272, 275–280, 282, 283
Sulfonyl aromatic polyamides, 82, 83
Sulfur, 124

SUBJECT INDEX

T

Tetraamines, 139–141, 188–190, 192, 193, 275–283
Thermal polymerization, 139–141, 151, 152, 159, 160, 177, 179, 188, 193, 197, 202, 274–283
Thermal stability, of aliphatic-aromatic polyamides, 84
  of aluminum–nitrogen polymers, 242
  of aluminum–oxygen polymers, 242
  of aromatic–aliphatic polyhydrazides, 88–90
  of aromatic copolyamides, 95
  of aromatic copolyamide fibers, 297–299
  of aromatic copolyimides, 174–176
  of aromatic heterocyclic polyhydrazides, 88–90
  of aromatic polyamide fibers, 292–297
  of aromatic polyamide papers, 304–314
  of aromatic polyamides, 94–97
  of aromatic polyhydrazide fibers, 296, 297
  of aromatic polyhydrazides, 88–90
  of boron–phosphorus polymers, 222
  of chelates of 8-hydroxyquinoline-formaldehyde polymer, 256, 257
  of cyclized dienes, 273
  definition, 1, 5, 23
  effect of environment, 22
  effect of structure, 18–22
  factors contributing to, 14–21
  of ferrocene-containing polymers, 261, 262
  of germanium–oxygen polymers, 243
  of ladder polymers from Diels-Alder reaction, 278, 279
  of ladder polyquinoxalines, 276, 277
  of metal acetylacetonate chelate polymers, 252
  of metal chelates of polyhydrazides, 258
  of metal thiopicolinamides, 260
  of organoarsenic polymers, 248, 249
  of organogermanium polymers, 246, 247
  of organotin polymers, 246, 247
  of phosphorus–nitrogen polymers, 224, 227–229
  of polyanhydrides, 99
  of poly(benzimidazo–benzophenanthrolines), 283
  of polybenzimidazole adhesives, 286
  of polybenzimidazole composites, 290–293
  of polybenzimidazole fibers, 299–302
  of polybenzimidazoles, 145–151
  of polybenzothiazole adhesives, 287
  of polybenzothiazoles, 157, 158
  of poly(benzoxazinones), 209
  of polybenzoxazole adhesives, 287
  of polybenzoxazoles, 154, 155, 158
  of polycarbonates, 103–106
  of polycyanoterephthalylidenes, 72, 73
  of polyesters, 113, 114
  of polyethers, 117–123
  of polyimidazopyrrolones, 280–282
  of polyimide adhesives, 286
  of polyimide coatings, 287–289
  of polyimide composites, 290–293
  of polyimide fibers, 299–302
  of polyimide films, 302–303
  of polyimide resins, 314, 315
  of polyimides, 163–176
  of poly-3,4-isoprene, 272
  of poly-1,2,4-oxadiazoles, 178
  of poly-1,3,4-oxadiazoles, 183–188
  of poly-1,3,4-oxadiazole fibers, 299–302
  of poly-p-phenylenes, 44–49
  of polyphenylsilsesquioxanes, 271
  of polyphthalocyanine chelates, 254, 255
  of polypyrazoles, 191, 192
  of polyquinoxaline adhesives, 287
  of polyquinoxalines, 194–197
  of polyquinazolinedione, 209
  of polysiloxanes, 229–234
  of polysulfides, 124–126
  of polysulfonates, 128
  of polysulfones, 130–132
  of polytetraazopyrene, 203, 204
  of poly-1,3,4-thiadiazole, 198, 199
  of poly-1,3,4-thiadiazole fibers, 299–302
  of polythiazoles, 201, 202
  of poly-N-phenyl-1,2,4-triazoles, 206, 207
  of poly-N-phenyl-1,2,4-triazole fibers, 299–302
  of poly-p-xylylenes, 61–69

of pyrolyzed polyacrylonitrile, 271
of silicon–oxygen–aluminum polymers, 235
of silicon–oxygen–germanium polymers, 241
of silicon–oxygen–tin polymers, 238
of silicon–oxygen–titanium polymers, 237
of tin–oxygen polymers, 243
Thermogravimetric analysis, of aliphatic–aromatic polyamides, 84
of aliphatic polyhydrazides, 88, 89
of aluminum–oxygen polymers, 242
of aromatic–aliphatic polyhydrazides, 88, 89
of aromatic copolyamides, 95
of aromatic copolyimides, 174–176
of aromatic polyamides, 94–96
of aromatic polycarbonates, 104
of aromatic polyesters, 114
of aromatic polyhydrazides, 88, 89
of aromatic polysulfonates, 129
of aromatic polysulfones, 130–132
of boron–nitrogen polymers, 217, 219
of boron–phosphorus polymers, 222
of chelates of 8-hydroxyquinoline–formaldehyde polymers, 257
description, 27
of ferrocene-containing polymers, 262
of ladder polymers from Diels-Alder reactions, 278, 279
of ladder polyquinoxalines, 276
measurement, 27, 28
of metal acetylacetonate chelate polymers, 252
of organoarsenic polymers, 248, 249
of organogermanium polymers, 246, 247

of organotin polymers, 224, 229
of phosphorus–nitrogen polymers, 145–151
of polybenzimidazoles, 283
of poly(benzimido–benzophenanthrolines), 154, 155, 158
of poly(benzoxazinones), 209
of polybenzothiazoles, 157, 158
of polyethers, 118–123
of poly(imidazopyrrolines), 281, 282
of polyimides, 163–168
of poly-1,3,4-oxadiazoles, 183–186
of poyl-p-phenylenes, 45–47
of polyphenylsilsesquioxanes, 271
of polypyrazoles, 191
of poly(quinazolinediones), 209
of polyquinoxalines, 194–197
of polysiloxanes, 231
of polytetraazopyrenes, 124–126
of poly-1,3,4-thiadiazoles, 203, 204
of poly-N-phenyl-1,2,4-triazoles, 198, 199
of polyureas, 206
of polyurethanes, 132
of poly-p-xylylenes, 63–65
of poly-p-xylylidene, 71
of silicon–oxygen–aluminum polymers, 235
of silicon–oxygen–tin polymers, 239
of silicon–oxygen–titanium polymers, 237
Thiophenol, 124
Tin–oxygen polymers, 243
Torsional braid analysis, 26

U

Ullman reaction, 116